P9-DMY-450

How To Know

POLLEN AND SPORES

How To Know

POLLEN AND SPORES

Ronald O. Kapp

Alma College

WM. C. BROWN COMPANY PUBLISHERS
Dubuque, Iowa

Pictured Key Nature Series

Library of Congress Catalog Card Number 73-89538

SBN 697–04848–9 (Paper)
SBN 697–04849–7 (Cloth)

THE PICTURED-KEY NATURE SERIES

How To Know The—

AQUATIC PLANTS, Prescott, 1969
BEETLES, Jaques, 1951
BUTTERFLIES, Ehrlich, 1961
CACTI, Dawson, 1963
EASTERN LAND SNAILS, Burch, 1962
ECONOMIC PLANTS, Jaques, 1948, 1958
FALL FLOWERS, Cuthbert, 1948
FRESHWATER ALGAE, Prescott, 1954, 1970
FRESHWATER FISHES, Eddy, 1957, 1969
GRASSES, Pohl, 1953, 1968
GRASSHOPPERS, Helfer, 1963
IMMATURE INSECTS, Chu, 1949
INSECTS, Jaques, 1947
LAND BIRDS, Jaques, 1947
LICHENS, Hale, 1969
LIVING THINGS, Jaques, 1946
MAMMALS, Booth, 1949
MARINE ISOPOD CRUSTACEANS, Schultz, 1969
MOSSES AND LIVERWORTS, Conard, 1944, 1956
PLANT FAMILIES, Jaques, 1948
POLLEN AND SPORES, Kapp, 1969
PROTOZOA, Jahn, 1949
SEAWEEDS, Dawson, 1956
SPIDERS, Kaston, 1952
SPRING FLOWERS, Cuthbert, 1943, 1949
TAPEWORMS, Schmidt, 1970
TREES, Jaques, 1946
WATER BIRDS, Jaques-Ollivier, 1960
WEEDS, Jaques, 1959
WESTERN TREES, Baerg, 1955

Other subjects in preparation

Printed in United States of America

INTRODUCTION

In the nineteenth century, when our understanding of the reproductive functions of spores and pollen grains was developing, some European geologists and paleobotanists were already reporting the occurence of fossil pollen grains in pre-Quaternary deposits and postglacial peat. However, the science of palynology* actually began in Scandinavia with the work of Lennart von Post, the first person to develop the methods and some of the applications of pollen analysis. His techniques were widely adopted after 1920 by phytogeographers and students of postglacial vegetation and climate.

Following World War II pollen analysis was applied to oil exploration and has since been used by geologists in paleontological work, primarily as a correlation tool. Archeologists, plant and animal ecologists, plant taxonomists, foresters, agricultural researchers, honey producers, aero-allergists and glaciologists have also recognized the potential contribution of palynology to these diverse fields. The broadened scope of application of palynology has produced a geometric growth in the number of workers and publications in the past few decades. Today many scientists, technicians, students, and hobbyists are seeking information to aid in the identification of pollen grains and spores.

Most palynologists in this country depend upon European keys and manuals as identification aids; fortunately there is sufficient circumpolar floristic similarity at the family and generic levels, in north temperate and arctic regions, to make these references useful. The absence of a North American manual, however, provides the impetus for preparing this contribution to the Pictured-Key Nature Series.

Confident identification of many pollen and spore types can only be accomplished by comparison of unknown specimens with a reference collection of microscope slides prepared from authoritatively identified plants. This manual can only serve as a general guide in identification; serious students of palynology should prepare a reference collection appropriate to their own needs. To facilitate such work, the laboratory techniques for preparation of modern pollen samples, as well as fossil sediments, are briefly described.

Very few pollen types can at the present time be identified with confidence to the species level; therefore these identification keys generally lead to genera and families. Where several pollen types occur in the same family or genus, I have attempted to key or illustrate the different types. Unfortunately, the morphological types of palyno-

---

*The term palynology refers to the science of pollen and spores with all of its ramifications. The word is derived from the Greek *palynos*, meaning "fine dust." It is sometimes convenient to refer to pollen grains and spores collectively as palynomorphs.

morphs often do not coincide with established taxonomic groups; this makes it necessary for the worker to be aware of the several taxa to which an unidentified specimen might belong. Since the primary objective of this handbook is to serve as an identification guide, the organization and keys are based on the most obvious morphological characteristics, not upon taxonomic groups like plant families.

Wind disseminated (anemophilous) pollen is most commonly encountered by palynologists; this includes pollen of many trees, a few shrubs, and certain herbaceous plants such as grasses, sedges, chenopods, and amaranths. (Examples of pollen of most of the aforementioned anemophilous plant groups are included in the manual.) The pollen of common, native tree genera of temperate and boreal North America are treated in the keys, as well as the other pollen and spore types routinely encountered by pollen analytical workers in this geographic area. While it has not been possible to include pollen and spores of uncommon or exotic plants because of limitations of space and inavailability of reference material, representatives of the common plant families have been included to provide clues for the identification of most morphological types. The Pictured-Key is limited almost exclusively to pollen and spores of vascular plants; however, spores of certain mosses and fungi, as well as resistant remains of algae and protozoa, are described in appropriate parts of the text or key.

Recognition and identification of palynormorphs requires careful microscopic examination. For comfort during many hours of work, a *binocular* compound microscope is most practical. The optical system should be of high quality, especially if fine structural details must be observed, and an oil immersion objective (950-1500 X magnification) is almost a necessity.

ACKNOWLEDGMENTS

Too many people to mention contribute to the preparation of a book such as this through suggestions, materials, encouragement and helpful criticism. However, certain persons gave indispensible aid. Rick Hall, a masterful biologist-artist, prepared all illustrations except those of the fungal spores; without his services the book would not have been attempted. Mrs. Darlene Helmich Southworth selected a representative sample of fungal spores for inclusion and prepared drawings of them. My wife, Phyllis, aided with illustrations and typed the manuscript. Dr. John McAndrews provided helpful advice and read the manuscript. Susan Walker aided in many ways, especially in reading proofs and in the preparation of the Index. Alma College provided equipment as well as funds for illustrating from its NSF Institutional Grant Fund.

To the persons and institutions enumerated, and the many unnamed contributors, deep gratitude is hereby acknowledged. Any inadequacies and errors are, however, my own responsibility.

Alma, Michigan — Ronald O. Kapp

CONTENTS

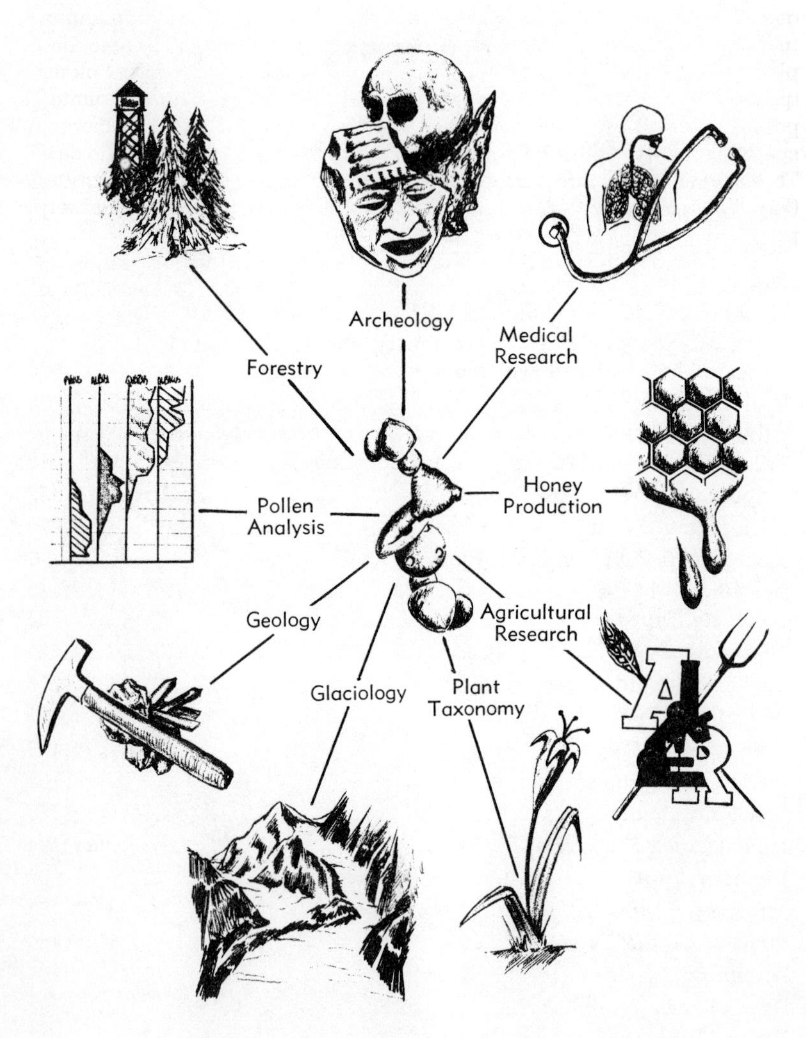

Some important applications of palynology.

FORMATION AND FUNCTION

Spores are produced by nonvascular plants (algae, fungi, mosses) as either thick-walled resting cells or single or multicellular propagules; in either instance, germination of the spore at a favorable time and place produces a new individual. In the primitive vascular plants (psilopsids, lycopsids, sphenopsids, ferns, and their extinct counterparts), spore production follows one of two patterns. In homosporous species, a single morphological type of spore is formed by meiosis. The spores germinate and grow into either bisexual gametophytes (Fig. 1), or into two different sexual plants, one of which ultimately produces sperms, the other eggs.

In heterosporous plants, meiosis results in two distinct kinds of spores: the male microspores and female megaspores. In some living

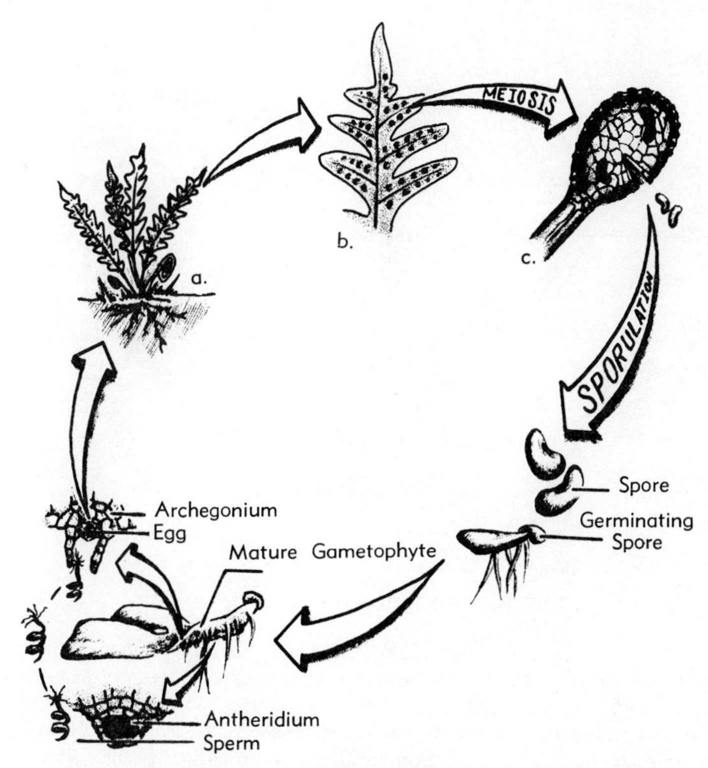

Fig. 1. Formation and function of spores in a homosporous fern. a, Mature fern sporophyte with aerial fronds and subterranean rhizomes and roots; b, portion of a frond showing clusters of sporangia or sori; c, a mature sporangium in which meiosis has occurred producing haploid spores.

and many extinct vascular plants, including many of the coal age ferns, lycopods, and sphenopsids, both the microspores and the much larger megaspores were shed from the parent plant. Both types of spores, often strikingly different even from the same species of plant, are frequently encountered by palynologists working in modern tropical floras or in geologically ancient sediments.

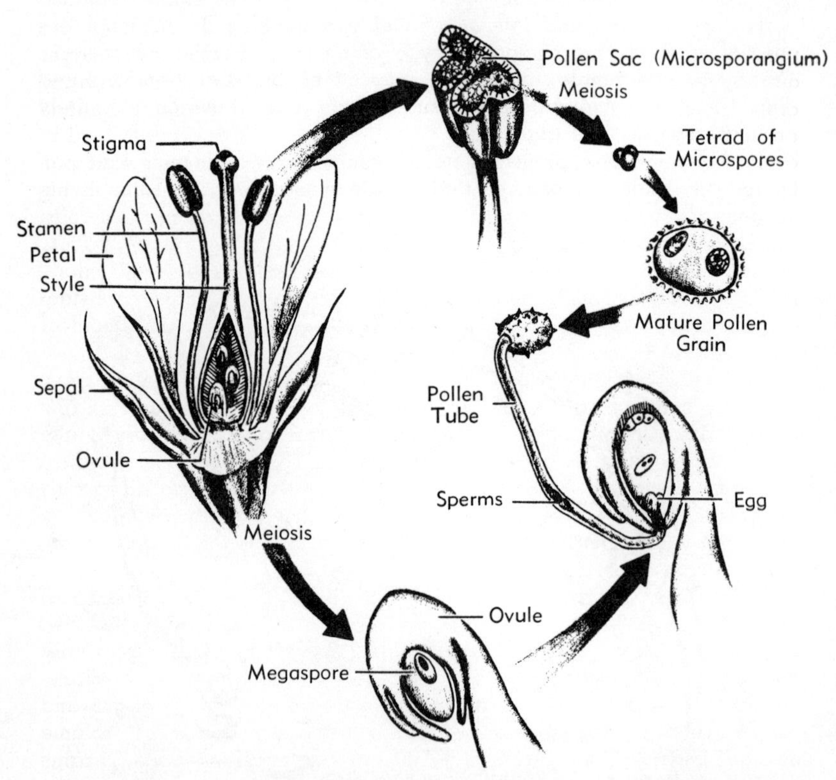

Fig. 2. Heterosporous spore production in a flowering plant. Flower parts are labelled and the male and female reproductive structures are enlarged to show where the microspores and megaspores are formed.

While seed plants are also heterosporous, the megaspore is retained within the ovule and there it matures into a female gametophyte or embryo sac containing an egg. The microspores typically mature into two-celled or binucleate pollen grains which are shed from pollen cones of gymnosperms, or from anthers of flowering plants. Pollination, usually by wind or by an animal agent, ultimately brings some of the pollen grains to the female part of the plant. The pollen grain then forms a pollen tube which either ruptures the pollen wall or emerges through an aperture in the wall of the grain. Eventually the pollen tube grows into the micropyle of the ovule where the sperms are released, permitting fertilization to occur (Fig. 2). The resultant embryo develops within the ovule; the ovule finally becomes a seed.

POLLEN AND SPORE STRUCTURE

Megaspores, microspores, and pollen grains (especially if they are produced by terrestrial plants) typically develop a resistant wall layer outside the usual cell wall. This wall layer is the exine. It is a waxy or resinous, chemically resistant layer which serves to minimize damage and prevent drying of the delicate cell. The exine is often constructed in a manner which aids in dissemination. Many anemophilous (wind pollinated forms) have walls modified to enhance bouyancy, while in entomophilous (insect pollinated) types, the presence of surface rods, spines and other sculptural features, serve to improve dissemination. Pollen and spores are among the most frequently preserved microfossils because of their widespread distribution and their persistant exines, which are remarkably resistant to chemical and biological degradation.

The exine provides the microscopic features which the palynologist uses for identification; these features include: 1. the size and shape; 2. the shape, number and arrangement of wall apertures; and 3. the structure and ornamentation of the exine itself. It has become standard procedure for most palynologists to chemically treat pollen grains by acetolysis (see page 11), thus permitting satisfactory examination of the exine characteristics. This renders pollen in comparable condition, whether fresh or dried, recent or fossil.

The SIZE of palynomorphs varies from about 5 microns (0.005 mm) to more than 200μ, but most are between 20 and 50μ. When the cell contents are alive, the size and shape may vary greatly, primarily due to changes in the osmotic balance of the protoplast. R. P. Wodehouse, a pioneer American palynologist, illustrated these changes and coined the term harmomegathy for the accommodation of volume change. He noted that the furrow membranes of many pollen grains swell with increases in internal volume; furthermore, the entire outline of the grain changes from angular to distended with increase in internal turgor pressure. Size is more constant in fossil palynomorphs and in those grains which have been prepared by acetolysis. Approximate sizes are useful in identification; they are given in the descriptions of each pollen type of the Pictured-Key. Attempts by palynologists to separate species within the genera *Betula* (birch) and *Pinus* (pine) on the basis of pollen size indicate that there are consistent average size differences between species but considerable overlap within the size ranges. It seems unlikely that individual pollen grains could be assigned to a species on the basis of size alone. The type of slide mounting medium used is also known to affect size; for example, mounting in glycerin jelly causes some swelling over a period of time, while the use of silicone oil avoids this problem.

In SHAPE, most palynomorphs are ellipsoids with rotational symmetry in which it is possible to identify the polar axis and the equatorial plane. In some cases the shape is spheroidal or subspheroidal, but it may be variously elongated (prolate or perprolate) or flattened (oblate or peroblate). Shape classes have been established which define the limits of these terms (Fig. 3). These classes are based on the ratio between length of the polar axis (P) and the equatorial diameter (E).

P/E Index	Shape Class
>2.0	Perprolate
1.33-2.0	Prolate
0.75-1.33	Subspheroidal
0.50-0.75	Oblate
<0.5	Peroblate

Example:

$$P/E = 3/2 = 1.5$$

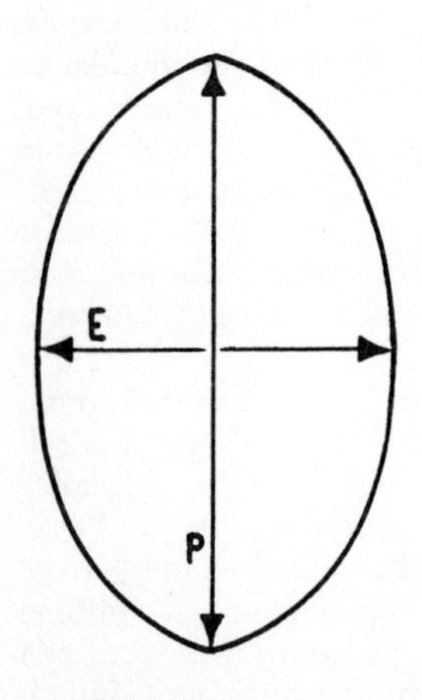

Fig. 3. The major shape classes of pollen grains and a diagram of a prolate grain showing the dimensions measured. Some workers subdivide the subspheroidal category into several additional shape classes.

Pollen grains and spores often depart from the ellipsoid shape. Spores frequently retain a pyramidal form resulting from developmental restrictions within a tetragonal tetrad; many fern spores attain a kidney shape. In polar view, pollen with protruding equatorial pores may assume varied shapes, depending upon the number of apertures: subtriangular, quadrate or hexagonal.

The triradiate scar of a spore identifies the proximal pole which was at the center of the original tetrad; the opposite pole was therefore distal with respect to the tetrad. The polar axis of a pollen grain similarly has a proximal and distal end; but, in many cases, it is virtually impossible to distinguish the proximal from the distal pole of the mature pollen grain. Monocolpate and monoporate angiosperm pollen grains have a single aperture at the distal pole.

The APERTURES of pollen grains and surface scars of spores provide the primary basis upon which this key is organized. Some pollen types show no evidence of a thin area in the exine and are said to be *inaperturate;* similarly, *alete* spores have no scar at the exine surface. *Monolete* spores have a single scar (which may somewhat resemble a furrow), while *trilete* spores are provided with a three-pronged scar (Fig. 4). Rupturing of the spore wall at the time of germination usually occurs at the scar.

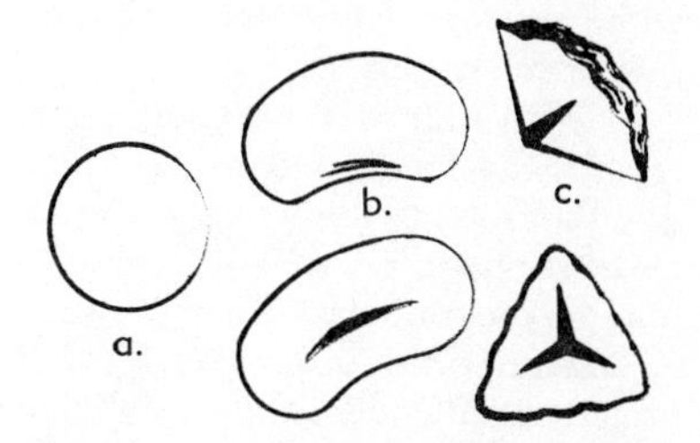

Fig. 4. The major spore types of vascular plants. a, alete; b, monolete, two views; c, trilete, two views.

True apertures are found in most pollen types and are either furrows or pores in the exine (Fig. 5). Recent electron microscopic studies in-

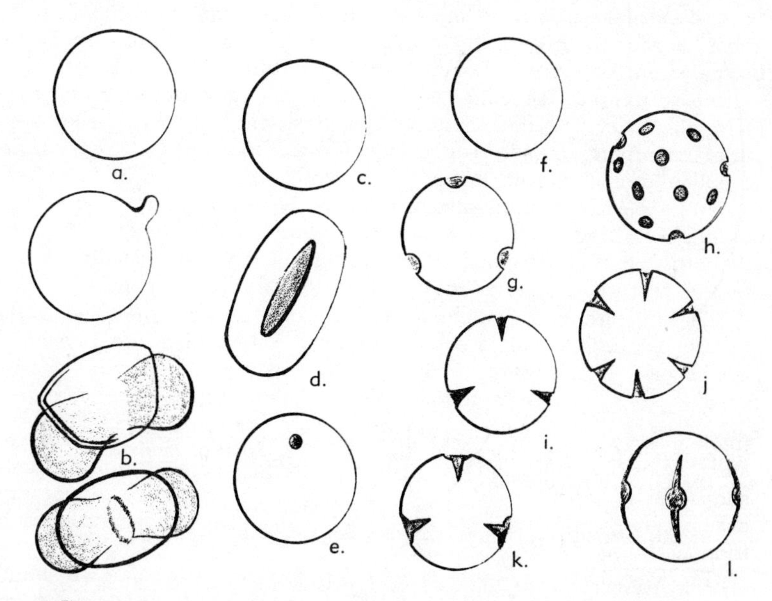

Fig. 5. Some common pollen types of seed plants. Gymnosperms: a, inaperturate; b, winged with an obscure furrow. Monocotyledons: c, inaperturate; d, monocolpate; e, monoporate. Dicotyledons: f, inaperturate; g-h, 3 to periporate; i-j, 3 to 6 colpate; k-l, colporate.

dicate that there is usually a thin layer of inner exine covering the apertures, although this is more readily observed in furrows than pores. Furrows are boat-shaped apertures which are more than twice as long as broad, while pores are circular to elliptical apertures. The furrow or pore membrane may be smooth and featureless, have a granular surface, or bear an operculum. The margin of the apertures may be even with the surrounding wall; or, if the adjacent wall is thinned or thickened, the pore is said to have an annulus (a bordered furrow is said to have a margo). In many dicotyledons the apertures of the pollen grains are compound, usually with a pore located in each furrow. Pollen grains with pores are described as porate; those with furrows (colpi) are colpate; and those with both furrows and pores are colporate.

The WALL of the pollen grain or spore is composed of two basic layers: the intine and exine. The intine is the cell wall which immediately surrounds the living protoplasm. It is partly cellulose, but appears to have a different chemical composition from most cell walls, with higher proportions of pectic substances, callose, and other polysaccharides. Pollen grains only retain their viability for short periods of time and after death the protoplasm and intine decay readily, leaving behind the outer wall layer, the exine.

Pollen exines are composed of poorly understood chemical compounds which have extraordinary resistance to degradation. The distinctive substances forming the exine have been called sporopollenins. These substances are poorly characterized chemically, but appear to be polymers of monocarboxylic or dicarboxylic fatty acids of high molecular weight. The exine withstands high temperatures and treatment with concentrated acids and bases. Resistance to decay, especially under anaerobic conditions, accounts for the widespread preservation of pollen grains and spores as microfossils; the latter even occur in Paleozoic rocks. Under aerobic conditions, however, pollen exines are suseptible to breakdown by enzymatic degradation by fungi and other microorganisms. Differential resistance to decay and conditions of optimal preservation have not been thoroughly investigated.

The structure and sculpture of the exine imparts the distinctive microscopic and submicroscopic morphology to pollen grains which permits their identification. Only recently the work of electron microscopists has clearly substantiated some of the morphological features which are equivocal under light microscopy. Several electron photomicrographs are reproduced in Figs. 470-479. Several systems of terminology have been developed by different authors with resultant confusion of terms and abundant synonomy and quasi-synonomy. An outline of the terminology that I have adopted follows.*

---

*The terminology of the various layers which may occur in walls of pollen grains and spores follows the system of Faegri and Iverson. In a specific pollen or spore type certain layers may be absent or impossible to resolve with light microscopy.

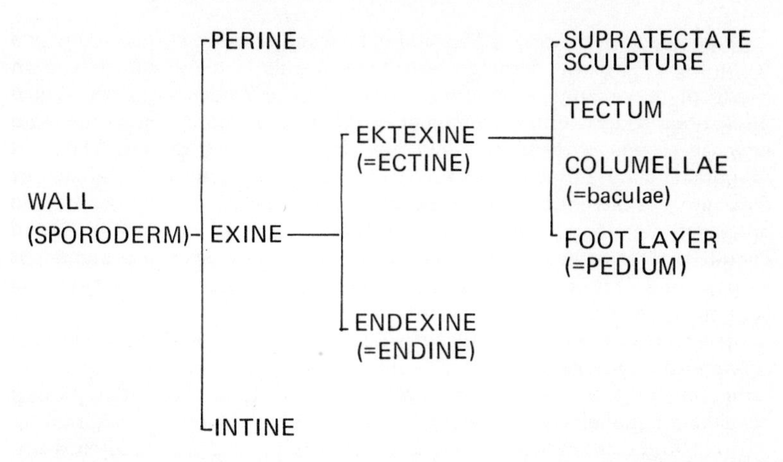

The endexine (endine) is structurally a rather homogeneous layer; any internal differentiation appears as horizontal layering. The ektexine (ectine) is peripheral to the endexine; it often exhibits radial orientation of the structural elements and stains more deeply with fuchsin b than does the endine. In sequence, the ektexine may be composed of a more or less continuous foot layer, radial columellae (syn. baculae), an outer more or less continuous tectum, and various supratectate sculpturing elements (Figs. 6, 470, 471).

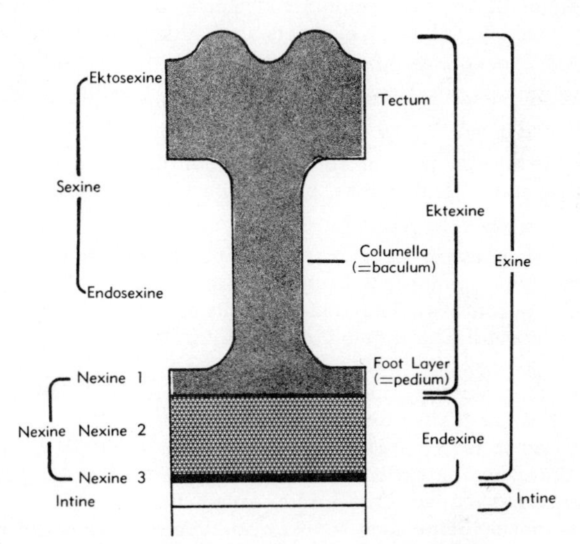

Fig. 6. Wall layers of a pollen grain comparing some proposed terms for the several layers. Terms on the left are basically those of Erdtman, terminology on the right that of Faegri and Iversen.

Palynologists usually differentiate between structural and sculptural features of the exine. Structural elements are those exine characteristics inside of the tectum; structural differences are most frequently based on various shapes and arrangements of the columellae. The columellae may be simple or branched elongate rods or low granules. Their arrangement may be scattered and irregular, continuous and regular, or they may form a reticulate, striate, or rugulate pattern. If the columellae are fused distally to form a more or less continuous tectum (Fig. 7) the condition is tectate; if there are numerous openings in the tectum, it is perforate tectate. Intectate forms lack an extensive or continuous tectum.

Sculptural elements are those which protrude beyond the outermost continuous layer of the exine. If the grain is tectate, then the sculptural elements are supratectate features of various types. Sculpturing of intectate pollen types refers to the features which lie external to the foot layer or endexine; in this instance structure and sculpture are based on essentially the same features. When the surface is smooth, the sculpturing condition is described as psilate. In some cases the surface may be pitted (foveolate) or grooved (fossulate).

The primary types of sculpturing elements, which are isodiametric in a surface view of the pollen wall, are shown in the upper part of Fig. 7. When the sculptural elements are horizontally elongated (in surface view), various kinds of surface patterns result (striate, reticulate, rugulate), as depicted in the lower part of Fig. 7.

The characteristics shown in Figure 7 are defined by the size, shape, and arrangement of the sculptural elements, as follows:

scabrate: small sculptural elements, less than 1μ in any dimension

verrucate: elements as broad as, or broader than, high, $>1\mu$ wide

gemmate: diameter of elements equal to, or greater than, height; constricted base, $\geqq 1\mu$ wide

baculate: elements post or rod-like, higher than broad, $\geqq 1\mu$ long

clavate: elements higher than broad with a constricted base or club-shaped tip, $\geqq 1\mu$ long

echinate: elements in form of pointed spines, sometimes called spinulate (with spinules) if between $1\text{-}3\mu$ long

rugulate: horizontally elongated elements in an irregular pattern

striate: horizontally elongated elements in a more or less parallel pattern

reticulate: horizontally elongated elements forming a net-like pattern of lacunae (holes) and muri (walls)

The exine layer of spores of lower vascular plants is structurally simpler than that of pollen; it lacks columellae, but is multilayered. It is sometimes called exosporium rather than exine, to emphasize these differences. Some such spores also have a rather loose wall layer termed the perine. It lies external to the exine and is readily lost in processing of samples. This outermost layer seems to be deposited by

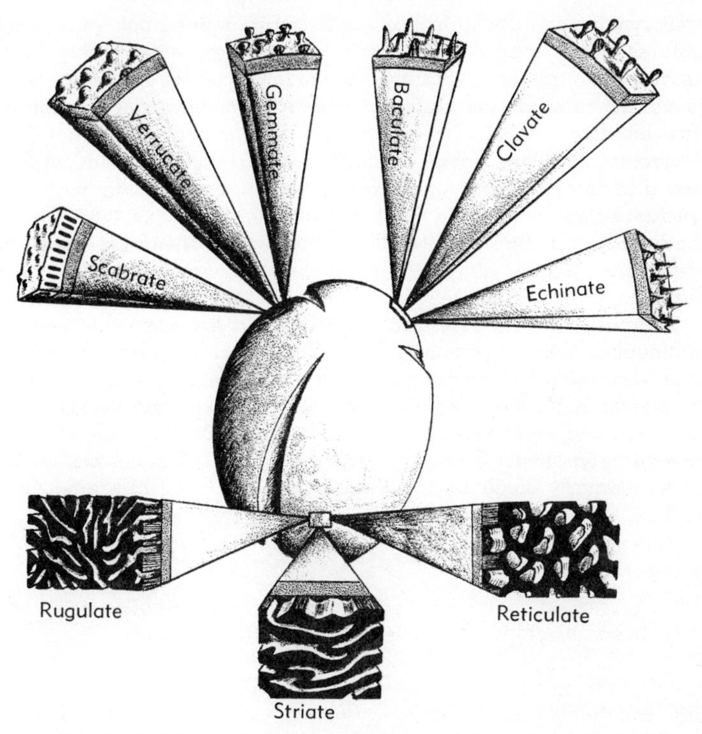

Fig. 7. Sculptural elements and patterns of the pollen exine.

the tapetal cells which surround the spores during development in the sporangium.

The characteristics described in the foregoing pages and figures are used throughout the pictured keys in this book; other specialized structures found in specific groups are described in the index-glossary. Many of the detailed features can be seen effectively only through oil immersion objectives at high magnification.

It must be emphasized that the positive identification of any spore or pollen grain finally depends upon comparison of the entity with pollen or spores from authoritatively identified reference specimens. Recent innovations in microscope accessories have greatly facilitated such direct comparisons. For this purpose, comparison bridges, which permit binocular split-field examination of objects on two separate microscopes, are being used in some palynological laboratories.

Detailed studies of exine structure require special microscopic equipment. For example, phase contrast microscopy and polarizing accessories may greatly aid in the visualization of exine stratification. Ultimately, detailed analysis of minute features of the pollen or spore wall may depend upon electron microscopy. Both carbon surface replica and ultra-thin section techniques have been successfully used in palynological studies; recently the scanning electron microscope has been used to study the surface topography of the exine in detail. (See Figs. 470-479).

Several workers have recently prepared very useful descriptive monographs describing pollen-spore morphology of certain plant groups, usually an analysis of a selected family. These, in addition to the photographically-illustrated pollen-spore floras of certain parts of the world, are of great aid to the student of palynology. Several of these publications are listed below among the selected references.

Two international periodicals are devoted exclusively to palynology and publish many taxonomic-morphological studies, as well as results of pollen analysis. *Grana Palynologica* (Almqvist and Wiksell Publishers, 26, Gamla Brogatan, Stockholm C, Sweden) was established in 1954; it is published mostly in English and German. *Pollen et Spores* (Laboratoire de Palynologie, 61, rue de Buffon, Paris V^{e}, France) has been published regularly since 1959, primarily in French, English, and German.

PREPARATION TECHNIQUES

Special techniques for collecting and processing pollen-rich materials have been developed by scientists in different fields of application. While it is impossible to describe all of the techniques and devices that have been perfected, several commonly-used pieces of apparatus and widely-adopted processing schedules are included. These will be adequate for the needs of most readers. For variations of the methods mentioned and more exhaustive description of techniques, the books of Brown (1960) and Kummel and Raup (1965) should be consulted.

Reference Slides

Permanent reference slides may be prepared in several ways. In any case, the cover glasses should be ringed to anchor the glass and to prevent drying of the mounting medium, or, the slides should be stored flat in dustproof trays or boxes. The slide labels should clearly indicate the origin of the reference material, preferably showing the full identification, locality, collector, and herbarium and the accession number of the specimen.

Allergists often must identify pollen which is fresh, having the cellular contents intact. The Wodehouse technique for preparing comparative reference material has been most frequently used by these specialists.

Wodehouse technique for reference pollen:

1. Place a small amount of pollen in the center of a microscope slide, add a drop of alcohol, and allow it to evaporate. A second or third drop may be necessary.

2. As the alcohol evaporates, an oily or resinous extract from the pollen is deposited on the slide. Remove the oily ring with cotton moistened in alcohol.
3. Add a drop of melted glycerin jelly stained with methyl green or basic fuchsin just before all the alcohol has evaporated. Stir with a needle to secure even distribution of the pollen.
4. Cover with a clean, heated No. 0 cover glass. It is important to use just enough glycerin jelly to fill the cover glass and to keep the mount thin. After a few trials one will learn the quantity of jelly to use.

Silicone oil may be used instead of glycerin jelly as a mounting medium, and has the advantage of eliminating size increases which often occur after a period of time in glycerin jelly. In using silicone oil, the above schedule is modified by: (a) staining in step 2, followed by a benzene wash, (b) adding a drop of silicone oil, then allowing the benzene to evaporate from the sample for several hours, and (c) covering with a cover glass.

Most pollen specialists examine both reference pollen, modern samples, and fossil materials only after a chemical treatment which removes the cellular contents and the cellulose wall (intine) of the palynomorph. The most widely used procedure is the Erdtman acetolysis technique. Chemical treatment may be followed by staining, but this is often unnecessary because the process itself enhances the contrast of the exine features. The processing may be followed by mounting in the medium of choice, but silicone oil is highly recommended. Wearing rubber gloves is advisable.

Erdtman acetolysis technique:

1. Place dried or fresh anthers or sporangia in a centrifuge tube and macerate in glacial acetic acid with a stirring rod. This moistens dry materials for maceration and dehydrates fresh specimens.
2. Centrifuge and decant off the acetic acid.
3. Carefully add about 10 ml. of cold acetolysis solution, a chemical mixture which degrades most organic tissues. (This is prepared fresh daily by slowly adding 1 part of concentrated sulfuric acid to 9 parts of acetic anhydride, with constant agitation.) Stir and leave stirring rod in tube while heating to boiling.
4. Heat in boiling water for one minute, stirring occasionally.
5. Centrifuge, and decant off the acetolysis mixture.
6. Wash with glacial acetic acid.
7. Wash twice with water (by successive stirring, centrifuging, and decanting).

Reference material may then be stored in glycerin jelly by: (a) washing in 50% glycerin, followed by (b) suspending in molten glycerin jelly. Reference slides may be made at once or from storage vials.

If silicone oil is used, the water-washed sample (step 7, above) is treated as follows: (a) wash with 95% alcohol, (b) wash with absolute alcohol and stain with fuchsin, if desired, (c) wash with benzene and decant, (d) add about 1 ml. benzene, transfer to a small vial, add silicone oil and leave open for evaporation for 24 hours*, and (e) add the appropriate amount of silicone oil to attain the desired concentration of pollen or spores. Slides may be made at any time; because high viscosity silicone oil is used (2000 to 6000 cps), the droplet spreads under a cover glass very slowly. If one wishes to roll the pollen grains (by pushing the edge of the cover glass), the cover glass should not be ringed; if this is not necessary, the slides may be sealed with clear fingernail polish or some other suitable ringing cement.

Sediment Sample Preparation

Pollen analysts and geologists often need to separate and concentrate pollen and spores from peats, soils, clay, marl, coal, and sedimentary rocks. There are special techniques and processes for each of these materials, and the references by Brown (1960) and Kummel and Raup (1965) should be consulted for details.

In each type of matrix the first objective of processing is to attain sufficiently fine maceration that pollen and spores are freed from other particles. Removal of the inorganic (mineral) fraction of samples and concentration of pollen must also be achieved.

Some common maceration and concentration techniques are:

1. *KOH maceration.* Peat samples may be boiled in 10% KOH (potassium hydroxide) prior to acetolysis. Powdered lignites may be soaked in 10% KOH at room temperature from a few hours to several days prior to acetolysis.

2. *Acid maceration.* Dilute (6%) HCl (hydrochloric acid) is commonly used as a first step in macerating lake sediments, marls, pulverized limestone, and soil samples. Calcareous sediments are removed efficiently and the samples are usually completely macerated following this treatment.

In many cases (lake or alluvial sediments, pulverized shales, clays, etc.), the siliceous component is subsequently removed by treatment with concentrated HF (hydrofluoric acid). The sample may be mixed with 2 or 3 times its original volume of HF in polyethylene beakers, and processed at room temperature for several days, or in a hot water bath for a few minutes to several hours. This is followed by several washes in water and glacial acetic acid prior to acetolysis.

Nitric acid is especially useful in macerating coals. Several versions of the Schulze technique are in common usage; the Schulze solution is a mixture (varying proportions) of nitric acid and potassium chlorate.

One method is as follows:

1. Mix 1 gram of powdered coal with 1 gram of dry potassium chlorate.

---

*The alcohol and benzene washes can be eliminated and one wash with tertiary butyl alcohol substituted.

2. Drop by drop (to prevent explosion), add 5 cc of concentrated nitric acid; this should be in good ventilation or in a fume hood.
3. Allow to stand several hours or overnight.
4. Decant top fluid and add 10% KOH; allow to stand overnight.
5. Centrifuge and wash twice in water.
6. Suspend in a mounting medium or make wet mounts with the residue of the last water wash.

In many cases, the oxidative treatment by acetolysis or Schulze solution may render the pollen and spores a very dark color. Bleaching is usually accomplished by chlorination. One simple procedure is to suspend the residue in a 33% sodium chlorate solution and then add 1 to 3 drops of concentrated HCl. Chlorine gas will be liberated. Immediately centrifuge, decant, and wash twice with water; long treatment will degrade or destroy pollen.

Many other concentration techniques have been employed. Some workers use flotation techniques to remove the lighter pollen-spore fraction from denser inorganic sediments (e.g., suspended in a concentrated solution of $ZnCl_2$, specific gravity about 2.0). In other laboratories, ultrasonic generators are used for maceration or deflocculation.

SAMPLING PROCEDURES

Most techniques for air and sediment sampling have been developed to meet the specific needs of a research problem. As a result, there are samplers which work on many different principles and numerous "improved designs" continue to be built as modifications of some of the "standard samplers." Only a few types of air and sediment samplers are under commercial production; most are custom-built from published or unpublished plans.

Air Sampling

Air samplers which are in common use range in complexity from gravity slide devices to motorized, directional, volumetric devices. In each instance, airborne particles, such as fungal spores and pollen, are impacted on an adhesive surface which is subsequently examined microscopically, usually after staining but often without chemical treatment.

The simplest collecting method is by gravity slides. A selected portion of a microscope slide (or cover glass) is streaked with an adhesive mounting medium such as glycerin or silicone and then exposed for a length of time in a horizontal position at the sampling point. Umbrella-type hoods and other shields are usually devised to protect the slide from rain.

The Rotobar Sampler has been used for many years by the aeroallergy laboratories at the University of Michigan. This device is designed to sample a known volume of air per hour by impact on a rotating bar on which is mounted an adhesive strip (Fig. 8). The aerodynamics of this device have been studied in some detail, resulting in design modifications which maximize the sampling efficiency for pollen and spore-sized particles.

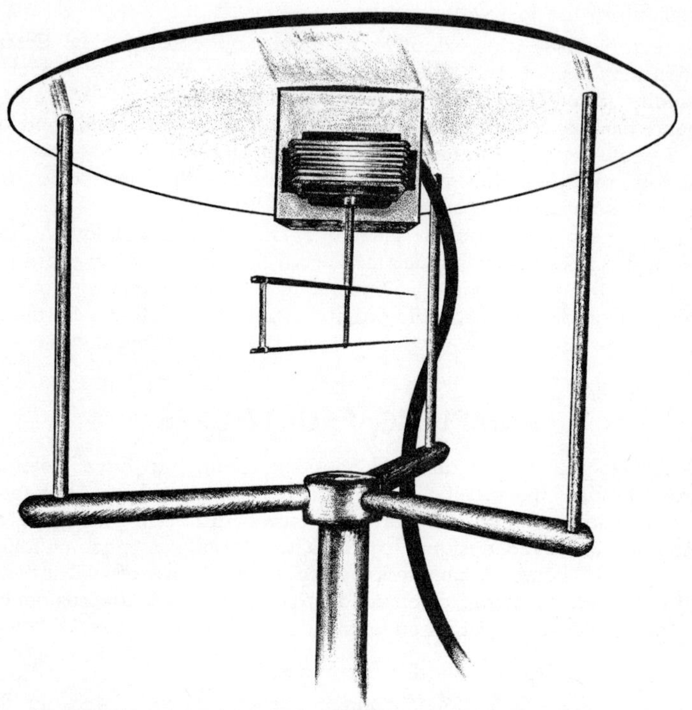

Fig. 8. The rotobar air sampler developed at the University of Michigan. An adhesive strip is mounted on the motor-driven vertical bar which turns at a rate appropriate to sample a known volume of air per hour.

A commercially-built and highly accurate continuous air sampler, a modification of the Hirst sampler, is now being marketed by the Burkard Mfg. Co. Ltd., (Rickmansworth, Hertfordshire, England); it is rapidly becoming the standard sampling device in research laboratories. The apparatus (Fig. 9) has a built-in motor drive and vacuum pump. A known volume of air (10 liters per min.) is impacted upon a transparent plastic tape which is mounted upon a clock work-driven drum. Interchangeable orifices can be obtained on special order to improve the trapping efficiency for particles in the range of 1 to 10μ in diameter. The apparatus samples continuously for periods up to seven days; the double-sided adhesive tape bearing the particles is later cut into daily (or hourly) segments which can be mounted directly on glass slides for counting.

In some cases it is important to sample the cumulative pollen rain of an entire season or for several recent years. In this case, sampling depends upon the day-by-day accumulation of pollen and spores which fall from the atmosphere and are preserved in favorable places. Pollen and spores are well preserved in bodies of water, especially if fungal and bacterial activity are limited. Some palynologists have suspended bottle-traps in lakes to catch the pollen settling

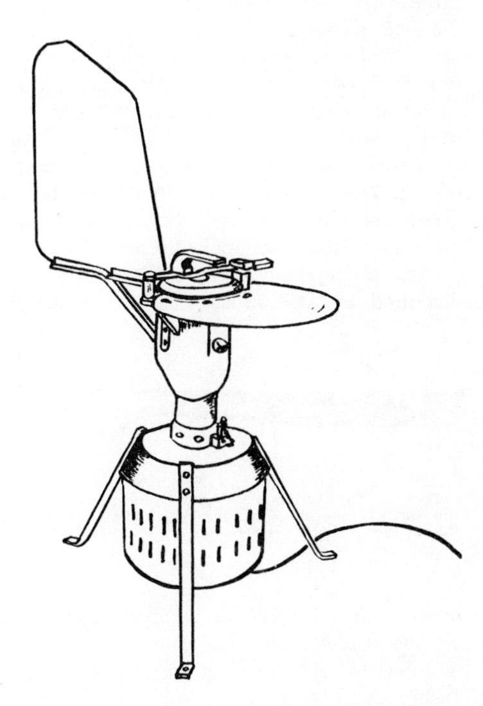

Fig. 9. The Burkard seven-day recording, volumetric spore trap. The windvane rotates the turret so that the oriface faces upwind.

through the water; others collect the recently-accumulated lake sediments for analysis of the modern pollen rain. Even cattle-watering tanks and open reservoirs may contain several years' pollen accumulation. Clumps of moss, duff at the surface of forest soils, lichen thalli, and even the terrestrial, gelatinous alga *Nostoc*, are known to be effective traps for recent pollen. Pollen and dust may be washed from moss polsters in alcohol or KOH; in other cases the sediments or lichen and *Nostoc* thalli may be subjected to acetolysis. Studies of the regional and local pollen rain have shown that different plant communities produce distinctive and recognizable mixtures of pollen and spores.

Sediment Sampling

Sediment samplers vary considerably depending upon the type of matrix which must be penetrated (Kummel and Raup, 1965). Oil drillers may retrieve rock cores from depths of several thousand feet with special rotary drills. Pollen analysis may then permit stratigraphic correlation of cores and palynological characterization of fossiliferous geologic strata.

From the earliest days of pollen analysis, there has been great interest in boring peat bogs and lakes. Lake and bog sediments often have very high pollen frequencies and continuous deposition during the postglacial period permits reconstruction of changes in pollen rain (and thus vegetation and habitat) at the site. The Hiller peat sampler was designed in Sweden and is now sold by A/B Borros, Solna, Sweden. It has a 0.5 meter long collecting chamber which is turned clockwise and pushed into the sediments to the appropriate depth; the chamber is opened and filled by twisting the appartus counter-clockwise, cutting a sediment sample. The device is then rotated counterclockwise, to close the chamber, and then pulled up. Successive lengths of rod may be added to deepen the bore hole.

The Livingstone sampler (Fig. 10) is a variant of a device originally designed by Dachnowsky. It is preferred for sampling in lakes and

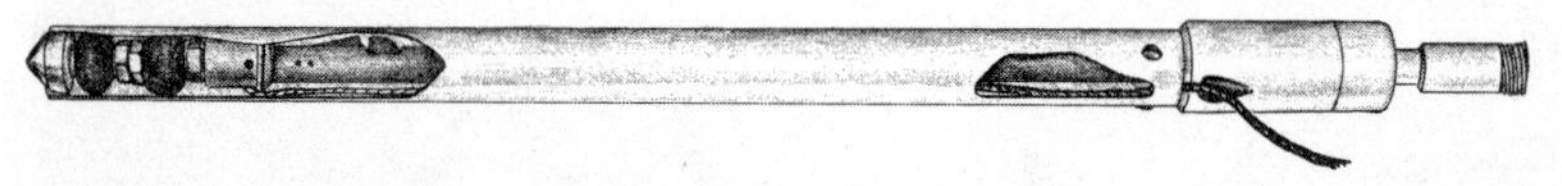

Fig. 10. The modified Livingstone sampler has a steel tube enclosing a retractable piston. To fill the chamber, the push-rod is withdrawn and engages a spring catch at the top of the tube; the empty tube is then pushed into the sediments while holding the piston stationary by means of its attached steel cable.

ponds and on occasions when the sediments are silty or slightly sandy. This is a piston-type sampler which remains closed while the sampler is lowered. At the desired depth, the push-rod is retracted and its spring catch engages at the top of the sample chamber. The tubular chamber is then pushed into the sediment while holding the piston in the same position by means of a steel cable. Up to one meter of sediment may be taken in each drive; the core may be extruded into storage tubes or placed in bottles. In all instances in which ancient sediments are being collected for pollen analysis, every precaution should be taken to prevent contamination of the sediments by modern pollen from the air, dust, or soil.

Heavily weighted, free-fall samplers have been designed for collecting sediment samples from very deep lakes.

BIBLIOGRAPHY

Anderson, Svend Th. 1960. Silicone oil as a mounting medium for pollen grains. Geological Survey of Denmark, Series 4, *4*(1):7-24.

Arms, Bernard C. 1960. A silica depressant method for concentrating fossil pollen and spores. Micropaleontology, *6*(3):327-328.

Arnold, Chester A. 1950. Megaspores from the Michigan coal basin. Contributions Museum Paleontol., Univ. of Mich., *8*(5) 59-111 (18 pls.).

Barnett, H. L. 1960. Illustrated genera of imperfect fungi. Burgess, Minneapolis. Pp. 218.

Bassett, John I. and P. W. Voesey. 1961. A new continuous pollen sampler. Canadian Journal of Plant Science, *31*:840-853.

Bassett, John I. 1959. Surveys of air-borne ragweed pollen in Canada with particular reference to sites in Ontario. Canadian Journal Plant Science, *39*:491-497.

Bhoj Raj and Saxena, M. R. 1966. Pollen morphology of aquatic angiosperms. Pollen et Spores, *8*(1):49-55.

Brown, Clair A. 1960. Palynological techniques. C. A. Brown, 1180 Stanford Ave., Baton Rouge, La. Pp. 188.

Cain, Stanley A. 1948. Palynological studies at Sodon Lake: I. Size-frequency study of fossil spruce pollen. Science 108:115-117.

Cain, Stanley and Louise G. Cain. 1948. Palynological studies at Sodon Lake. II. Size-frequency studies of pine pollen, fossil and modern. American Journal Botany 35:583-591.

Chanda, Sunirmal. 1965. The pollen morphology of Droseraceae with special reference to taxonomy. Pollen et Spores, 7(3):509-528.

Clausen, Knud. E. 1962. Size variations in pollen of three taxa of Betula (1). Pollen et Spores, *4*(1):169-174.

Colinvaux, P. A. 1964. Sampling stiff sediments of an ice-covered lake. Limnology and Oceanography, *9*(2):262-264.

Cushing, Edward J. 1961. Size increase in pollen grains mounted in thin slides. Pollen et Spores, *3*(2):265-274.

Dahl, A. Orville. 1952. Electron microscopy of ultra-thin frozen sections of pollen grains. Science, *116*(3):465-467.

Dahl, A. Orville and John Rowley. 1956. Modifications in design and use of the Livingstone piston sampler. Ecology, *37*(4):149-151.

Davis, Margaret. 1963. On the theory of pollen analysis. American Journal of Science, *261*:897-912.

Davis, Margaret B. and John C. Goodlett. 1960. Comparison of the present vegetation with pollen spectra in surface samples from Brownington Pond, Vermont. Ecology, *41*(2)346-357.

Echlin, Patrick. 1968. Pollen. Scientific American, 218(4):80-90.

Erdtman, G. 1943. An introduction to pollen analysis. The Ronald Press, New York. Pp. 239.

——— 1952. Pollen morphology and plant taxonomy I. Angiosperms. The Chronica Botanica Co., Waltham, Mass. Pp. 539.

——— 1957. Pollen and spore morphology/plant taxonomy II. The Ronald Press, New York. Pp. 151.

——— 1965. Pollen and Spore Morphology/Plant Taxonomy III. Gymnospermae, Bryophyta Text. Almqvist-Wiksell, Stockholm.

——— 1966. A propos de la stratification de l'exine. Pollen et Spores, *8*(1):5-7.

——— 1966. Sporoderm morphology and morphogenesis. A collection of data and suppositions. Grana Palynologica, *6*(3):317-323.

Erdtman, G., B. Berglund, and J. Praglowski. 1961. An introduction to a Scandinavian pollen flora. Vol. I. Almqvist and Wiksell, Stockholm. Pp. 92 plus 74 plates.

Erdtman, G., J. Praglowski, and S. Nilsson. 1963. An Introduction to a Scandinavian pollen flora. Vol. II. Almqvist & Wiksell, Stockholm. Pp. 89 plus 58 plates.

Faegri, K. and Johs Iversen. 1964. Textbook of Pollen Analysis. Munksgaard, Copenhagen. 237 Pp.

Funkhouser, John W. and William R. Evitt. 1959. Preparation techniques for acid-insoluble microfossils. Micropaleontology, *5*(3):369-375.

Graham, Alan. 1962. The role of fungal spores in palynology. Journal of Paleontology, *36*(1):60-69.

Gray, J. 1962. Fossil pollen and archaeology. Archaeology, *15*(1):16-26.

Hanson, H. P. 1949. Pollen contents of moss polsters in relation to forest composition. Amer. Midl. Nat., 42:473.

Harrington, J. B. 1959. High-efficiency pollen samplers for use in clinical allergy. The Journal of Allergy, *30*(4):357-375.

Helmich, Darlene E. 1963. Pollen morphology in the maples (*Acer* L.). Papers Mich. Acad. Sci., Arts, Letters, *48*(1962 meeting): 151-164.

Helsop-Harrison, J. 1962. Origin of Exine. Nature 195:1069.

——— 1968. Pollen wall development. Science, 161:230-237.

Hyde, H. A. and K. F. Adams. 1958. An Atlas of Airborne Pollen Grains. Macmillan & Co., London. Pp. 112.

Irwin, Henry and Barghoorn, Elso S. 1965. Identification of the pollen of maize, teosinte and *Tripsacum* by phase contrast microscopy. Botanical Museum Leaflets, Harvard University, Volume 21, No. 2. Pp. 37-57.

Kosanke, R. M. 1950. Pennsylvanian spores of Illinois and their use in correlation. Ill. Geol. Surv. Bull. 74.

Kremp, G. O. W. 1965. Morphologic Encyclopedia of Palynology. University of Arizona Press, Tucson. Pp. 263.

Kremp, G. O. W. and W. Spackman, editors. Catalog of fossil spores and pollen. The Pennsylvania State University, University Park, Pa. (Periodical).

Kummel, B. and D. Raup, editors. 1965. Handbook of Paleontological Techniques. Freeman, New York. Pp. 852.

Larson, Donald A., John J. Skvarla and C. Willard Lewis, Jr. 1962. An electron microscope study of exine stratification and fine structure. Pollen et Spores. *4*(2):232-246.

Lee, H. 1964. A modified method of coal maceration and a simple technique for slide preparation. Micropaleontology, *10*(4):486-490.

Leopold, Estella B. and Richard A. Scott. 1958. Pollen and spores and their use in geology. Smithsonian Report for 1957. (Publication 4322:303-323.)

Livingstone, D. A. 1955. A light-weight piston sampler for lake deposits. Ecology, *36*(1):137-139.

Maloney, Norma A. 1961. Comparative Morphology of spores of the ferns and fern allies of Minnesota. PhD thesis, Univ. of Minnesota, 1961.

Martin, P. S. 1969. Pollen analysis and the scanning electron microscope. *Scanning Electron Microscopy,* 1969.

Martin, P. S. and C. M. Drew. 1969. Scanning electron photomicrographs of Southwestern pollen grains. *Journal of the Arizona Academy of Science,* 5:170.

Martin, Paul S. and Jane Gray. 1962. Pollen analysis and the Cenozoic. Science, *137*(3524):103-111.

McAndrews, J. H. and A. R. Swanson. 1967. The pore number of periporate pollen with special references to *Chenopodium*. Review of Paleobotany and Palynology, *3*:105-117.

McClymont, J. 1955. Spore studies in the Musci, with special reference to the genus Bruchia. Bryologist, 58(4):287-306.

McDonald, J. E. 1962. Collection and washout of airborne pollens and spores by raindrops. Science, *135*:435.

McVaugh, Rogers. 1935. Studies on the spores of some northeastern ferns. American Fern Journal, *25*(3):73-85.

Meeuse, B. J. D. 1961. The story of pollination. The Ronald Press, New York. Pp. 243.

Nair, P. K. K. 1964. Advances in palynology. National Botanic Gardens, Lucknow, India. Pp. 438.

Ogden, J. G. III. 1960. Recurrence surfaces and pollen stratigraphy of a postglacial raised bog, Kings County, Nova Scotia. American Journal of Science, *258*:341-353.

——— 1965. Pleistocene pollen records from eastern North America. The Botanical Review, *31*(3):481-504.

Oliphant, E. M., Staplin, Pocock and Jansonius. 1960. Palynological techniques for sediments. (Revision of note in Micropaleontology, *6*(3):320-331.)

Pittenger, T. H. and E. F. Frolik. 1951. Temporary mounts for pollen abortion determinations. Stain Technology, *26*(3):181-184.

Potter, L. D. and J. Rowley. 1960. Pollen rain and vegetation, San Augustin Plains, New Mexico. Botanical Gazette, *122*(1):1-25.

Rowley, J. R. 1959. The fine structure of the pollen wall in the Commelinaceae. Grana Palnologica, *2*(1):3-31.

Rowley, J. R. and A. O. Dahl. 1956. Modifications in design and use of the Livingstone piston sampler. Ecology, *37*(4):849-851.

Selling, O. H. 1947. Studies in Hawaiian pollen statistics. Part II. The pollens of the Hawaiian phanerogams. Bishop Museum, Honolulu. Pp. 430 plus 58 plates.

Steeves, Margaret Wolfe and Elso S. Barghoorn. 1959. The pollen of *Ephedra*. Journal of the Arnold Arboretum, *40*(3):221-254.

Stix, Erika. 1960. Pollen-morphologische untersuchungen an compositen. Grana Palynologica, *2*(2):41-104 (21 pl.).

Thanikaimoni, G. 1966. Pollen morphology of the genus Utricularia. Pollen et Spores, *8*(2):265-284.

Ting, William S. 1961. On some pollen of California Umbelliferae. Pollen et Spores, *3*(2):189-199.

——— 1966. Pollen morphology of Onagraceae. Pollen et Spores, *8*(1): 9-36.

Ting, William S., Charles C. Tseng, and Mildred E. Mathias. 1964. A survey of pollen morphology of Hydrocotyloideae (Umbelliferae) (1). Pollen et Spores, *6*(2):479-514.

Tsukada, Matsuo. 1963. Pollen morphology and identification I. Eucaesalpinieae. Pollen et Spores, *5*(2):239-284.

——— 1963. Pollen morphology and identification. II. Cactaceae, Pollen et Spores, *6*(1):45-84.

Tsukada, M. 1967. Chenopod and Amaranth pollen: electron microscopic determination. *Science* 157:81.

Urban, J. B. 1961. Concentration of palynological fossils by heavy liquid flotation. Oklahoma Geology, Notes *21*:191.

Waterman, Ann H. 1960. Pollen grain studies of the Labiatae of Michigan. Webbia, *15*(2):399-415.

Whittmann, Gerda and D. Walker. 1965. Towards simplification in sporoderm description. Pollen et Spores, 7(3):433-456.

Whitehead, Donald R. 1961. A note on silicone oil as a mounting medium for fossil and modern pollen. Ecology, *42*(3):591.

——— 1963. Pollen morphology in the Juglandaceae, I: pollen size and pore number variation. Journal of the Arnold Arboretum, *44*(1): 101-110.

——— 1965. Pollen morphology in the Juglandaceae, II: Survey of the family. Journal of the Arnold Arboretum, *46*(4):369-410.

Wilson, L. R. and G. J. Goodman. 1963. Techniques of Palynology Part I. Collection and preparation of modern spores and pollen. Oklahoma Geology Notes, *23*(7):167-171.

Wilson, L. R. 1949. The correlation of sedimentary rocks by fossil spores and pollen. Journal Sedimentary Petrology, *16*:110.

Wodehouse, R. P. 1935. Pollen Grains. McGraw-Hill, New York (Reprinted, 1959, Hafner, New York). Pp. 574.

——— 1945. Hayfever plants. The Chronica Botanica Co., Waltham, Mass. Pp. 245.

PICTURED KEY TO MAJOR SPORE AND POLLEN GROUPS*

1a Pollen grains and spores solitary and free, not united into groups 2

1b Pollen grains and spores united in groups of two or more..............21

2a Walls with vesiculate (wing-like) extensions (Fig. 11b) or with an irregular warty surface (Fig. 11a)..............
............*Vesiculate* type, page 34

a. Vesiculate—Tsuga type.

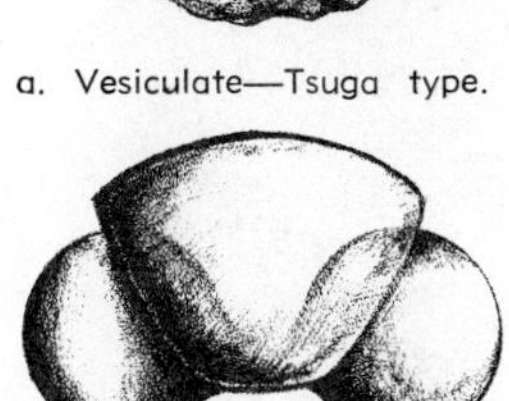

b. Vesiculate—Bisaccate type.

Fig. 11. Vesiculate pollen types, a, Tsuga type; b, Bisaccate type.

2b Walls with various aperture arrangements, but without wings or irregular warty surface.......................3

3a Pollen grains polyplicate, with numerous sharp-crested meridional ridges (Figure 12)......................*Polyplicate* type, page 39

3b Pollen grains without sharp-crested, longitudinal ridges; aperture various, including types with several longitudinal furrows and rounded meridional ridges..4

4a Grains inaperturate or with a trilete scar........................5

4b Grains with various types and numbers of apertures.............6

a. Side view.

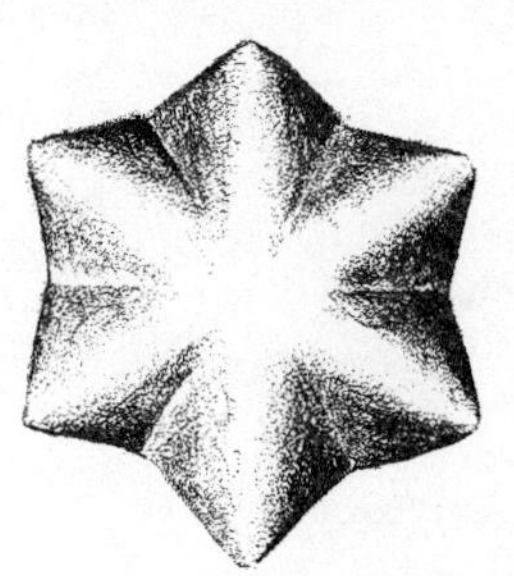

b. End view.

Fig. 12. Polyplicate pollen type. a, Side view; b, End view.

*The interpretation of some characteristics, especially wall structure and sculpture may sometimes vary among different observers. Insofar as possible such types are keyed in several sections of the key. If identification is not achieved, try other sections of the key or examine reference pollen suggested by the manual.

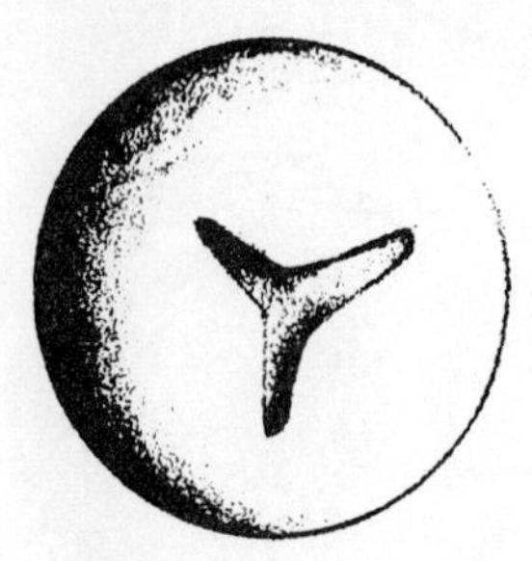

a. Trichotomocolpate pollen type.

b. Trilete spore type.

Fig. 13. a, Trichotomocolpate pollen type with three fused apertures; b, Trilete spore type with tetrad scar.

5a Trilete scar present, on proximal face of spherical or pyramidal spores, or with three furrows joined at one pole (Figure 13)............................*Trilete* or *Trichotomocolpate* type, page 41

Fig. 14. Inaperturate type.

5b Trilete scar absent; spore or pollen grain inaperturate, but sometimes torn open or folded to give appearance or one or more irregular furrows (Figure 14)...*Inaperturate* type, page 52

6a Aperture solitary, either more or less round pores or elongated furrows or scars..........7

6b Apertures two or more, either furrows or pores, or both..8

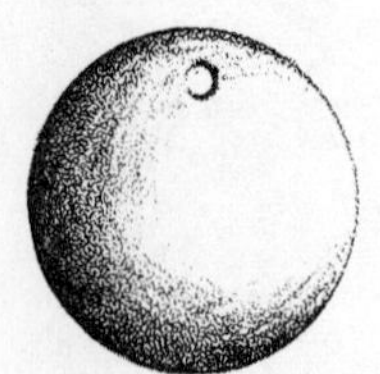

Fig. 15. Monoporate type.

7a Single aperture, a pore, sometimes indistinct (Figure 15)..........*Monoporate* type, page 70

7b Single aperture a furrow, at least twice as long as broad, or a furrowlike scar (Figure 16)...... .....*Monocolpate* and *Monolete* types, page 76

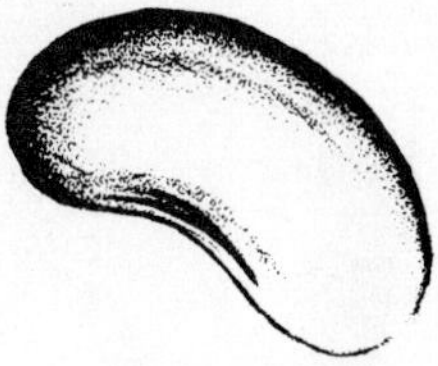

8a Furrows present; pores, if any, associated with the furrows.................................9

8b Furrows absent, all apertures poroid (round or less than twice as long as broad), or grains with ridges enclosing large lacunae..............17

9a Pores or transverse furrows never associated with the furrows.............................10

9b Pores or transverse furrows occur with the furrows 14

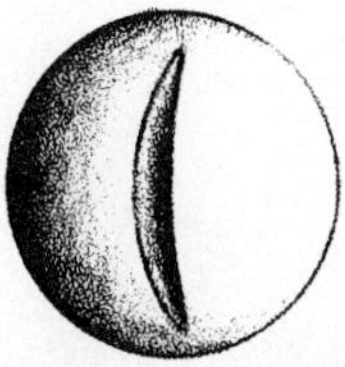

a. Monolete type.

b. Monocolpate type.

Fig. 16. a, Monolete scar, common among fern spores; b, Monocolpate aperture, especially common among monocotyledons.

10a Furrows fused, sometimes forming spiral or irregular aperture (Figure 17).................. ....*Syncolpate* type, page 90

10b Furrows free.............................11

11a Furrows, two or three......................12

11b Furrows, four or more.....................13

a.

b.

Fig. 17. a, furrows fused into a spiral pattern; b, three long furrows, fused at one or both poles.

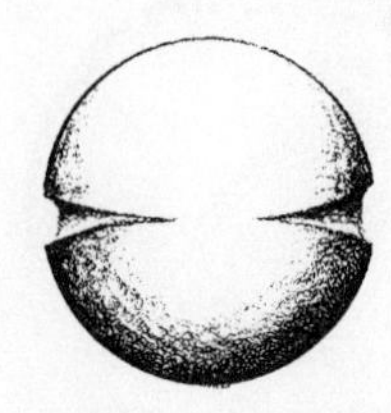

Fig. 18. Dicolpate type.

12a Furrows, two (Figure 18) .... *Dicolpate* types, page 93

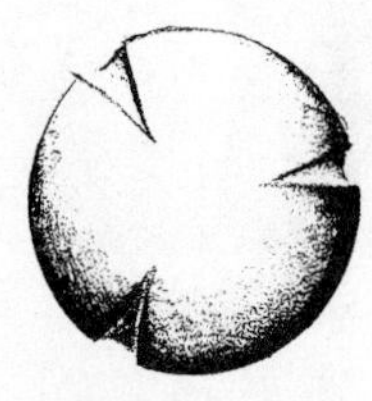

Fig. 19. Tricolpate type.

12b Furrows, three (Figure 19) .... *Tricolpate* types, page 94

Fig. 20. Stephanocolpate type.

13a Arrangement of furrows meridional (Figure 20) .... *Stephanocolpate* types, page 117

Fig. 21. Pericolpate type.

13b. Arrangement of furrows scattered or regular over entire surface of grain, not all meridional (Figure 21) .... *Pericolpate* type, page 121

14a All furrows with associated pores or transverse furrows 15

14b Some furrows with associated pores or transverse furrows; other, pseudocolpi, lacking pores (Figure 22) .... *Heterocolpate* type, page 122

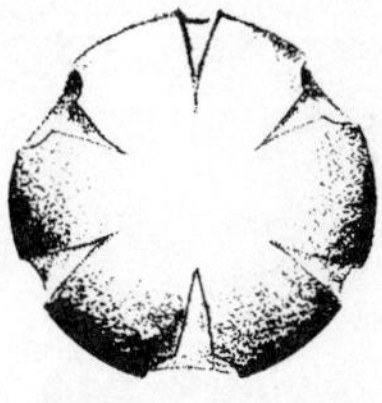

Fig. 22. Heterocolpate type.

15a Colpi, three, with either pores or transverse furrows (Figure 23) .... *Tricolporate* type, page 123

15b Colpi, four or more, with either pores or transverse furrows 16

Fig. 23. Tricolporate type.

16a Apertures centered on the equator of the grain (meridional) (Figure 24) .... *Stephanocolporate* type, page 163

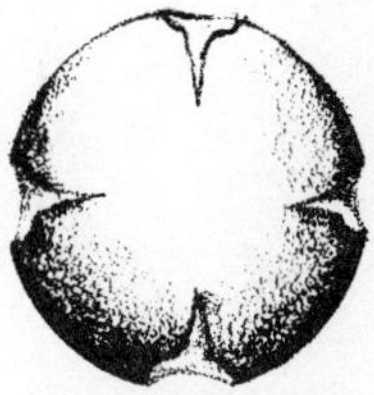

Fig. 24. Stephanocolporate type.

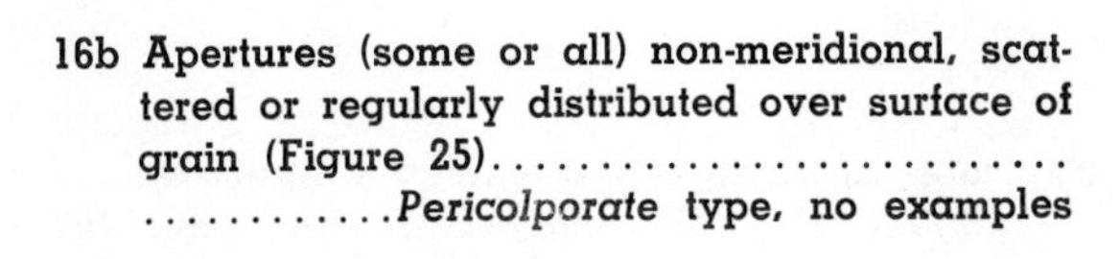

16b Apertures (some or all) non-meridional, scattered or regularly distributed over surface of grain (Figure 25) .... *Pericolporate* type, no examples

Fig. 25. Pericolporate type.

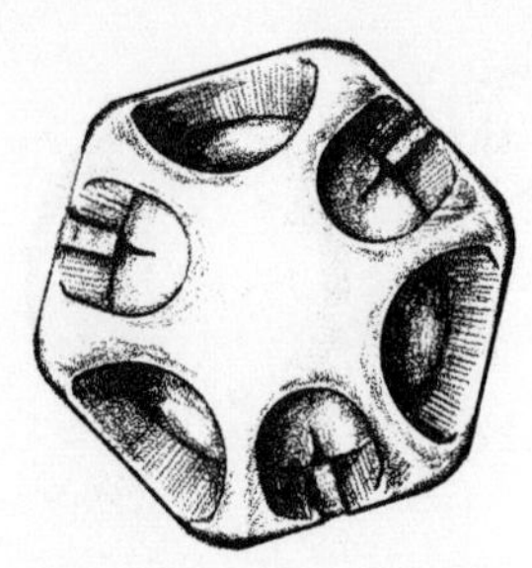

Fig. 26. Lophate and Fenestrate type.

17a Lacunae, surrounded by muri, present, often in complex geometric patterns, some usually containing apertures (Figure 26)................................
Lophate and *Fenestrate* types, page 164

17b Lacunae absent, bearing poroid apertures only..........................18

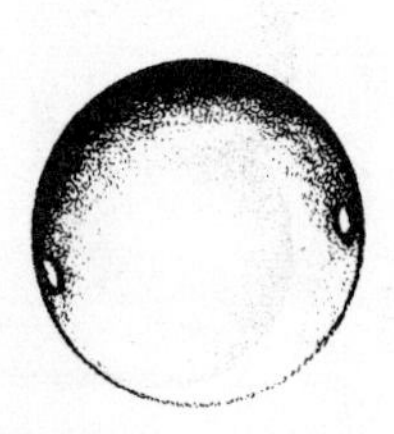

Fig. 27. Diporate type.

18a Pores, two (Figure 27)........................
......................*Diporate* type, page 167

18b Pores, three or more.......................19

Fig. 28. Triporate type.

19a Number of pores, three, usually on the equator (Figure 28)..........*Triporate* type, page 168

19b Number of pores four or more.............20

Fig. 29. Stephanoporate type.

20a Pores restricted to the equator (Figure 29).....
..............*Stephanoporate* type, page 180

20b Pores irregularly or regularly arranged on surface, not restricted to the equator; in heteropolar types pores may be restricted to one hemisphere (Figure 30)........................ ....*Periporate* type, page 185

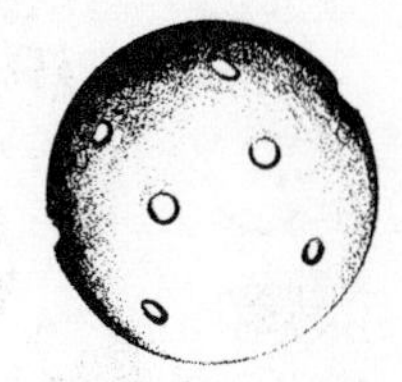

Fig. 30. Periporate type.

21a Pollen or spores united in groups of two (Figure 31) ....*Dyad* type, page 196

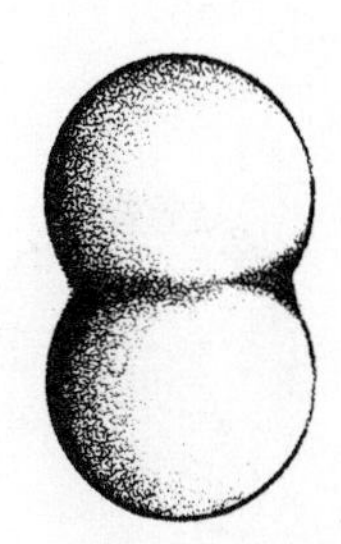

21b Pollen or spores united in groups of four or more 22

Fig. 31. Dyad type.

22a Spores in aggregations of four spores or pollen grains (sometimes three after detachment of one grain, or apparently three because one is obscured by others) (Figure 32)......*Tetrad* type, page 197

Fig. 32. Tetrad type.

22b Aggregations of more than four spores or pollen grains; geometrically regular or irregular (includes pollinia and certain multicellular aggregations which may be confused with pollen)..........*Polyad* type, page 203

Fig. 33. Polyad type.

PALYNOMORPHS WITH UNUSUAL SYMMETRY AND ENTITIES WHICH MIGHT BE CONFUSED WITH POLLEN

Palynologists frequently encounter organisms or parts of organisms which are in the same size range as palynomorphs and which may be confusing. Many of these are tests, or cysts, of Protozoans; they may also be diatom remains. Furthermore, some pollen and spore types have such unusual symmetry that their relationship may be uncertain. The following illustrations include examples of bodies of these types. These are generally resistant to destruction by desiccation and/or decay and may be encountered in air or sediment samples.

Unusual Palynomorphs

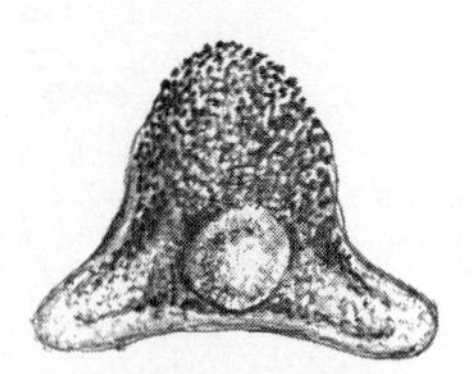

Fig. 34. *Aquillapollenites pulvinus* Stanley (Pollen et Spores, 3(2):327-352 Coll. Cretaceous sediments of Montana. Relationship unknown. Redrawn from: Stanley, E. A. 1961. The fossil pollen genus *Aquillapollenites. Pollen et Spores,* Vol. III(2):331 (Pl. 1).

Aquillapollenites

SIZE: 32-38μ, Equatorial: 16-20μ.

RANGE: Upper Cretaceous of Rocky Mts. and Siberia.

The broadly rectangular body has three equatorial, radially-arranged, wing-like projections. A fourth wing lies at right angles to the other three and forms the polar axis. The exine of the body and "dorsal" wing are structurally and sculpturally different from the equatorial wings.

The genus *Aquillapollenites* has proved to be an important plant microfossil in the upper Cretaceous rocks of the Rocky Mountains and adjacent plains of the United States and southern Canada.

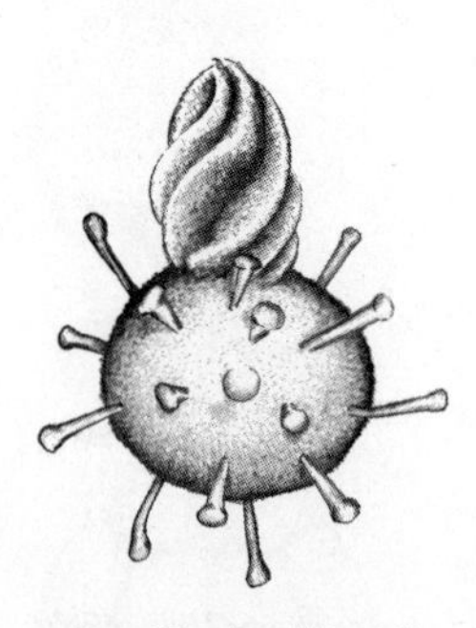

Fig. 35. *Arcellites disciformis* Miner Coll. Cretaceous sediments of Iowa. Megaspore of the HYDROPTERIDACEAE (fern). Redrawn from: Hall, J. W. 1967. Cover photograph. *Bioscience,* Vol. 17(8).

Arcellites

SIZE: Spore about 230μ (body appendages excluded), overall length about 530μ.

RANGE: Cretaceous Dakota formation Central United States.

Originally thought to be a freshwater rhizopod, *Arcellites* (and also *Molasporites* and *Balmeisporites*) are believed to be the megaspores of aquatic hydropterid ferns. The six-winged, twisted appendage apparently served as a float for the spore; it may have been a receptive surface on which microspores lodged and germinated.

Algae

Pediastrum

SIZE: Colonies vary in size, up to about 125μ diameter.

RANGE: The genus is worldwide in distribution and ranges in age from Cretaceous (or earlier?) to the present.

These colonial green algae are distinguished by disc-shaped colonies composed of a variable number of cells (usually 8-64). The cells are essentially concentric in arrangement, but the outer row usually differs from the inner ones by bearing 1-3 projections. The cells may be closely packed or may have intervening interstices. The fossil specimen illustrated is similar to *P. boryanum*, a modern species in North America.

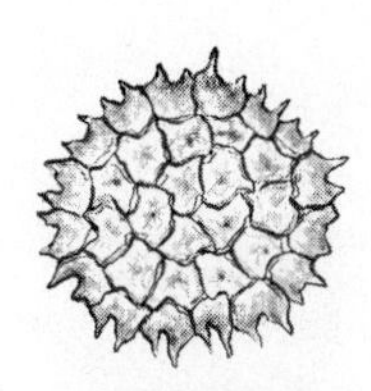

Fig. 36. *Pediastrum* sp. Coll. Tertiary of California. Chlorophyta, HYDRODICTYACEAE. Redrawn from: Wilson, L. R. and W. S. Hoffmeister. 1953. Four new species of fossil *Pediastrum*. *Amer. Jour. Sci.*, Vol. 251:753-760 (Fig. 1).

Fossil *Pediastrum*

SIZE: 69 to 155μ, colony diameter.

The accompanying drawing of a fossil specimen is included because of the similarity of the colony to modern *P. clathratum* and to *P. simplex* var. *duodenarium*. This type of colony has large interstices among the cells.

Fig. 37. *Pediastrum kajartes*. Coll. Cretaceous of Sumatra. Chlorophyta, HYDRODICTYACEAE. Redrawn from: Evitt, W. R. 1963. Occurrence of freshwater alga *Pediastrum* in Cretaceous marine sediments. *Amer. Jour. Sci.*, Vol. 261:890-893 (Fig. 4).

Diatoms

Diatoms are in the golden algae, Chrysophyta. The siliceous walls of diatoms are very resistant to degradation and might be classed with pollen by the novice. The walls (frustules) are of one of two symmetry patterns. The order Centrales (Example *Coscinodiscus lacustris* Grun., Fig. 38) is radial in symmetry (valve view) and pill-box-shaped. The order Pennales includes diatoms in which the shells have basically bilateral symmetry in valve view. This is a very common fresh-water form; two species of the genus *Navicula* (Fig. 39) are illustrated as examples of this type. (Illustrations from Prescott, G., 1964. How to Know the Fresh-Water Algae, Figures 391 and 443.)

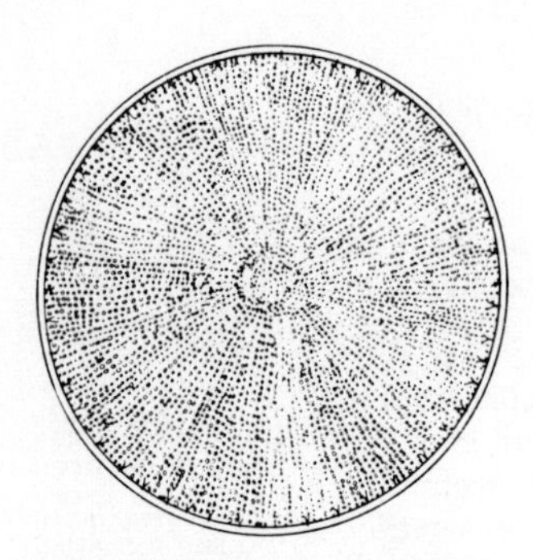

Fig. 38. *Coscinodiscus lacustris* Grun. From Prescott, G. W. *How to Know Freshwater Algae*, Figs. 391 and 443. © H. E. Jaques, 1954. Used by permission of the Wm. C. Brown Company Publishers.

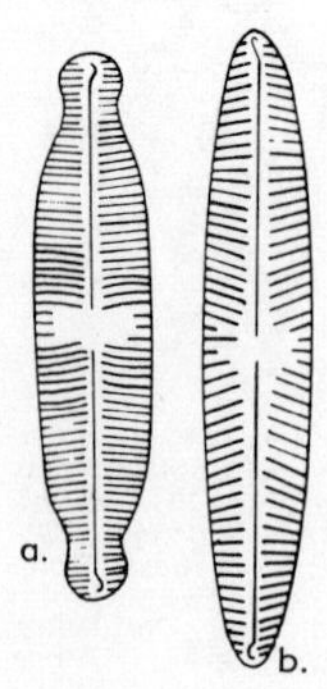

Fig. 39. a. *Navicula Petersenii* Hustedt. b. *Navicula digitoradiata* forma *minor* Foged. Valve view. From Prescott, G. W. *How to Know Freshwater Algae*, Figs. 391 and 443. © H. E. Jaques, 1954. Used by permission of the Wm. C. Brown Company Publishers.

Protozoa

Dinoflagellate

SIZE: 50μ

RANGE: Marine, coastal Pacific.

The thecae of dinoflagellates are composed of cellulose; they do not fossilize, but may be encountered with pollen in air or modern sediment samples. In life, one transverse flagellum extends around the girdle and a second longitudinal flagellum extends down the vertical sulcus. Such armored dinoflagellates are almost ubiquitous in aquatic sediments and frequently produce resistant cysts or tests which may be preserved with fossil pollen. The paleoecology of the group is poorly known.

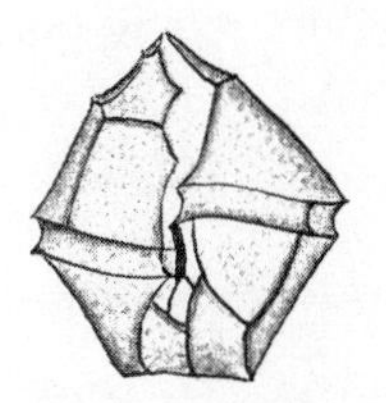

Fig. 40. *Gonyaulax polyhedra* Stein. Coll. in U.S. DINOFLAGELLATA. Redrawn from: Evitt, W. R. 1964. Dinoflagellates and their use in petroleum geology. Pp. 65-72 In: *Palynology in Oil Exploration*, Soc. of Economic Paleontologists and Mineralogists (Text-figure 2).

Hystrichosphere

SIZE: Body about 45μ, 85μ including projections.

RANGE: Unknown.

Hystrichospheres have recently been shown to be cysts of dinoflagellates. Some of these cysts or tests rather faithfully reflect the pattern of plates of the motile dinoflagellate theca, others are spiny spheres with the expanded or forked spine tips corresponding to the pattern of the thecal plates. Frequently the hystrichosphere has an operculum which itself bears a few spines; it may be detached from the test and be found alone.

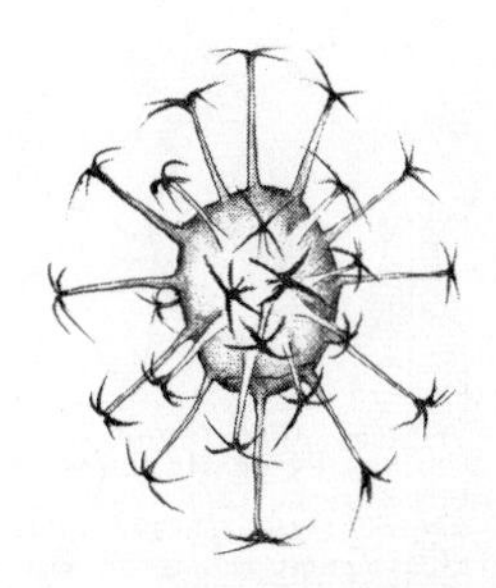

Fig. 41. *Hystrichosphaerium recurvatum* (White). Coll. Upper Cretaceous; DINOFLAGELLATA (cyst). Ibid. (Plate 1, Fig. 11).

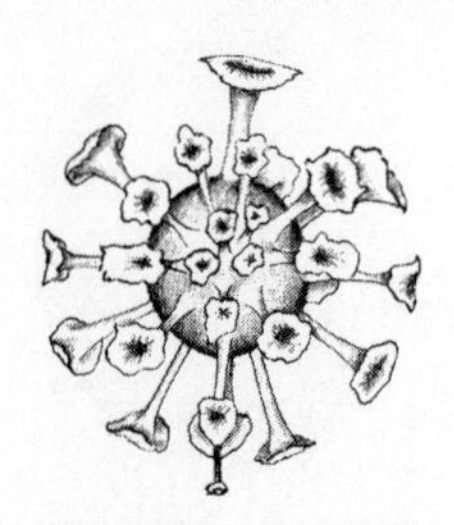

Fig. 42. *Hystrichosphaeridium tubiferum* (Ehrbg.). Coll. Redbank Formation, Teritary, New Jersey; DINOFLAGELLATA (cyst). Ibid. (Plate 1, Fig. 17).

Hystrichosphere

SIZE: Body about 35μ, 80μ including projections.

RANGE: Unknown.

The cyst or test of this dinoflagellate has conically-expanded tips on the processes. Dinoflagellate fossil remains are known in the fossil record since Permian time.

Fig. 43. *Globorotalia menardii multicamerata* Cushman and Jarvis. Coll. Atlantic Ocean Core; FORAMINIFERIDA. Redrawn from: Ericson, D. B., M. Ewing, and G. Wollin. 1963. Cover photograph to accompany: Pliocene-Pleistocene Boundary in deep-sea sediments. *Science*, Vol. 139. © American Association for the Advancement of Science; used by permission.

Foraminifera

SIZE. About 800μ.

RANGE: Ocean waters. (The species illustrated is Caribbean and Atlantic, Miocene to Pleistocene.)

There are about 30 species of marine planktonic Foraminifera which secrete tests of calcium carbonate. Upon death these settle to ocean floor and are incorporated into the sediments. Such remains are stratigraphically important. Recently certain species have been shown to reflect temperature changes by reversal of the direction of shell coiling as the ocean waters cooled at the beginning of the Pleistocene. A key for identification of species of Foraminifera in the families Globigerinidae and Globoratalíidae was published in 1967 by the Conseil Permanent International Pour L'Exploration de la Mer, Charlottenlund Slot, Denmark.

Testaceous Rhizopod

SIZE: 90μ diameter.

RANGE: Unknown; habitat: moist rapidly growing peat moss (*Sphagnum*).

The tests of sphagnicolous rhizopods may be recovered with pollen from peat bog cores. They are generally well preserved and may in some cases be identified to species.

Fig. 44. *Amphitrema wrightianum*. Coll. Massachusetts (postglacial); SARCODINA (testaceous rhizopod).

Arcella

SIZE: Various species range from 30μ to nearly 150μ in maximum dimension.

The genus *Arcella* secretes a smooth chinous test which is usually hemispherical with a single opening through which the amoeboid animal emerged. Some species, however, have marginal spines on the test. *A. artocera* lives in growing peat moss and may therefore be encountered in bog cores with pollen and spores. The testaceous rhizopod *Difflugia* differs in that the test appears to be constructed of grains of sand.

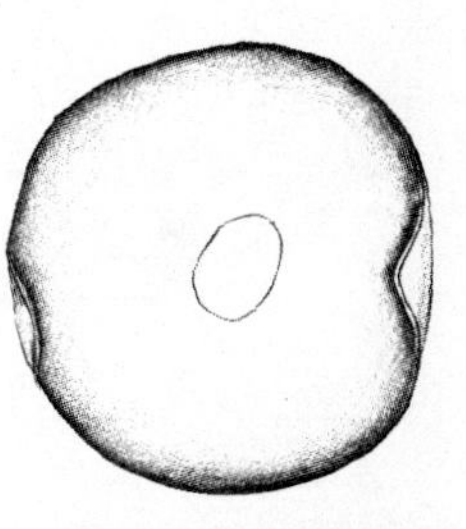

Fig. 45. *Arcella artocera*. Coll. Massachusetts (postglacial); SARCODINA (testaceous rhizopod).

Testaceous Rhizopod

SIZE: 42μ x 160μ.

RANGE: Unknown; habitat: moist *Sphagnum*.

The cylindrical test is constricted near the middle. Unlike pollen and spores, rhizopod tests lack wall stratification.

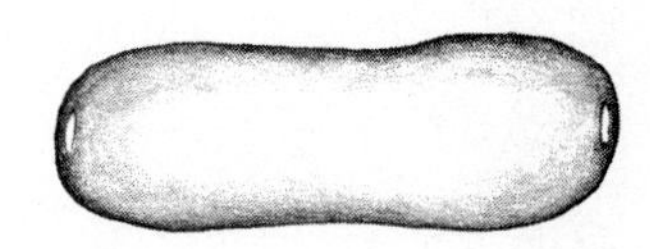

Fig. 46. *Ditrema flavum*. Coll. *Massachusetts* (postglacial); SARCODINA (testaceous rhizopod).

VESICULATE TYPES

In North America, pollen of this type is restricted to the genus *Podocarpus* and many, but not all, genera in the PINACEAE. Most vesiculate types are winged (bisaccate as in pine, spruce, fir); hemlock (*Tsuga*) pollen is included in this group because it has a vesiculate "frill" surrounding the grain.

1a Pollen grains with large blunt and undulating projections irregularly covering the surface of the pollen grain, sometimes appearing as a "fringe" bordering a large depression; greater than 40μ diameter (Fig. 47)...*Tsuga*

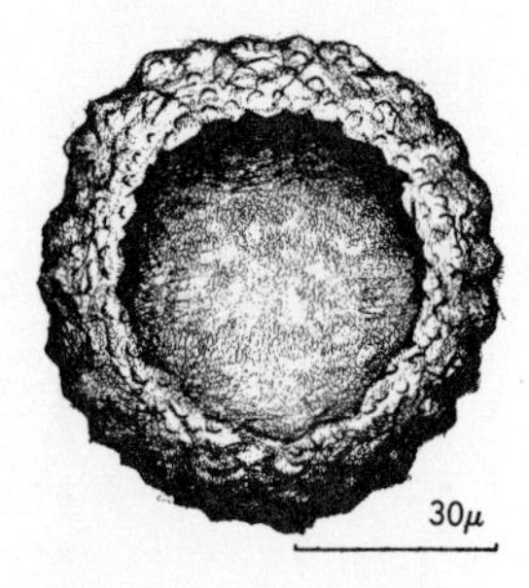

Fig. 47. *Tsuga canadensis* (L.) Carr. Coll. Michigan; PINACEAE.

Eastern Hemlock

SIZE: Diameter approximately 70μ to 80μ.

RANGE: Southeastern Canada, northeastern U.S. and Appalachian Mts.

Grain essentially spherical in polar view and discoid in lateral view with a large circular indentation usually apparent on the distal face. Ektexine separated from inner wall layers and developed into a covering of convolutions. Detailed electron microscopic studies show minute spines on the exine of most species of *Tsuga*. *Tsuga mertensiana* (Mountain Hemlock of Pacific Northwest and Alaska) has small wings and unlike other species of *Tsuga* lacks fine surface spinules.

1b Pollen grains with two large wings (bladders)..................2

2a Wings attached to body (cap) of grain along an indistinct transitional zone. Exine of body with a granular surface which changes gradually to a reticulum on the bladders (Fig. 48 and 49) .... ***Picea***

White Spruce

SIZE: Body 80μ long, 60μ wide; overall length 125μ.

RANGE: Canada, Alaska, Northeastern U.S. and Black Hills of S. Dakota and Wyoming.

Bladder reticulum becomes finer close to the body. The thick exine of the cap has a fine and uniform verrucate texture. Size frequency analysis shows that among eastern North American species, *Picea rubra* and *P. glauca* are consistently larger than *P. mariana*. The size ranges overlap, however, making identification of individual grains difficult.

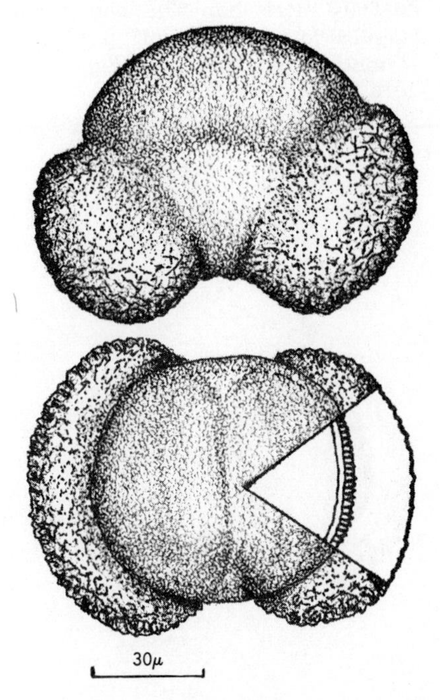

Fig. 48. *Picea glauca* (Moench) Voss Coll. Michigan; PINACEAE.

Black Spruce

SIZE: Body about 65μ long, about 50μ wide; overall length 85μ.

RANGE: Alaska, Canada and northeastern U.S.

Size frequency measurements demonstrate that Black Spruce produces the smallest pollen of the eastern North American spruce species. Its pollen, like other spruces, tends to fold when dry; the distal face, which bears an indistinct aperture, becomes very concave and the bladders close together.

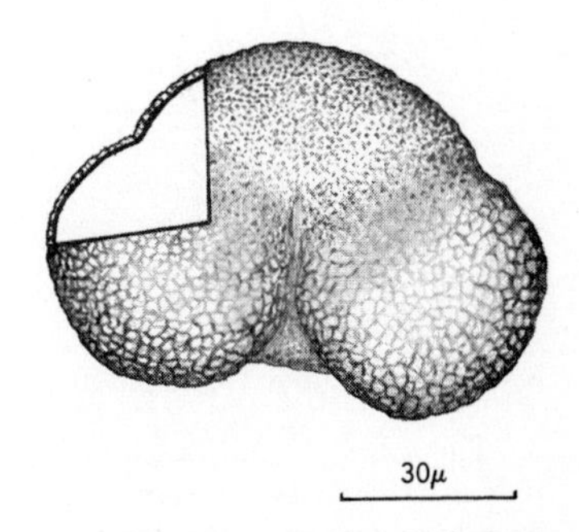

Fig. 49. *Picea mariana* (Mill) BSP.

2b **Wings loosely attached to the body of the grain, the line of attachment distinct and usually forming an indentation or distinct angle. Bladder reticulum begins abruptly at point of attachment to body. . 3**

3a **Body more than 90μ long, exine of body with a faint rugulate or substriate "fingerprint" pattern, exine thick, but with distinct thinning at proximal pole (Fig. 50 and 51). . . . . *Abies***

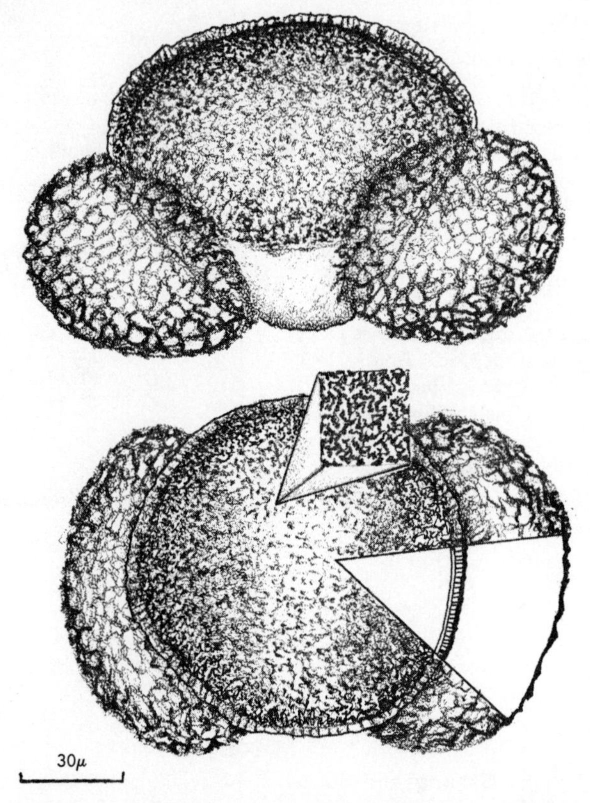

Fig. 50. *Abies balsamea* (L.) Mill Coll.; Arn. Arb.; PINACEAE

Balsam Fir

SIZE: Body (90-) 125μ long, 110μ wide, 95μ high; overall length up to 190μ.

RANGE: Eastern Canada and northeastern U.S.

The exine of the body of fir pollen is very coarse in texture and thick. It has a swirled, finely rugulate surface pattern which is distinctive. The bladders appear to be loosely attached to the body.

Subalpine Fir

SIZE: Body 105 to 110μ long; overall length 160 to 165μ.

RANGE: Southeastern Alaska and central Yukon and Rocky Mountains to Arizona and New Mexico.

It will probably be very difficult to separate pollen of the several western Firs on the basis of size. Detailed morphological studies are not available. Subalpine Fir pollen has a very thick exine along the top (proximal side) of the body. Note the thinning of the exine at the proximal pole, Fig. 51, a distinctive characteristic of *Abies* pollen.

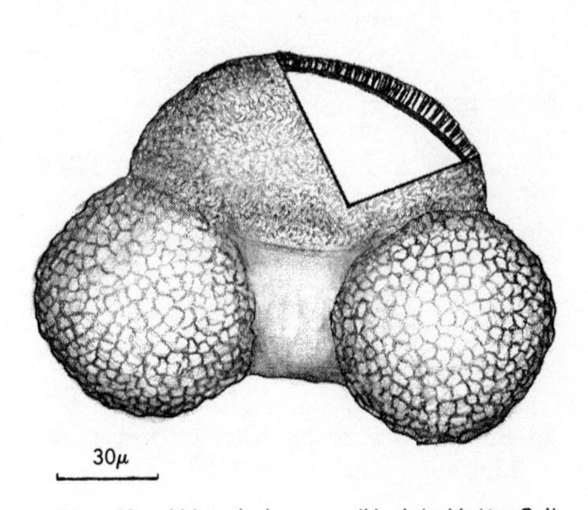

Fig. 51. *Abies lasiocarpa* (Hook.) Nutt. Coll. Arizona; PINACEAE.

3b Body less than 65μ long, exine of body verrucate or scabrate with granular appearance..4

4a Body usually less than 40μ long, width often only 1/2 or 2/3 the width of bladders. (*Podocarpus dacrydioides* is larger, 50μ, and has three bladders.) (Fig. 52 and 53)..................*Podocarpus*

Reich's Podocarp

SIZE: Body 26.5μ x 32μ; overall dimensions 32μ x 56μ.

RANGE: Mexico.

Exine surface granular or indistinctly verrucate on the distal and lateral faces of body. Bladders coarsely reticulate with indistinctly parallel ridges on the distal faces. The furrow on the distal wall of the body is often distinct.

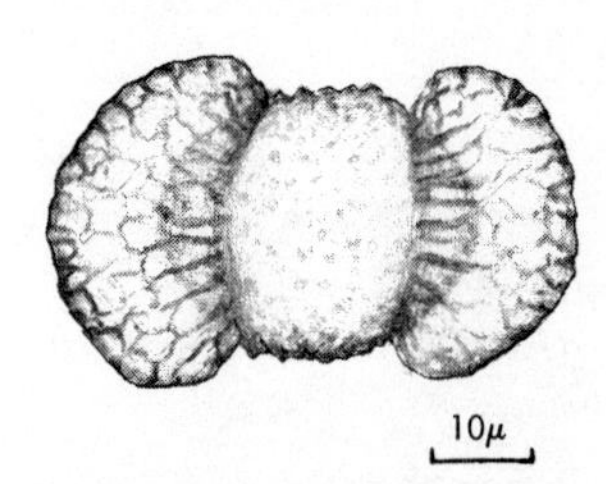

Fig. 52. *Podocarpus reichii* Coll. Mexico; PODOCARPACEAE.

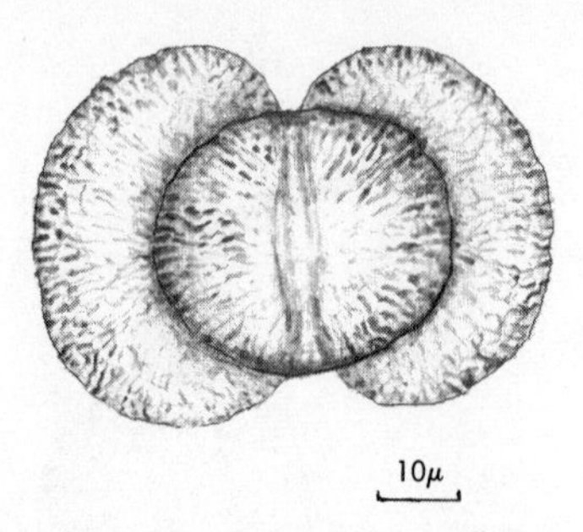

Fig. 53. *Podocarpites radiatus* Brenner Coll. Maryland; Cretaceous.

Fossil Podocarp

SIZE: Overall length about 65μ, body width about 25μ.

RANGE: Fossil from Potomac Group, Cretaceous of Maryland.

Members of this fossil genus are apparently ancestral to modern *Podocarpus*. *P. radiatus* has ridges on the bladders which radiate from the line of distal attachment.

4b Body more than 40μ long, typically as wide or wider than the bladders (Fig. 54 and 55) .... *Pinus*

Size frequency statistics may be of some value in identification of pine pollen if large populations of grains are available and measurements can be made of several dimensions. Variations in both bladder morphology and size, and overlapping characteristics, make separation at the species level difficult.

The aperture area (leptoma) at the distal pole of the pollen grains of many species of the subgenus Haploxylon is "studded" with small, circular, slightly thickened flecks (Erdtman, 1965). These "belly warts" (see Fig. 55 have been helpful in separating *Pinus strobus* pollen from other species in the postglacial sediments of the northeastern United States. Such morphological characteristics are more promising than size statistics for species determination.

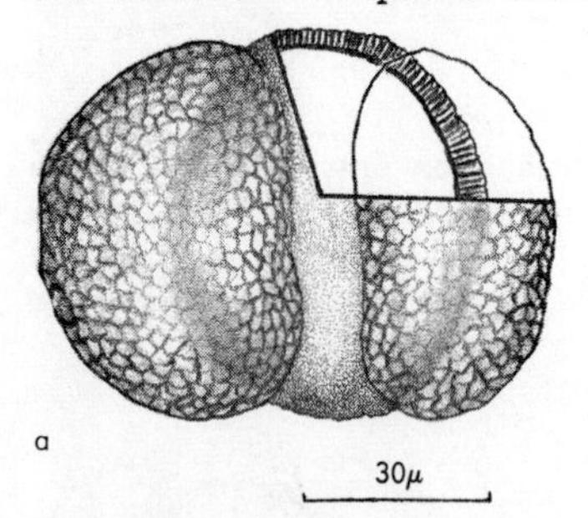

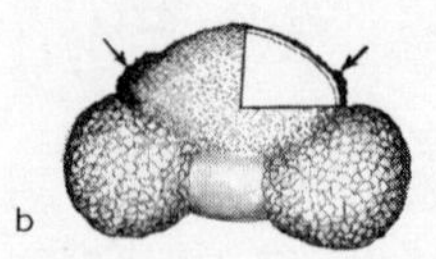

Fig. 54. *Pinus resinosa* Ait. a, distal view; b, lateral view showing exine crest. Coll. Illinois; PINACEAE.

Red (Norway) Pine

SIZE: Body width 35 to 40μ, length 40 to 48μ; overall, 80μ.

RANGE: Northeastern U.S. and southeastern Canada.

Size frequency statistics indicate that Red Pine is intermediate in size between Jack Pine (*P. banksiana*) and White Pine (*P. strobus*), but is most similar to the smaller Jack Pine. Red Pine frequently has an exine crest or "scallop" on the body adjacent to the point of bladder attachment; this is visible in side view.

Pinyon Pine

SIZE: Body 35μ wide, length about 50μ, overall 70μ.

RANGE: New Mexico, northeastern Arizona, southeastern Utah, and southwestern Colorado.

Pinyon Pine pollen overlaps in size with *Pinus contorta, P. monophylla,* and *P. flexilis;* these are markedly smaller than *P. ponderosa.* The specimens examined show a very distinct lateral line between the proximal and distal polar areas. Between the bladders (distal face) the exine of the body is thinner than on the proximal face and ornamented with fleck-like verrucae.

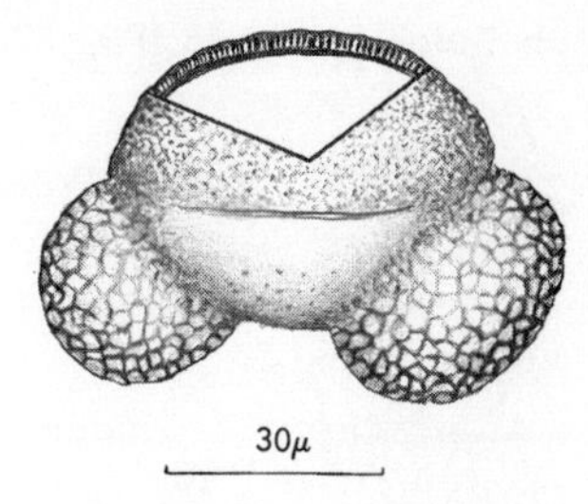

Fig. 55. *Pinus edulis* Engelm. Coll. Colorado; PINACEAE.

POLYPLICATE TYPES

Pollen of this type is restricted to the Gymnosperm genera *Welwitschia* (southwest Africa) and *Ephedra,* in the Gentales. *Ephedra,* the Mormon Tea, is represented by several species in the southwestern United States; its distribution also includes arid regions of South America and the Old World. Pollen of *Ephedra* has also been found in geologic sediments of Cretaceous and Tertiary age, but it is reported to be still more ancient. Steeves and Barghoorn (1959) present an analysis of *Ephedra* pollen in which four morphological types are recognized. Only two basic types are distinguished in this key, those with branched and those with straight furrows.

1a Furrows branched (Fig. 56 and 57) (*Ephedra nevadensis* type)....2

1b Furrows straight or obscure, not branched (Fig. 58 and 59) (*Ephedra torreyana type*)....................................3

2a Ridges 5 to 9 (Fig. 56)............................*Ephedra viridis*

SIZE: Polar axis 61μ; equatorial 34μ.

RANGE: Southwest Colorado, Utah, Nevada, Arizona and California.

The grains have straight, smooth ridges with thin crests. The ridges are broad-triangular in cross section having a broad rounded crest about 3.5μ wide. *Ephedra viridis* may be indistinguishable from *E. cloyeyi, E. coryi,* and *E. funera* which are also found in southwestern United States.

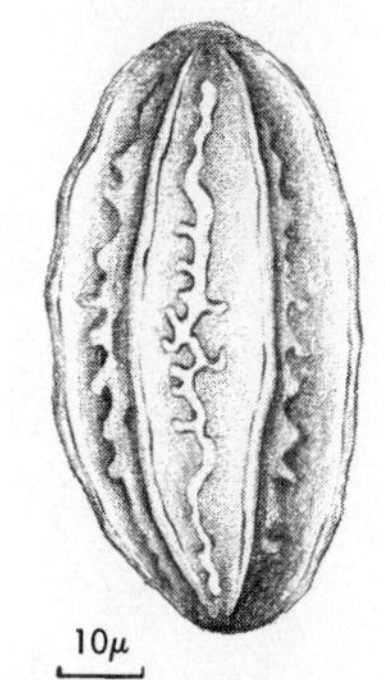

Fig. 56. *Ephedra viridis* Cov. Coll. Calif.; EPHEDRACEAE.

2b Ridges 11 to 15 (Fig. 57)........................*Ephedra nevadensis*

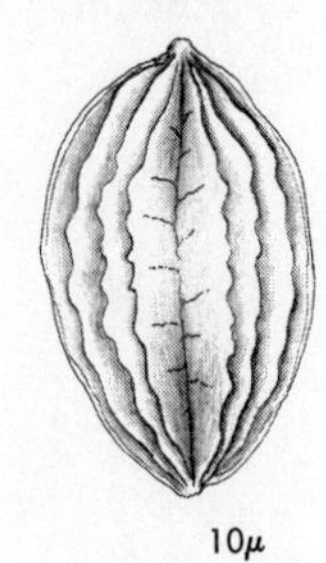

Fig. 57. *Ephedra nevadensis* S. Wats. Coll. Arizona; EPHEDRACEAE.

SIZE: Polar axis about 35μ, equatorial diameter 22μ.

RANGE: Oregon and California to Utah and Arizona.

Ridges vary from 11 to 15 in number (average 13). The colpi in the centers of the furrows are undulate. The ridge crests are thin (0.5μ) and either straight or undulate.

3a In polar view, ridges low, gently-rounded, or semi-circular; ridges straight, 11 to 17 in number (Fig. 58).........*Ephedra trifurca* Torr.

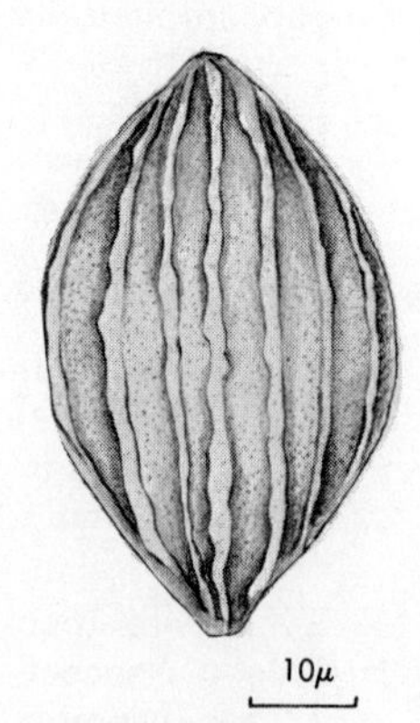

Fig. 58. Ephedra *trifurca* Torr. Coll. New Mexico; EPHEDRACEAE.

SIZE: Polar axis 53μ, equatorial diameter 29μ.

RANGE: Southwestern Texas to southern California and adjacent Mexico.

This species should be distinguishable from all other native species. The ridges are gently rounded (in cross section), about 2.8 to 5μ wide, and average 13.7 in number.

3b In polar view, ridges high and triangular, narrowing to a sharp crest (Fig. 59) .... *Ephedra antisyphilitica* Berland.

SIZE: Polar axis 59μ, equatorial diameter 43μ.

RANGE: Arid regions of western Texas to southern Colorado.

Pollen of this species is among the largest of the *Ephedra* pollen of the Torreyana type. The ridges are highly undulate to nearly straight. In southwestern North America its pollen will be most frequently confused with that of *E. californica* and *E. pedunculata.*

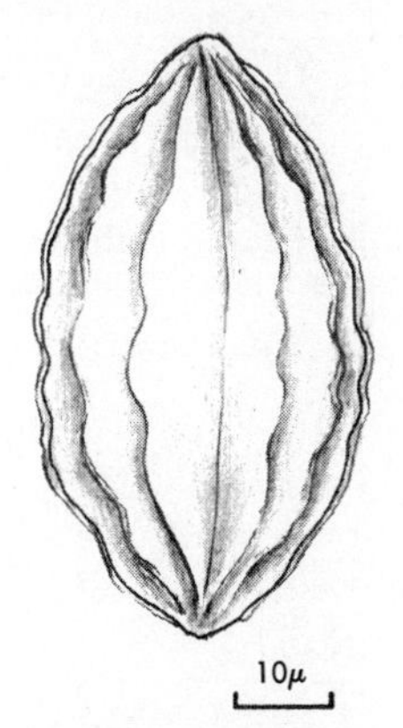

Fig. 59. *Ephedra antisyphilitica* Berland. Coll. Texas; EPHEDRACEAE.

TRILETE SPORES AND TRICHOTOMOCOLPATE POLLEN

Pollen of the trichotomocolpate type (three furrows joined at one pole) is encountered in the coconut, *Cocos nucifera.* It may therefore be collected occasionally in subtropical North America where coconut has been introduced.

Trilete spores are produced by a few mosses, some of the ferns, and many of the lycopsids (Microphyllophyta). These groups have a long fossil history and fossil spores from these groups and their extinct relatives have been recovered from sediments as old as the Devonian age (400 million years ago). In some places coal seams may be composed primarily of carbonized remains of megaspores! The wall is sometimes termed perine to emphasize structural distinctiveness from the pollen exine; the terms exine and perine are used interchangeably here.

The heterosporous vascular plants which do not produce seeds, usually shed both the large megaspores and small microspores, the latter in much larger numbers. The characteristics of the megaspore and microspore from a particular species may be either rather similar or very different in exine morphology.

This key includes examples of trilete spores of *Sphagnum* (peat moss), *Lycopodium* (club moss) and the fern families Ophioglossaceae, Osmundaceae, and Polypodiaceae (all homosporous). Heterosporous groups included are from the Selaginellaceae and Isoetaceae. The several fossil spores included only provide a small sample of the diversity of the trilete spores of these extinct forms. In many cases these fossil spores are from plants whose modern relationships are unknown; many are from completely extinct groups.

1a Spores very large, longest axis greater than 250μ, sometimes 1000μ or larger (megaspores)..2

1b Spores smaller, usually 75μ or less (microspores)...............4

2a Spines present, straight or curved (Fig. 60)................*Isoetes*

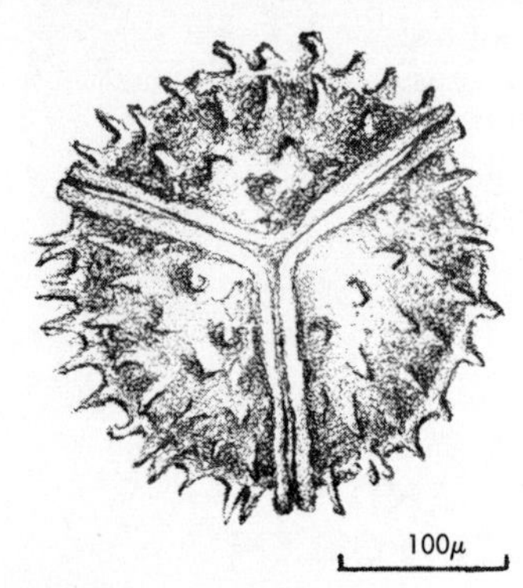

Fig. 60. *Isoetes echinospora* Dur.; var. *braunii* (Dur.) Engelm. ISOETACEAE (megaspore).

Quillwort

SIZE: Variable 250-750μ.

RANGE: Greenland to Alaska, south to New England, Utah, California.

Prominent trilete ridge (crest) on proximal face. Apiculate, curved, or blunt spines throughout. Megaspores of this species differ from other species of quillwort in which the megaspores may be reticulate or nearly smooth (see Fernald, 1950, Gray's Manual of Botany, 8th edition, for drawings).

2b Spines absent...3

3a Triradiate crest prominent extending to equatorial rim (Fig. 61).... ....*Triletes triangulatis Zerndt* (Fossil)

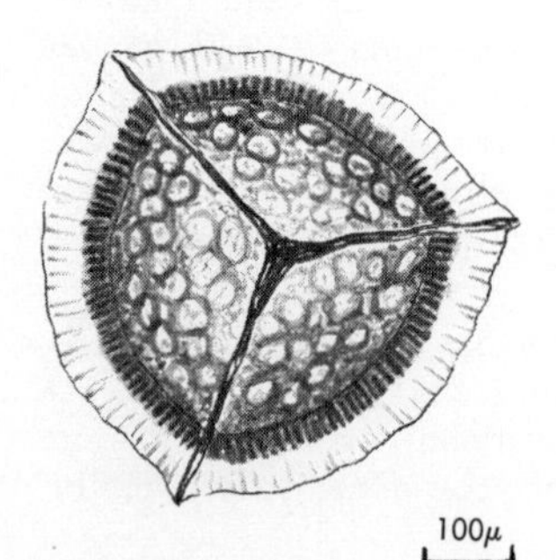

Fig. 61. *Triletes triangulatis* Zerndt. Redrawn from: Arnold, C. A. 1950. Megaspores from the Michigan coal basin. *Contrib. Mus. Paleontol. Univ. Michigan,* Vol. VIII(5):59-111 (Plate VI, Fig. 1).

SIZE: 500 to 600μ.

RANGE: Pennsylvanian age strata, common at several Michigan localities.

The long dark triradiate scar or crest extends to the equatorial limb. Margin of the spore has a "frill" with radial structure (exosporium?) and the endospore wall apparently has circular thin spots of varying size.

3b Triradiate scar extends only one-third distance from proximal pole to equator (Fig. 62)............ *Triletes mamillaris* Bartlett (Fossil)

SIZE: From less than 1.5 mm to slightly more than 2 mm.

RANGE: Frequently recovered from Pennsylvanian strata in Michigan.

Trilete scare with irregular marginal thickenings. Outline of spore round to flattened or subtriangular. Exine often folded and studded with large gemmae.

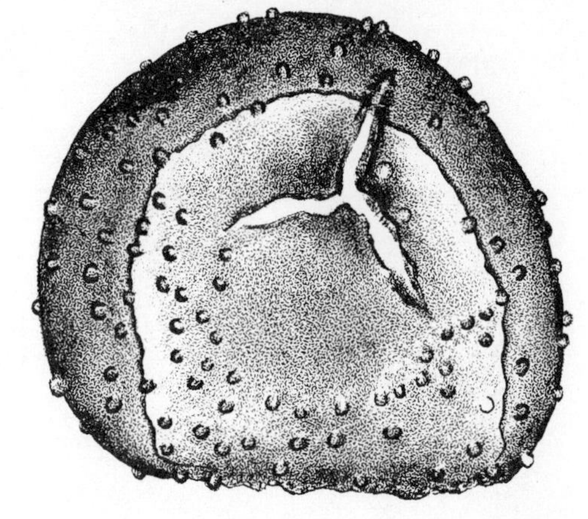

Fig. 62. *Triletes mamillaris* Bartlett. Ibid. (Plate VII, Fig. 1).

4a Exine psilate and smooth, sometimes folded....................5

4b Exine spiny, reticulate, pitted, or roughened..................8

5a Trilete scar tapering to pointed tips, often more or less open.....6

5b Trilete scar slender throughout.............................7

6a Age: Cretaceous fossil (Fig. 63)...*Cyathidites minor* Cooper (Fossil)

SIZE: (23-) 38 (-52)μ

RANGE: Lower to Upper Cretaceous

Subtriangular with rounded angles. Scar open, exine with two distinct layers. The concave sides and well rounded corners distinguish the genus from others. Relationship uncertain, but apparently a leptosporangiate fern, probably in the Cyatheaceae.

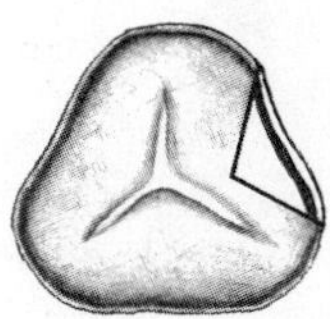

Fig. 63. *Cyathidites minor* Cooper. Redrawn from Oltz, 1968, Cretaceous of Montana.

6b Age: Recent or fossil (Fig. 64)........................*Sphagnum* sp.

Peat Moss

SIZE: About 30μ

RANGE: North Temperate and Boreal areas

Smooth subtriangular spore wall (a) sometimes with a distinct exosporium and endosporium (b). In some species the arms of the trilete scar may only extend one-third the distance from proximal pole to the equator.

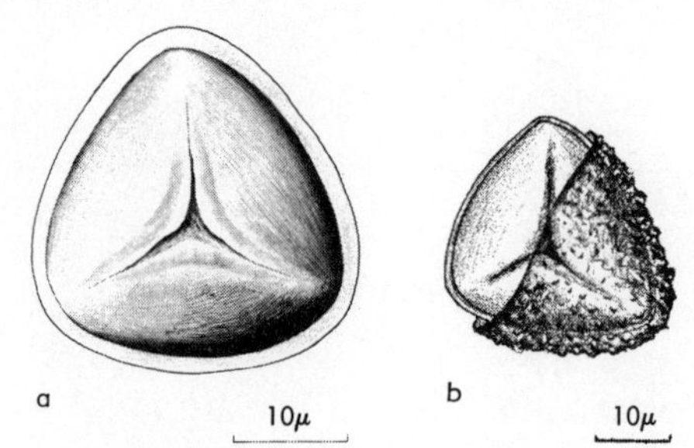

Fig. 64. *Sphagnum* sp. Coll. Michigan. SPHAGNACEAE.

7a Surface pattern appears to have short folds of various lengths and with irregular orientation (Fig. 65)................................ ....*Hamulatisporis hamulatus* Krutzsch (Fossil)

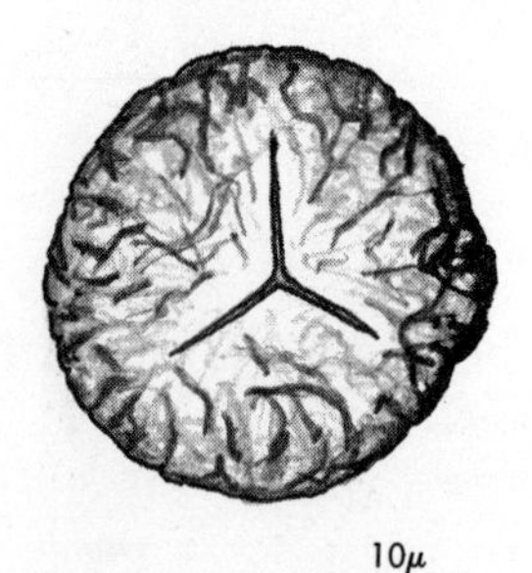

Fig. 65. *Hamulatisporis hamulatus* Krutzsch. Redrawn from Stanley, 1965 (Bull. of Amer. Palentology, *49*, Pl. 29) Cretaceous of South Dakota.

SIZE: Equatorial diameter 27-34μ

RANGE: Upper Cretaceous and Tertiary, Europe and North America

Psilate surface irregularly folded forming a "rugulate" pattern. Trilete scar slender and extending about one-half distance of pole to the equator.

7b Surface pattern with appearance of several concentric subtriangular thickenings (Fig. 66).. ......*Polycingulatisporites reduncus* (Fossil) Playford and Dettman

SIZE: About 50μ

RANGE: Cretaceous

The more-or-less concentric thickenings of the wall may be due to a consistent and distinctive pattern of compression of the spore during flattening.

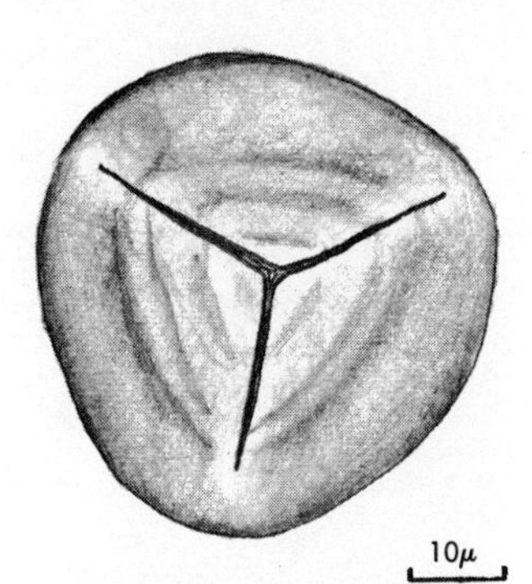

Fig. 66. *Polycingulatisporites reduncus* Playford and Dettman. Redrawn from Oltz, 1968. Cretaceous of Montana.

8a Perine with long pointed spines on the distal face of the grain (Fig. 67)............................*Selaginella kraussiana* A. Br.

SIZE: Tetrad diameter 80.5μ, Equatorial width 48μ, excluding spines

RANGE: Introduced

Spore with smooth wall except for small scattered gemmae and large straight or curved pointed spines. *Selaginela densa* microspores are not echinate, but have a dense granular or verrucate sculpture.

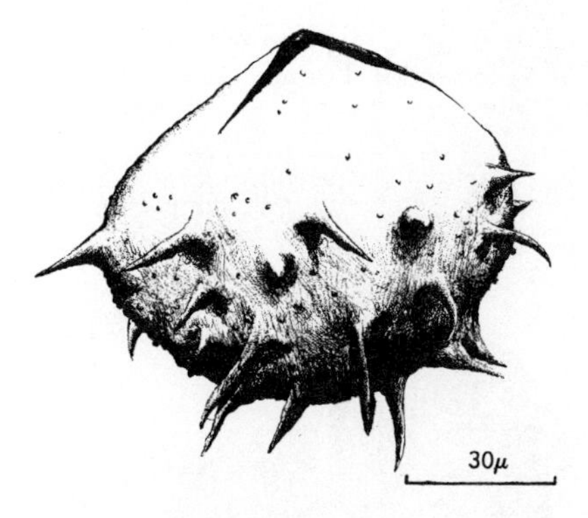

Fig. 67. *Selaginella kraussiana* A. Br.; SELAGINELLACEAE. Cultivated University of Michigan Botanical Gardens.

8b Perine reticulate, verrucate, striate, sometimes with a loose perisporium 9

9a Surface pattern reticulate...................................10

9b Surface otherwise...12

10a Reticulum coarse with straight-walled muri (walls of reticulum) (Fig. 68).................... ***Zlivisporis blanensis*** **Pacltova (Fossil)**

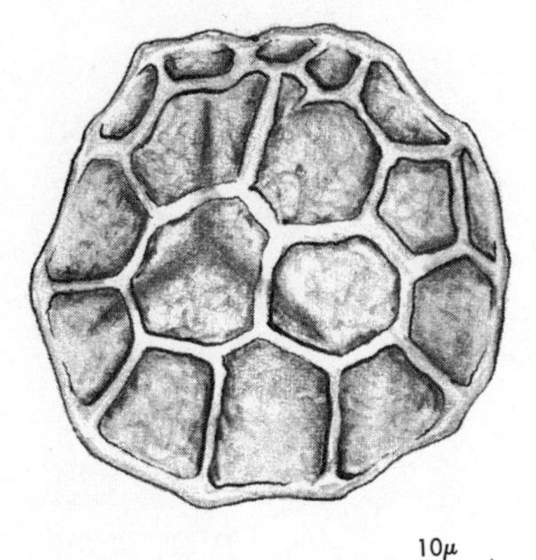

Fig. 68. *Zlivisporis blanensis* Pacltova. Redrawn from Oltz, 1968. Cretaceous from Montana.

SIZE: Diameter 36-45μ

RANGE: Cretaceous fossil

Trilete scar thin and extending less than half way from pole to equator. Possibly related to modern Lycopodiaceae.

10b Reticulum finer with the muri more or less winding, lumina with irregular bounaries..11

11a Trilete scare with conspicuously thickened border (Fig. 69).......
.... ***Ophioglossum engelmanni*** **Prantl**

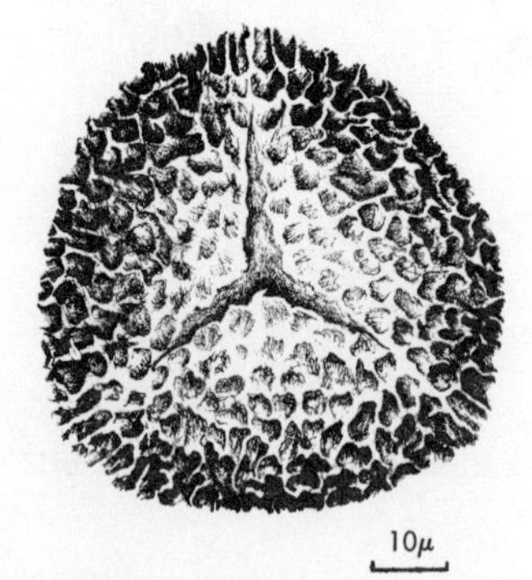

Fig. 69. *Ophioglossum engelmanni* Prantl. Coll. Texas; OPHIOGLOSSACEAE.

Adder's—tongue fern

SIZE: Diameter 50-60μ

RANGE: Florida to Arizona and Mexico north to Kansas, Ohio and northern Virginia

Wall reticulate on both proximal and distal faces. Muri tapered to sharp crests.

11b Trilete scar lacks conspicuously thickened border (Fig. 70, 71, 72) .... *Lycopodium* (in part)

Bristly Clubmoss

SIZE: Maximum dimension about 40μ

RANGE: Widespread in north temperate regions; circumboreal

Wall coarsely reticulate on the distal face, smooth on the proximal face where the trilete scar is located.

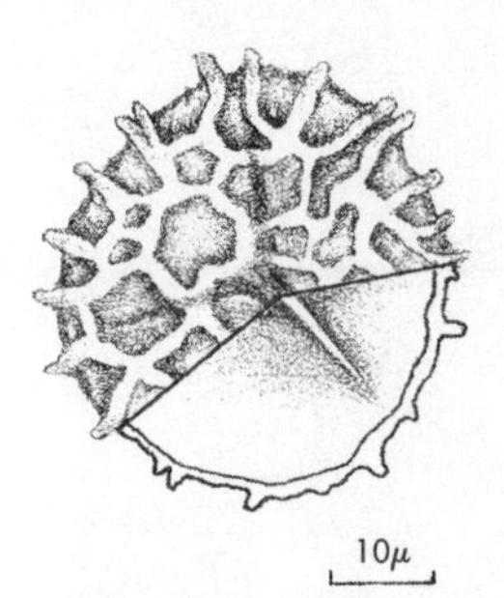

Fig. 70. *Lycopodium annotinum* L. Coll. W. Virginia; LYCOPODIACEAE.

Running Clubmoss

SIZE: 40 to 45μ maximum dimension

RANGE: Widespread in north temperate regions, circumboreal

Reticulum finer than in *Lycopodium annotinum* and present on both proximal and distal faces.

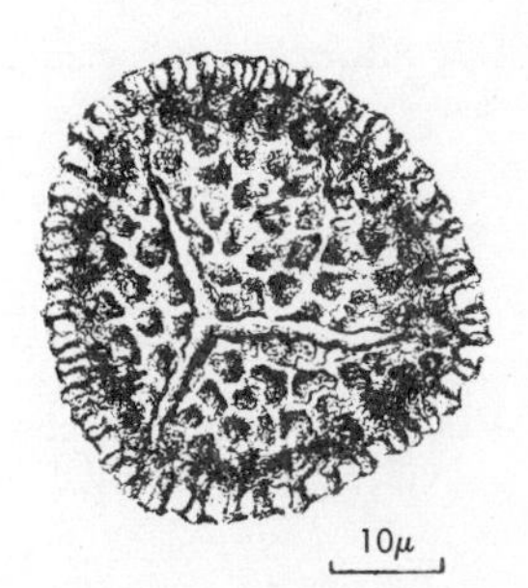

Fig. 71. *Lycopodium clavatum* L. Coll. W. Virginia; LYCOPODIACEAE.

Bog Clubmoss

SIZE: 45-50μ

RANGE: Widespread in U. S., Canada, and Eurasia

Reticulum irregular, sometimes nearly rugulate; proximal face smooth.

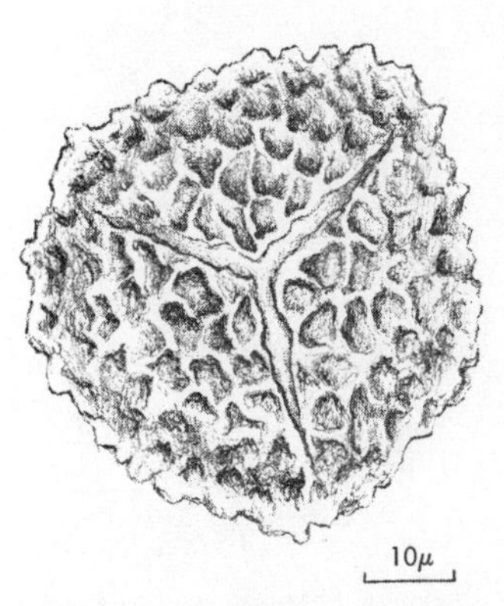

Fig. 72. *Lycopodium inundatum* L. Coll. Maine; LYCOPODIACEAE.

12a Surface pitted (Fig. 73)........................*Lycopodium* (in part)

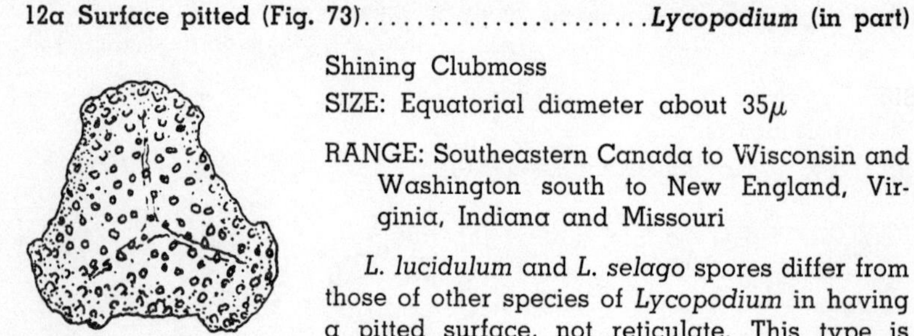

Fig. 73. *Lycopodium lucidulum* Michx. Coll. Minnesota. LYCOPODIACEAE.

Shining Clubmoss

SIZE: Equatorial diameter about 35μ

RANGE: Southeastern Canada to Wisconsin and Washington south to New England, Virginia, Indiana and Missouri

L. lucidulum and *L. selago* spores differ from those of other species of *Lycopodium* in having a pitted surface, not reticulate. This type is also hexagonal in outline rather than rounded-triangular.

12b Surface otherwise....................................13

13a Persporium present and loose..............................14

13b Perisporium absent (often present, but ephemeral) or, if present, closely associated with endospore.............................15

14a Perispore smooth (Fig. 74)...*Botrychium biternatum* (Sav.) Underw.

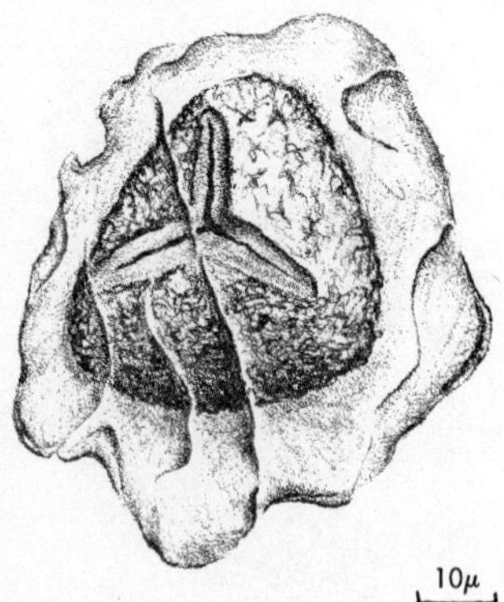

Fig. 74. *Botrychium biternatum* (Sav.) Underw. Col. Alabama; OPHIOGLOSSACEAE.

SIZE: Endospore 30 to 36μ; perispore to 53μ maximum dimension

RANGE: Southeastern United States

Surface of endospore unevenly thickened, subrugulate. Scar margin often thickened and smooth.

14b Perispore granular-roughened (Fig. 75)..........................
..............................*Selaginella densa Rydb.* (microspore)

SIZE: Perispore to 110μ, endospore about 50 to 55μ

RANGE: British Columbia and Idaho east to Alberta, south to New Mexico and Arizona

Perispore persistent and thick, its surface coarsely verrucate. Endospore broadly triangular, scar with narrow thickened margin.

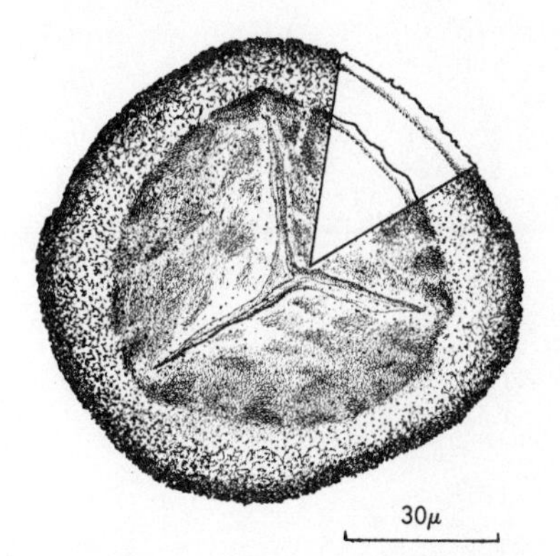

Fig. 75. *Selaginella densa* Rydb. (Microspore). Coll. Colorado; SELAGINELLACEAE.

15a Exine with rib-like thickenings (Fig. 76).........................
......*Appendicisporites tricornatus* Weyland and Greifeld (Fossil)
(Fern spores produced by the Schizaeceae will also key to this point.)

SIZE: 35-45μ

RANGE: Cretaceous of Europe and North America

Heavy ridges more or less obscure the trilete scar. Angles of the spore thickened into knob-like projections. Thickened ridges 2-3μ wide lie parallel to the equator. An exinal thickening about 4μ wide borders the trilete scar.

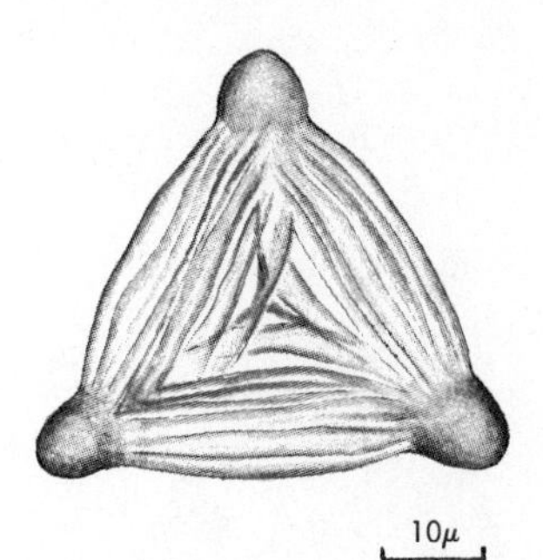

Fig. 76. *Appendicisporites tricornatus* Weyland and Greifeld. Coll. Potomac Group (Cretaceous) of Maryland.

15b Exine with various patterns of verrucate or gemmate sculpturing...
..16

16a Perine roughened and irregularly verrucate, closely appressed to endospore (Fig. 77)................*Pteridium aquilinum* (L.) Kuhn

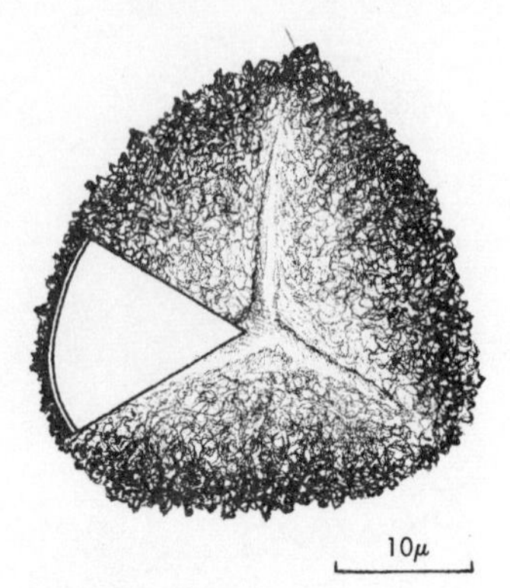

Fig. 77. *Pteridium aquilinum* (L.) Kuhn. Coll. Michigan; POLYPODIACEAE.

Bracken Fern

SIZE: 25-32μ

RANGE: Circumboreal

Exine surface verrucate, but with varying degrees of coarseness. Sometimes exinous material appears to be very loosely attached or exfoliating.

16b Perine absent, or if present coarsely warty....................17

**17a Shape rigidly triangular (Fig. 78)...............................
...............................*Cryptogramma aristicoides* R. Br.**

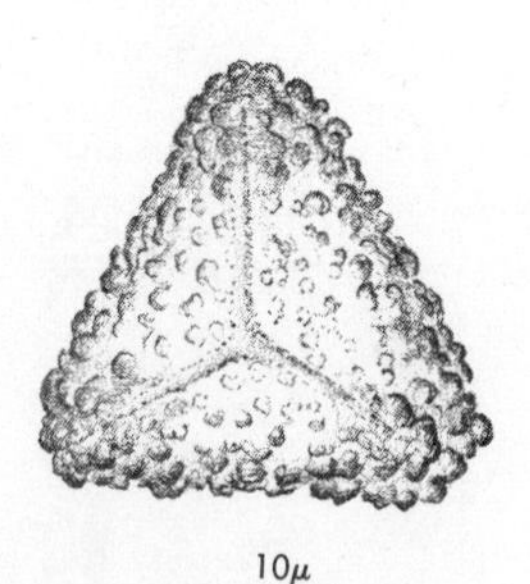

Fig. 78. *Cryptogramma aristicoides* R. Br. Coll. Washington; POLYPODIACEAE.

Rock-brake

SIZE: About 35μ

RANGE: Keewatin to Alaska, northeast Asia south to Great Lakes, and mountains to New Mexico and California.

Perine absent (?) or closely appressed to endospore. Surface very coarsely verrucate or gemmate. (The spores of *Adiantum pedatum* will also key to this point; its spores have concave sides.)

17b Shape subtriangular or rounded...............................18

18a Exine coarsely warty and folded, spongy in appearance (Fig. 79, 80) .... ***Botrychium* (in part)**

Grape-fern

SIZE: About 45μ

RANGE: Florida to eastern Texas, north to Maryland, Indiana, Missouri

Surface irregularly warty due to the thick spongy perisporium which is closely appressed to the endospore. Border of scar conspicously thickened.

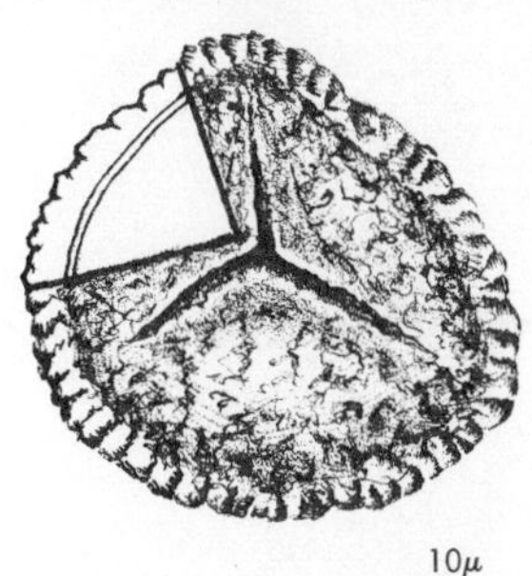

Fig. 79. *Botrychium tenuifolium* Underw. Coll. Arkansas; OPHIOGLOSSACEAE.

Rattlesnake Fern

SIZE: About 35μ

RANGE: Widespread throughout U.S. in deciduous or mixed woods.

Surface with large "bubble-like" projections. Scar with an inconspicuous narrow marginal thickening.

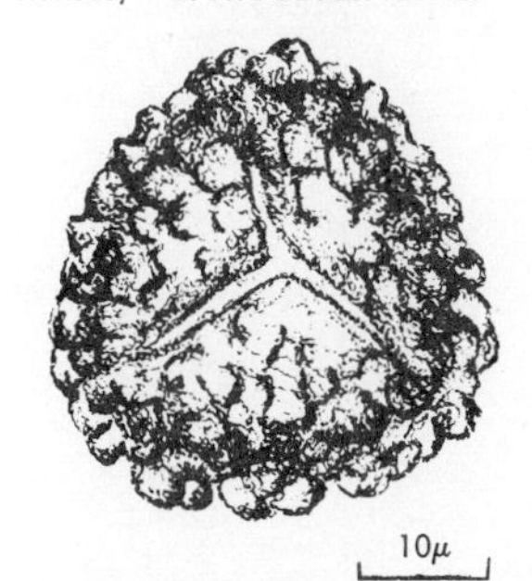

Fig. 80. *Botrychium virginianum* (L.) Sw. Coll. Michigan; OPHIOGLOSSACEAE.

18b Exine relatively thin with gemmate, clavate or baculate processes (Fig. 81, 82, 83) .... ***Osmunda***

Royal Fern

SIZE: 50 to 60μ

RANGE: Circumboreal and widespread in North America, south to Mexico

Spores of the royal fern may be distinguished from other species of *Osmunda* by their larger size and by the presence of baculae on the distal face.

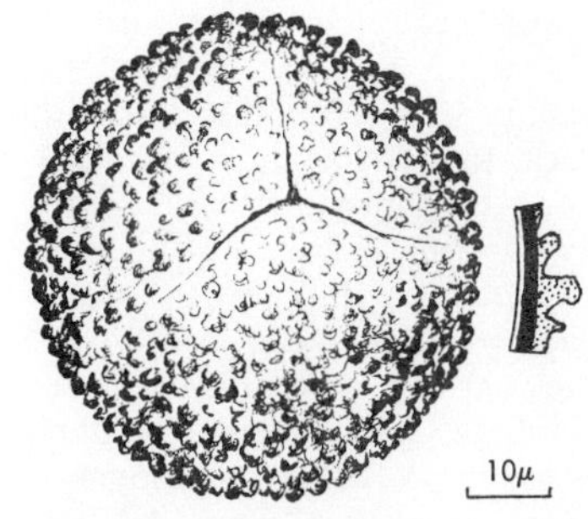

Fig. 81. *Osmunda regalis* L. Coll. Michigan; OSMUNDACEAE.

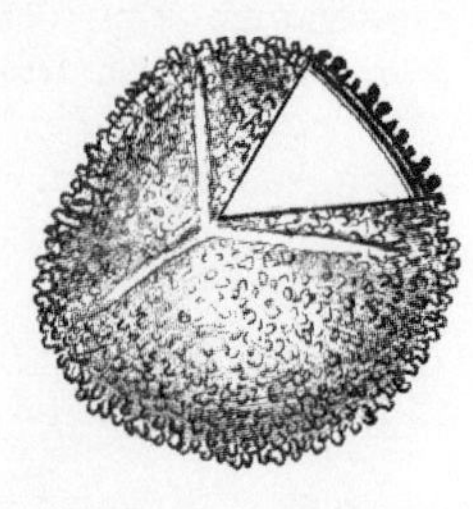

Fig. 82. *Osmunda claytoniana* L. Coll. Michigan; OSMUNDACEAE.

Interrupted Fern
SIZE: About 35μ

RANGE: Newfoundland to southeastern Manitoba, south to Georgia and Arkansas in uplands.

Spores of the interrupted fern have numerous clavate sculptural elements scattered among the baculae.

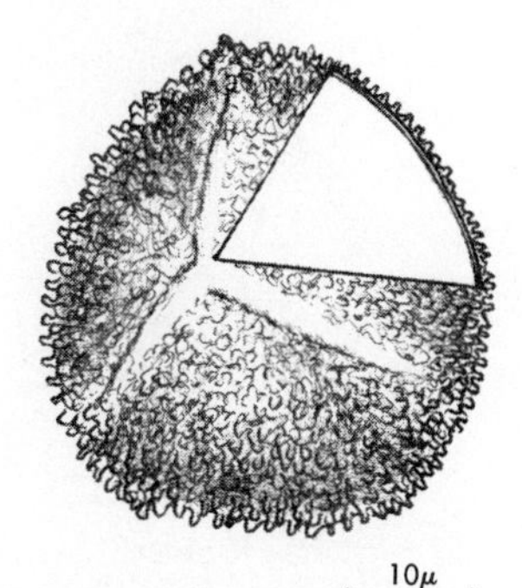

Fig. 83. *Osmunda cinnamomea* L. Coll. Michigan; OSMUNDACEAE.

Cinnamon Fern
SIZE: About 35μ

RANGE: Southeastern Canada to Minnesota south to Gulf of Mexico

The surface sculpturing of *Osmunda cinnamomea* differs from other species; most elements are verrucae and gemmae, a few short baculae are also found.

INAPERTURATE POLLEN AND SPORES

Inaperturate spores or pollen grains are produced by every major plant group; thus, they are among the most difficult to identify. Some bacteria produce endospores which are inaperturate and extremely small (approximately 1μ). Many species of algae produce thick-walled, resistant zygospores; among these types are members of the Conjugales and Oedogoniales. All groups of fungi produce spores with resistant walls; these may be the product of meiosis (slime mold spores, ascospores, basidiospores), asexual conidiospores, aeciospores and teliospores (rusts), or zygospores. (Some fungal spores are multicellular and are keyed with the dyad or polyad types; a few are monoporate.) Among mosses and liverworts, inaperturate spores are the most common type; a few are monocolpate, however. Among vascular plants, the horsetails (*Equisetum*), many gymnosperms, some monocots, and a few dicots produce inaperturate spores or pollen.

This key will only permit identification of representative types of spores of fungi and bryophytes, but is not inclusive of all species. The reader should assume that the genera and species keyed in these groups might be confused with several other species which are not illustrated.

1a Size less than 25μ (usually 20μ or smaller)....................2

1b Size greater than 25μ....................32

2a Maximum dimension greater than 15μ....................3

2b Maximum dimension less than 15μ....................8

3a Wall coarsely reticulate (Fig. 84).........*Tilletia caries* (DC.) Tul.

SIZE: About 20μ

Shape is spheroidal, reticulum very coarse and distinct, probably distinguishable from all other inaperturate types. *Tilletia* is a basidiomycete fungus which parasitizes wheat causing the condition known as "stinking smut." It is of almost universal occurrence.

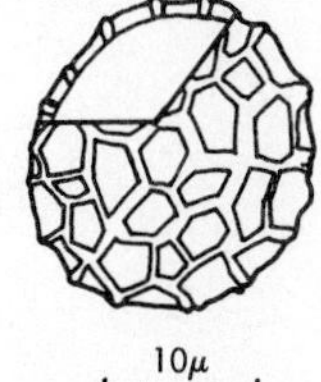

Fig. 84. *Tilletia caries* (DC.) Tul.

3b Wall psilate, verrucate, reticulate or striate....................4

4a Surface psilate, or sometimes coarsely striate or rugulate (Fig. 85) ....*Ascobolus* (ascospore)

SIZE: 12 x 23μ

Many species of *Ascobolus* are dung fungi, some very narrowly host specific. Wall psilate, but with very minute punctations, sometimes coarsely striate. Shape prolate, elliptical. Color purple when fresh.

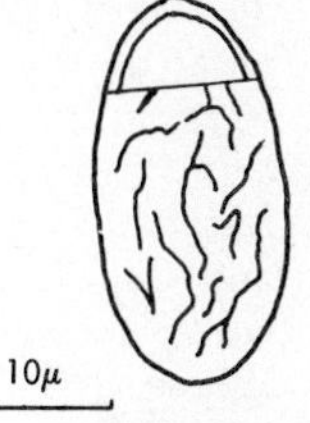

Fig. 85. *Ascobolus* sp.; PEZIZACEAE (Ascomycete fungus).

4b Surface verrucate (or may appear slightly reticulate or striate)....5

5a Shape prolate, elliptical, wall thick, tectate (?) (Fig. 86).......... ....*Dicranum scoparium* Hedw.

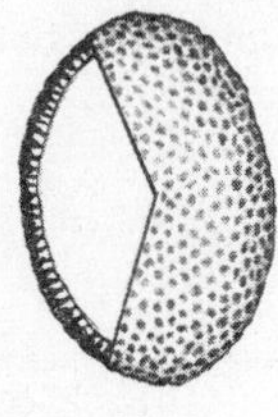

10μ

Fig. 86. *Dicranum scoparium* Hedw. Coll. Michigan; DICRANIACEAE (Moss).

SIZE: 13 x 18μ

RANGE: Widespread in North Temperate latitudes

Shape usually prolate, occasionally somewhat spherical. Surface fine reticulate (oil!).

5b Shape spherical or, if otherwise, very coarsely verrucate-gemmate 6

6a Verrucae very coarse, shape variable (Fig. 87)................... ....*Ulota crispa* (Hedw.) Brid.

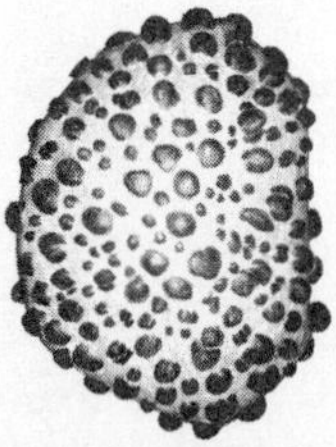

10μ

Fig. 87. *Ulota crispa* (Hedw.) Brid. Coll. Michigan; ORTHOTRICHACEAE (Moss).

SIZE: About 20μ, maximum dimension

RANGE: Quebec to North Carolina and Minnesota; Alaska

Very large "blotchy" verrucae and gemmae with smaller projections scattered among them. Shape varies from almost spherical to angular or kidney-shaped.

6b Verrucae finer, often irregularly spaced (like Cupressaceae !).....7

7a Sculpture pattern with scattered verrucae only (Fig. 88)........... ....*Hypnum imponens* Hedw.

SIZE: 18 to 19μ diameter

RANGE: Quebec to British Columbia, south to California and Georgia

Sculpture of scattered verrucae and gemmae; approximately spherical or sometimes angular-flattened. One hemisphere of the spore usually more finely and densely verrucate than the other.

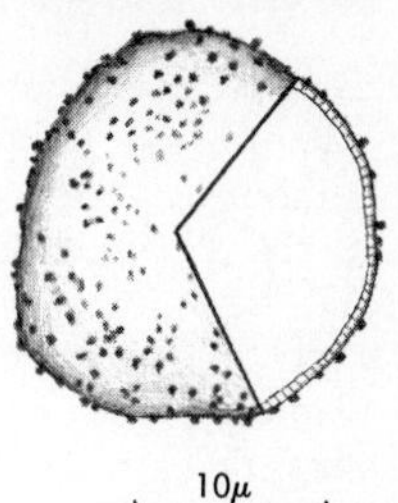

Fig. 88. *Hypnum imponens* Hedw. Coll. Michigan; HYPNACEAE. (Moss)

7b Sculpture pattern verrucate on one hemisphere, usually more-or-less striate on the other (Fig. 89).........*Didymodon rubellus* Bry. Eur.

SIZE: About 15 to 16μ

RANGE: Widespread

Shape approximately spherical. Some verrucae are elongated and nearly baculate in form. Wall thick and apparently "subtectate," at least on one face.

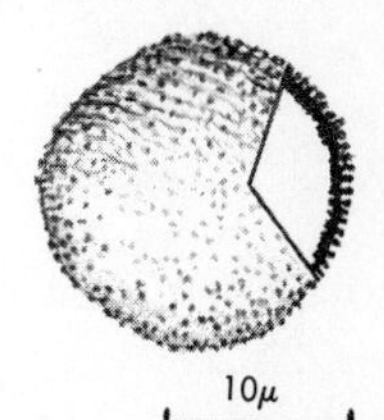

Fig. 89. *Didymodon rubellus* Bry. Eur. Coll. Michigan; POTTIACEAE (Moss).

8a Wall sculpturing absent, psilate...............................9

8b Wall sculpturing verrucate, baculate, echinate, striate, or reticulate 20

9a Shape spherical or slightly elliptical............................10

9b Shape elliptical to angular...12

10a Spores sometimes with tiny apiculus (hilum) (Fig. 90)........... ....*Lycoperdon* sp. (basidiospore)

Puffball

SIZE: 4 to 5μ

Shape spheroid. Small projection present on some spores. Psilate, but very slightly roughened; some species of *Lycoperdon* are more roughened. Fresh spores brown.

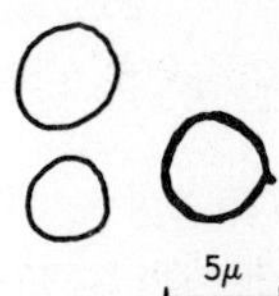

Fig. 90. *Lycoperdon* sp.; LYCOPERDACEAE (Basidiomycete fungus).

10b Spores smooth, not apiculate.................................**11**

11a Size 4 to 5μ (Fig. 91)..............***Penicillium* sp. (conidiospores)**

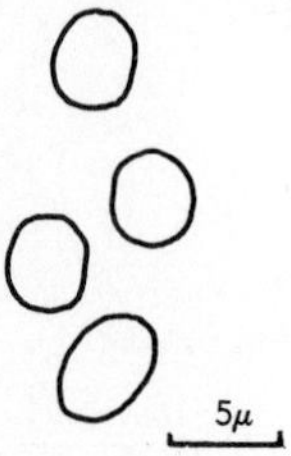

Fig. 91. *Penicillium* sp.; ASPERGILLACEAE (Ascomycete fungus).

Penicillium

SIZE: 4 to 5μ

Shape varies from spheroidal to ellipsoidal. Fresh spores light colored, but not hyaline. Very similar to conidiospores of *Aspergillus*.

11b Size 7 to 8μ (Fig. 92)...............***Polytrichum commune* Hedw.**

Hairy-cap Moss

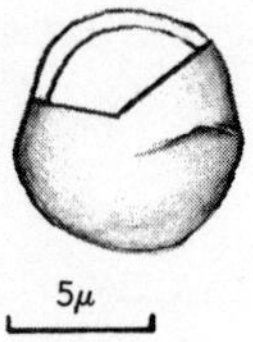

SIZE: (6.5 to) 7 to 8μ

RANGE: Worldwide

Spherical shape, but often collapsed or folded. Very similar to some fungal spores. This species is very similar to *Polytrichum juniperinum*, although the latter species has a slightly thinner exine.

Fig. 92. *Polytrichum commune* ..Hedw. Coll. Michigan; POLYTRICHACEAE (Moss).

12a Spore with two circular ridges, "pulley-wheel" shape (Fig. 93)... ....***Aspergillus***

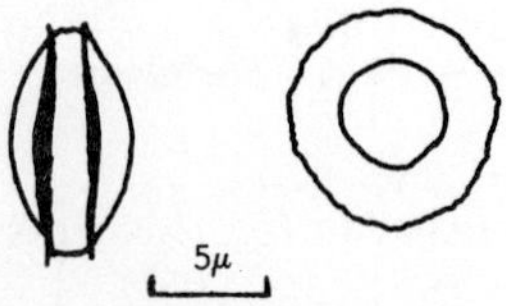

SIZE: 9 to 10μ diameter, breadth about 5μ

Surface psilate to scabrate; light-colored when fresh.

Fig. 93. *Aspergillus glaucus* (ascospore).

12b Spore elliptical, angular, or variable in shape................**13**

13a Hilum or apiculum present.......................................**14**

13b Hilum or distinct apiculum absent...............................**18**

14a Shape elliptical to prolate.......................................**15**

14b Shape angular (Fig. 94). . . . . *Rhodophyllus sericatum*

SIZE: Up to 10μ

Shape elliptical to angular. Fresh spores light colored (pinkish).

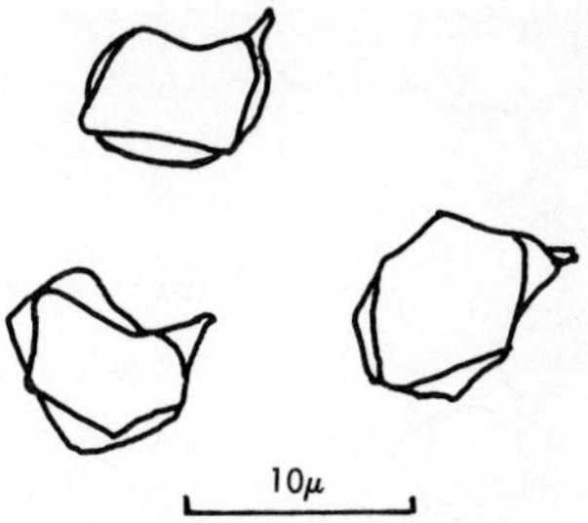

Fig. 94. *Rhodophyllus sericatum;* RHODOPHYLLACEAE (basidiospore).

15a Color of fresh spores dark brown (Fig. 95). . . . . *Agaricus campestris*

Field Mushroom

SIZE: 5 to 7μ

Subprolate or elliptical in shape.

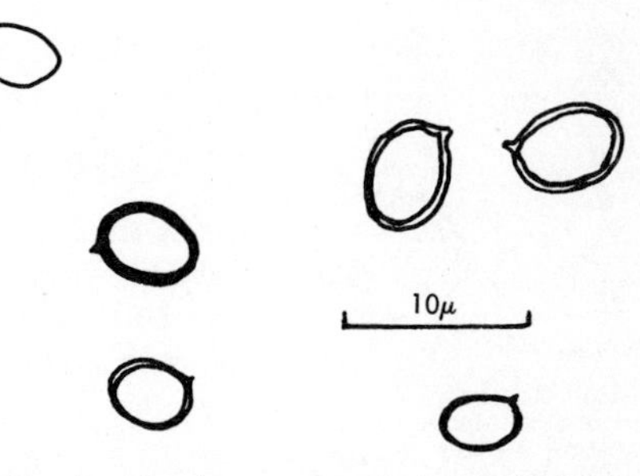

Fig. 95. *Agaricus campestris* Fries; AGARICACEAE.

15b Color of fresh spores light-colored. 16

16a Shape perprolate, slightly curved at apiculus (Fig. 96). . . . .
. . . . . *Boletus spectabilis* Peck (basidiospore)

Bolete

SIZE: 10 to 13μ long

Shape (as illustrated) very distinctive for the genus *Boletus*. Fresh spores yellowish to light brown.

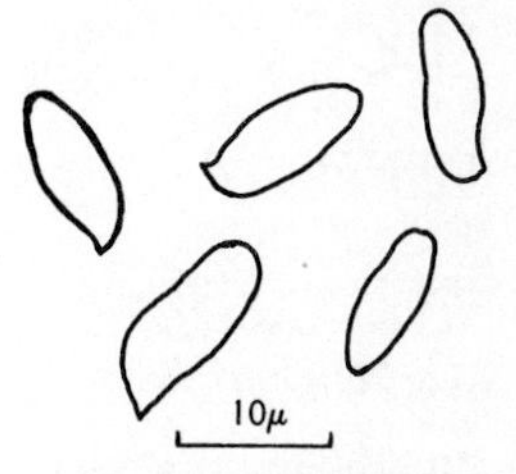

Fig. 96. *Boletus spectabilis* Peck; BOLETACEAE (basidiospore).

16b Shape subprolate to prolate (elliptical)........................17

17a Wall thick, surface smooth (Fig. 97)................*Lepiota* sp.

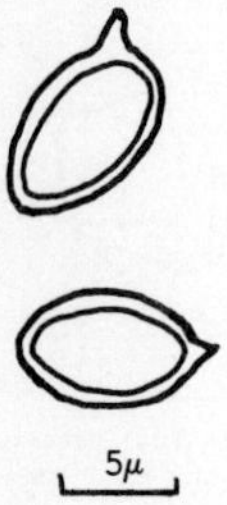

Parasol Mushroom

SIZE: 8 to 9μ

Hilum (apiculus) either terminal or subterminal. Wall surface completely smooth.

5μ

Fig. 97. *Lepiota* sp.; LEUCOPRINACEAE (basidiospores).

17b Wall thin, surface more-or-less scabrate (Fig. 98)...*Botrytis* sp. (conidiospore)

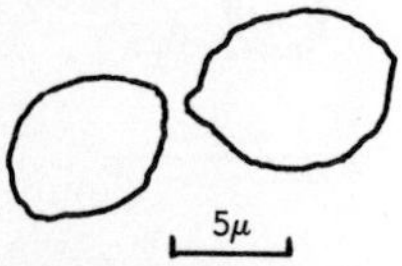

SIZE: Maximum dimension 7 to 9μ

Wall very thin, probably destroyed by acetolysis. Spores hyaline and gray. The fungal parasite *Botrytis* produces blight, a serious disease in some plants.

5μ

Fig. 98. *Botrytis* sp.; MONILIACEAE (Deuteromycete fungi, Fungi Imperfecti).

18a Spores simple, shape variable (Fig. 99)......*Phoma* (basidiospores)

SIZE: 3 to 5μ

Shape variable, wall featureless. Hyaline. Probably not separable from several other non-descript hyaline fungal spores. Species of *Phoma* are serious plant parasites, sometimes attacking commercial crops.

5μ

Fig. 99. *Phoma* sp.; SPHAEROPSIDACEAE (Deuteromycete fungi; Fungi Imperfecti).

18b Spores pseudoseptate or with vacuoles.......................19

19a Shape ellipsoidal, divided in two by a pseudoseptum (Fig. 100). . . . . *Caloplaca* (ascospore)

SIZE: 7 x 15μ

Shape perprolate, broad crosswall constricted at center and incompletely dividing the spore into two cells. Spores colorless. Probably indistinguishable from some other ascospores.

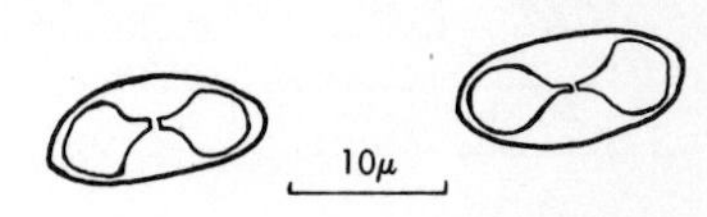

Fig. 100. *Caloplaca* sp. (ascospore).

19b Shape irregular, spherical to prolate, sometimes constricted (Fig. 101). . . . . *Sporobolomyces roseus* Kl. & vanN.

SIZE: 8 to 10μ

Conidiospores with characteristic vacuoles (Fig. 101, v.), these probably destroyed by acetolysis. Hyaline. Conidia of this genus may be distinct from other hyaline conidiospores.

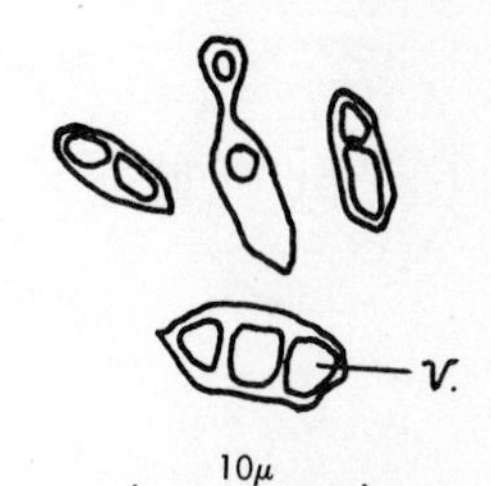

Fig. 101. *Sporobolomyces roseus* Kl. & van N.; SPOROBOLOMYCETACEAE (Deuteromycete fungi; Fungi Imperfecti), conidia.

20a Spores reticulate. 21

20b Spores verrucate, striate, echinate, or baculate 22

21a Shape prolate (Fig. 102). . . . .*Peziza* (Aleuria) *aurantia* Pers.

Orange-peel Peziza

SIZE: 7 x 13μ

Muri thin (compare *Tilletia*). Some species of *Aleuria* lack the reticulum.

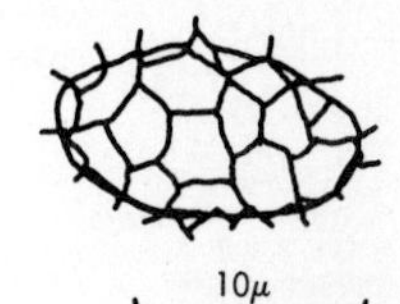

Fig. 102. *Peziza* (Aleuria) *aurantia* Pers., PEZIZACEAE (ascospore).

21b Shape spherical (Fig. 103)......................... *Lycogala* sp.

Slime Mold

SIZE: About 5μ

Spheroidal shape with fine reticulum. Probably distinguishable from other genera.

Fig. 103. Lycogala sp., MYXOMYCETACEAE (resting spore).

22a Sculpture verrucate, echinate, or baculate.................... 23

22b Sculpture faintly striate (Fig. 104)...................... *Rhizopus*

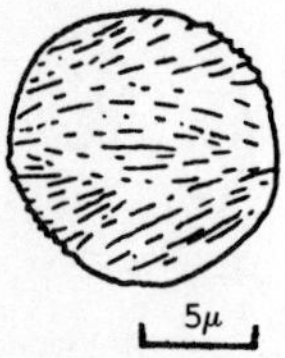

Black Bread Mold

SIZE. About 11μ

Shape spheroidal with flattened faces. Striations faint. Distinct from other genera.

Fig. 104. *Rhizopus* sp., *MUCORALES* (conidiospore).

23a Wall sculpture echinate or baculate......................... 24

23b Wall sculpture verrucate................................. 26

24a Shape ovate, apiculate (Fig. 105)................. *Russula* spp.

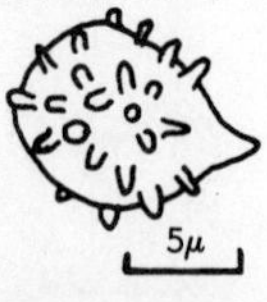

SIZE: 6 x 10μ

Shape slightly ellipsoid, hilum area extended as an apiculum. Surface blunt echinate. Fresh spores light-colored. This form appears to be distinctive for the genus *Russula*.

Fig. 105. *Russula paludosa* Britz., RUSSULACEAE (basidiospore).

24b Shape spheroidal... 25

25a Sculpture echinate (Fig. 106)..........................*Scleroderma*

Puffball

SIZE: About 10μ diameter

Spheroidal shape with numerous spines. Fresh spores yellowish brown. This form appears to be distinctive for the genus *Scleroderma*.

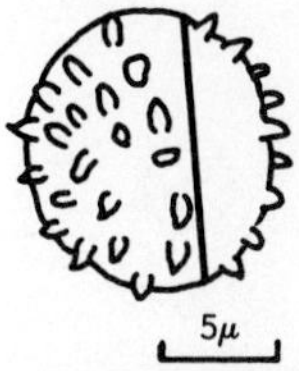

5μ

Fig. 106. *Scleroderma* sp., SCLERODERMACEAE (basidiospore).

25b Sculpture baculate (Fig. 107)....*Diphyscium foliosum* (Hedw.) Mohr.

Moss

SIZE: 6 to 8μ

RANGE: Northeastern United States

The very large coarse baculae of this spore are unique among the mosses examined. Shape spheroidal to ovate.

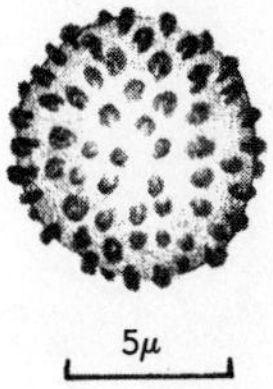

5μ

Fig. 107. *Diphyscium foliosum* (Hedw.) Mohr. Coll. Michigan; DIPHYSIACEAE.

26a Spores with stalk, apiculum, or distinct scar...................27

26b Spores without projections or scars..........................29

27a Shape spheroidal with a distinctive pedicellate hilum (Fig. 108) ....*Calvatia gigantea* (Pers.) Lloyd

Giant Puffball

SIZE: About 5μ, excluding pedicel

Wall surface scabrate or minutely verrucate. Stalk (pedicel) about 1μ long. This type appears to be distinctive for the genus *Calvatia*.

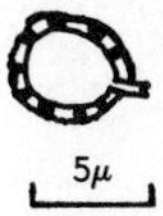

5μ

Fig. 108. *Calvatia gigantea* (Pers.) Lloyd; LYCOPERDACEAE (basidiospore).

27b Shape ovate or ellipsoidal...............................28

28a Attachment scar (hilum) a narrow apiculus (Fig. 109)...*Amanita muscaria* (Fr.) S. F. Gray

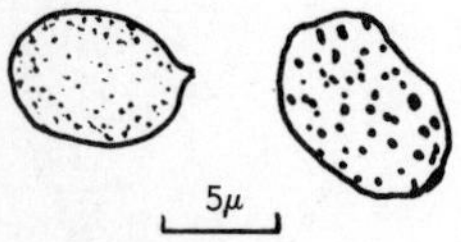

Fly Agaric

SIZE: About 8μ long

Shape prolate, irregularly elliptical. Surface scabrate or minutely verrucate. Fresh spores light-colored. This type appears to be typical for spores of the genus *Amanita*.

Fig. 109. *Amanita muscaria* (Fr.) S. F. Gray; AMANITACEAE (basidiospore).

28b Attachment scar is broad, flat, and terminal (Fig. 110).....................................*Fomes applanatus* (Pers. ex Wallr.) Gill.

Bracket Fungus

SIZE: About 9μ

Shape prolate; minutely verrucate. The distinctive attachment scar visible at base of the hyaline wall. Fresh spores yellow-brown. This spore-type appears to be characteristic of the species of *Fomes* which are sometimes placed in the genus *Ganoderma*.

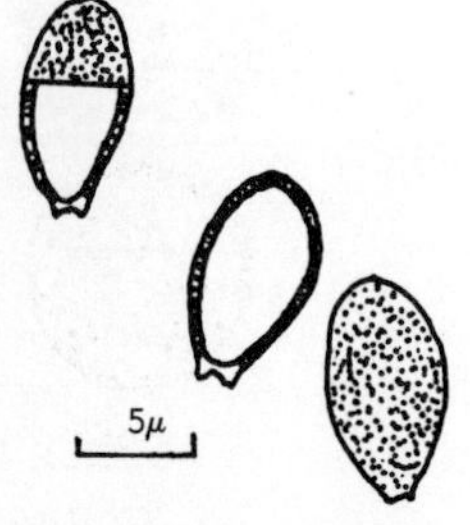

Fig. 110. *Fomes applanatus* (Pers. ex Wallr.) Gill.; POLYPORACEAE (basidiospore).

29a Shape prolate (occasionally subspheroidal) (Fig. 111)..*Cladosporium* sp.

SIZE: 15 to 16μ long

Shape prolate (9 x 15μ) to subspheroidal. Attachment scars indistinct, not projecting, one at each end of spore. May occur singly or in branched chains. Surface distinctly verrucate. Fresh conidia are brown in color.

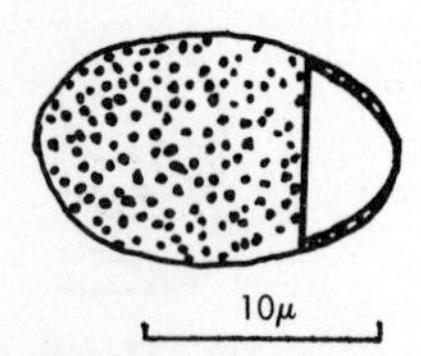

Fig. 111. *Cladosporium* sp.; DEMATIACEAE (Deuteromycete fungus, Fungi Imperfecti, condiospore).

29b Shape spheroidal..30

30a Size 11 to 13μ (Fig. 112).. ....*Plagiothecium denticulatum* (Hedw.) Bry. Eur

SIZE: 11 to 13μ diameter

RANGE: Woodlands, Canada to Georgia and Colorado

Shape spheroidal to sub-ovate. Verrucae coarse and scattered. Wall thin.

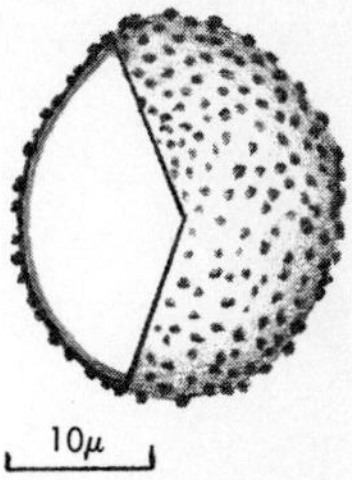

10μ

Fig. 112. *Plagiothecium denticulatum* (Hedw.) Bry. Eur. Coll. Michigan; HYPNACEAE (Moss).

30b Size less than 10μ...31

31a Verrucae very minute (Fig. 113)..................*Didymium* sp.

Slime Mold

SIZE: About 9μ

Inaperturate, minutely verrucate, spheroidal.

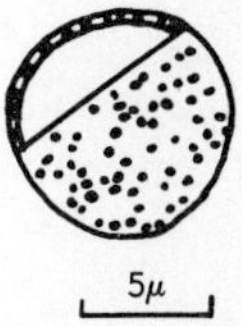

5μ

Fig. 113. *Didymium* sp.; PHYSARALES (resting spores).

31b Verrucae coarse (Fig. 114)..........*Ustilago maydis* (DC.) Corda.

Corn Smut

SIZE: About 7.5μ

Shape spheroidal, inaperturate; fresh spores brown in color. *U. maydis* spores have walls of uniform thickness. *U. nuda* very similar but wall of one hemisphere conspicuously thickened. *U. tritici* is similar but has finer verrucae.

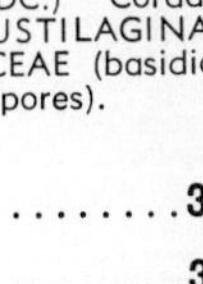

5μ

Fig. 114. *Ustilago maydis* (DC.) Corda.; USTILAGINACEAE (basidiospores).

32a Size greater than 50μ...33

32b Size less than 50μ...36

33a Surface verrucate (Fig. 115)...*Conocephalum conicum* (L.) Dumort.

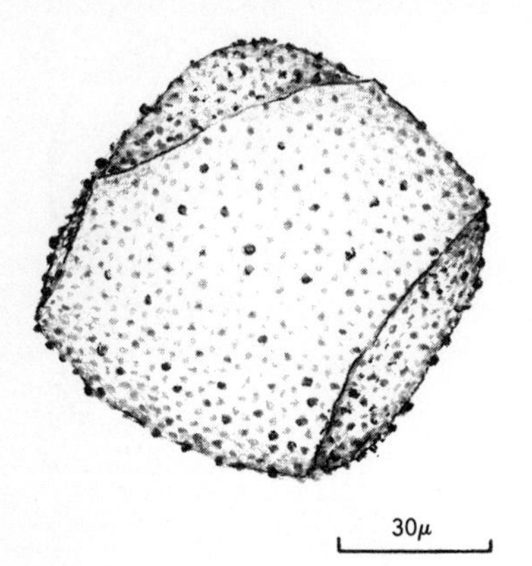

Fig. 115. *Conocephalum conicum* (L.) Dumort. Coll. Indiana; MARCHANTIACEAE.

Liverwort

SIZE: 90 to 110μ

RANGE: North America and Europe

Spore large, thin-walled usually with irregular folds. Wall thin and ornamented with numerous evenly-distributed verrucae and fewer large gemmae.

33b Surface psilate..34

34a Diameter usually greater than 110μ (Fig. 116).....................
...........................*Pseudotsuga menziesii* (Mirb.) Franco

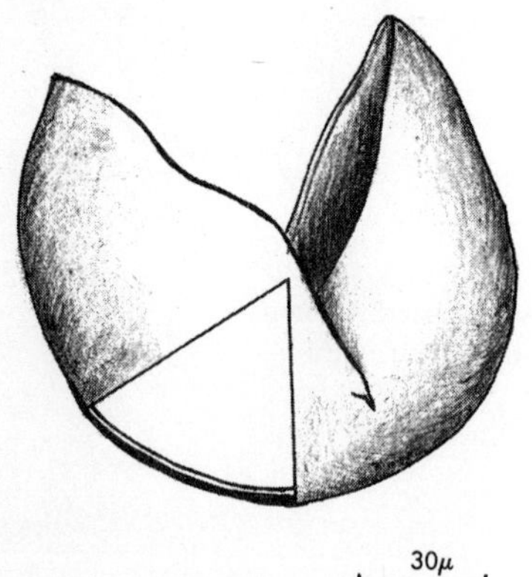

Fig. 116. *Pseudotsuga menziesii* (Mirb.) Franco. Coll. Montana; PINACEAE.

SIZE: About 110 to 115μ

RANGE: British Columbia to western Texas, Arizona and northern Mexico

Spheroidal in shape but often torn open or folded. Psilate.

34b Diameter 50 to 85μ..35

35a Fresh spore with elators (Fig. 117)..........*Equisetum hyemale* L.

SIZE: About 80 to 82μ

RANGE: Widespread throughout the United States and Canada

Living spores bear four strap-like, hygroscopic elaters. Wall often shows outer perisporium closely adhering to endospore. Often difficult to separate from *Larix*.

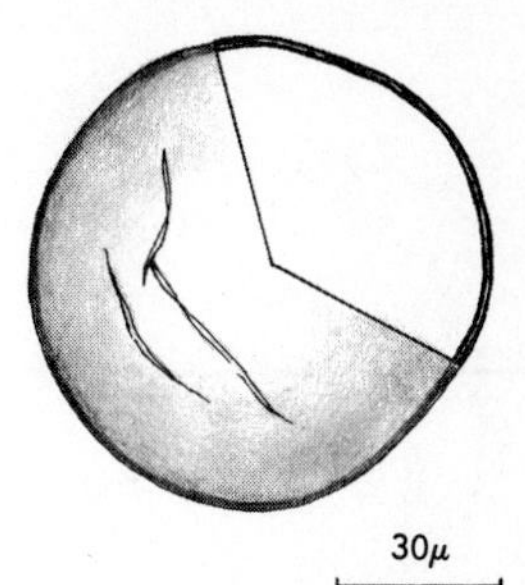

Fig. 117. *Equisteum hyemale* L. Coll. Texas; EQUISETACEAE.

35b Fresh pollen without elaters (Fig. 118)...*Larix laricina* (DuRoi) Koch

Larch (Tamarack)

SIZE: Variable, range 60 to 90μ

RANGE: Labrador to Alaska south to New England, West Virginia, northern Indiana and northeast British Columbia.

Exine thin, psilate and inaperturate. Often ruptures to form a furrow-like area.

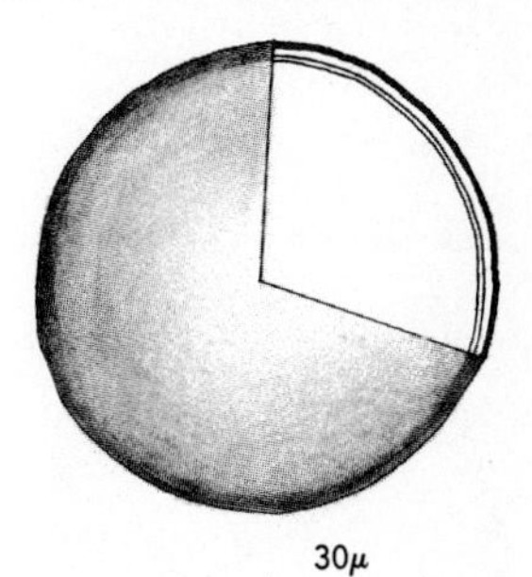

Fig. 118. *Larix laricina* (DuRoi) Koch. Coll. Michigan; PINACEAE.

36a Surface psilate or scabrate..............................37

36b Surface verrucate, reticulate, gemmate, or clavate.............39

37a Shape subtriangular, obscure pores (Fig. 119)...................................*Eleocharis obtusa* (Willd.) Schultes

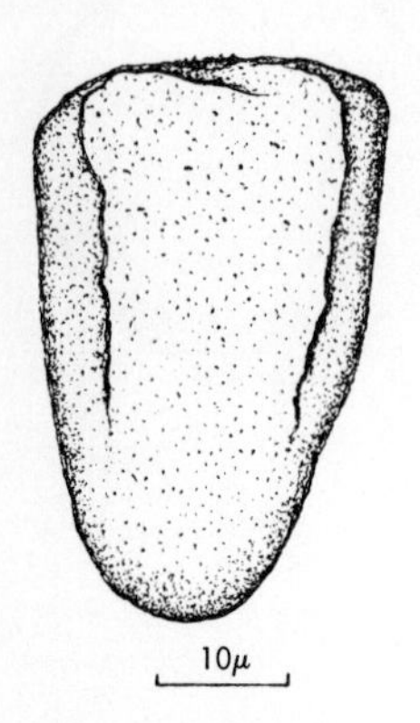

Fig. 119. *Eleocharis obtusa* (Willd.) Schultes. Coll. Michigan; CYPERACEAE.

Blunt Spikerush

SIZE: About 28 x 42μ

RANGE: Southeastern Canada to Minnesota and British Columbia, south to northern Florida and eastern Texas, also northern California

Shape subtriangular, but often irregularly folded, a single, obscure terminal pore may be seen with difficulty at broad end of pollen grain. Pores around middle of grain obsolete or obscure. (Sedge pollen is usually 3 to 4 porate.) Surface has a fine granular appearance due to columellae; wall thin.

37b Shape spheroidal.......................................38

38a Exine tectate, with fine columellae (Fig. 120)...................................*Torreya taxifolia* Arn.

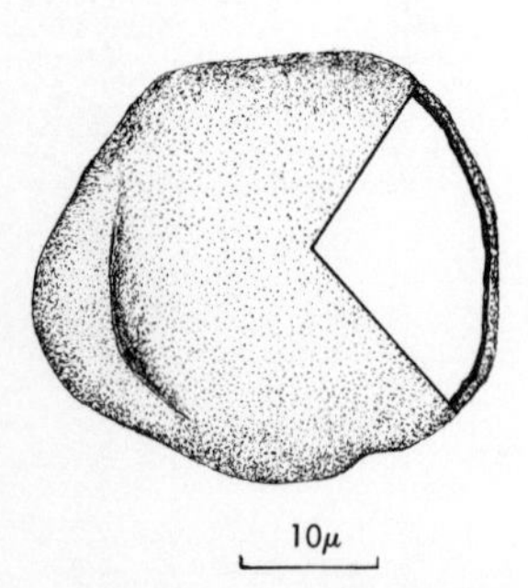

Fig. 120. *Torreya taxifolia* Arn. Coll. Florida; TAXACEAE.

Stinking Cedar

SIZE: About 32μ

RANGE: Northern Florida and adjacent Georgia

Inaperturate, indistinctly tectate. Spheroidal but outline irregular because of folds. Exine indistinctly tectate, appears granular due to subdued columellae.

38b Exine intectate (Fig. 121)........................*Equisetum* spp.

Field Horsetail

SIZE: 35 to 40μ diameter

RANGE: Throughout southern Canada to Alaska south to Virginia, Nebraska and California

Spherical, but often folded or torn. Exine may appear two-layered due to close-fitting exosporium. Four strap-like elaters present when fresh; lost in acetolysis or fossilization. Size in genus apparently varies from 30 to 75μ.

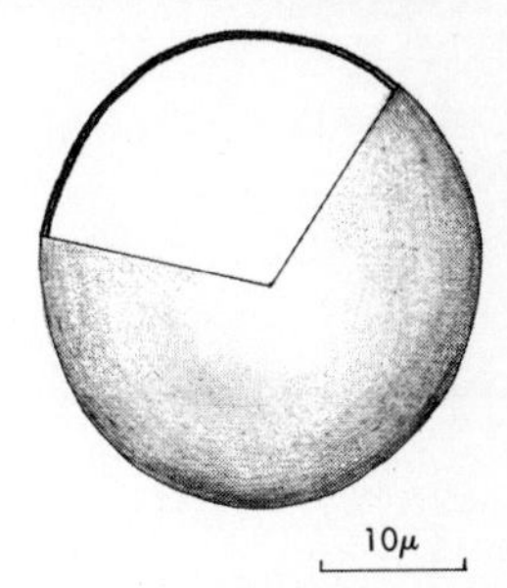

Fig. 121. *Equisetum arvense* L. Coll. Michigan; EQUISETACEAE.

39a Verrucae, or other sculptural elements, regularly spaced, usually dense and abundant; subtectate reticulum present in *Potamogeton* 40

39b Verrucae irregularly spaced, usually sparsely scattered (Fig. 122, 123, 124, 125)....................................*Cupressaceae*

Cypress

SIZE: About 28μ diameter

RANGE: West Coastal North America

Verrucae widely spaced, irregular. Spherical but often wrinkled or folded.

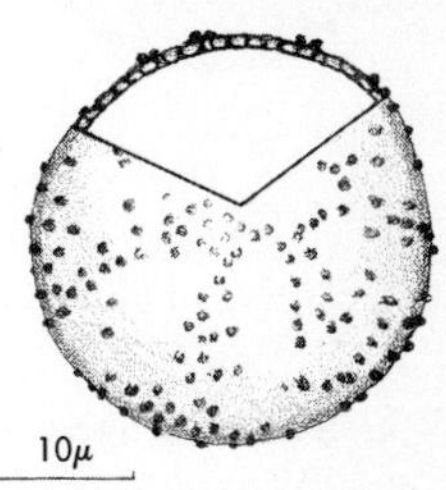

Fig. 122. *Cupressus macnabeana* Murray Coll. California; CUPRESSACEAE.

Yellow Cedar (Alaska Cypress)

SIZE: About 34μ

RANGE: Coastal Oregon to Alaska

Spheroidal, a single "sulcoid" tear is frequent. Intectate with numerous, dense but irregularly-spaced verrucae. Some sculptural elements approach shape of gemmae and baculae.

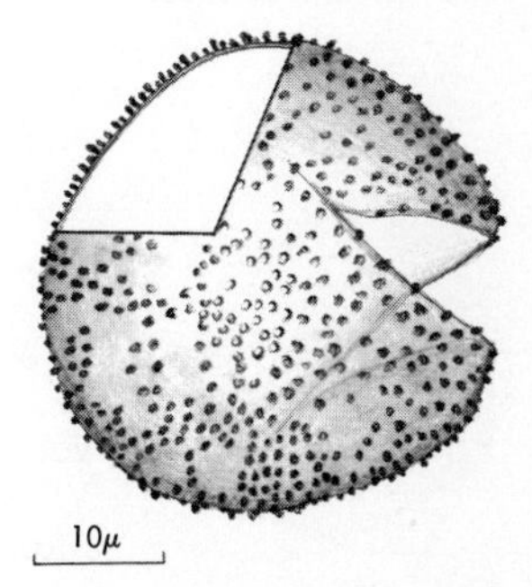

Fig. 123. *Chamaecyparis nootkanensis* (Lamb.) Spach. Coll. Washington; CUPRESSACEAE.

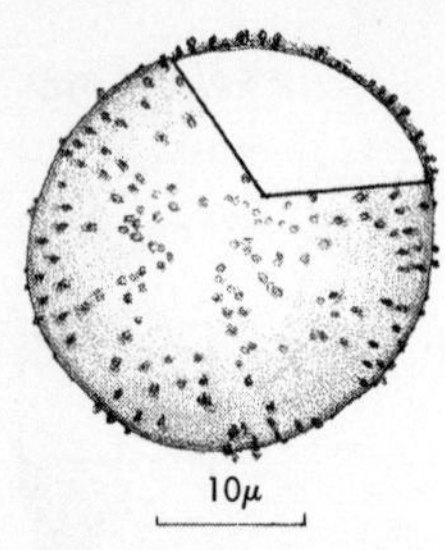

Fig. 124. *Juniperus scopularum* Sarg. Coll. Utah; CUPRESSACEAE.

Juniper

SIZE: About 27 to 28μ

RANGE: Alberta to British Columbia, south to Texas and Arizona

Spheroidal; inaperturate, but occasionally with a "dimple" area suggesting an aperture. Sparsely scattered verrucae and gemmae.

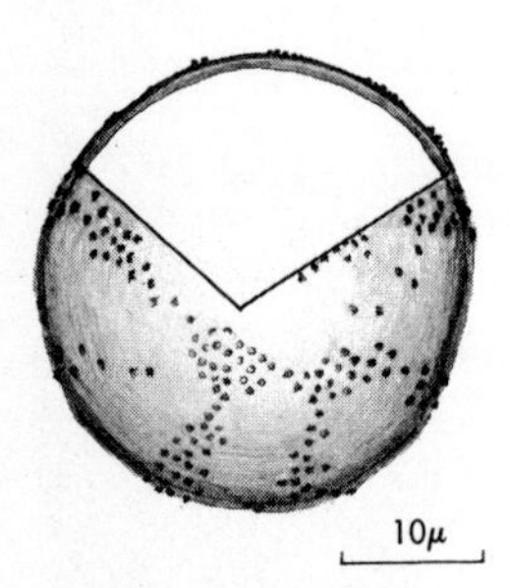

Fig. 125. *Thuja occidentalis* L. Coll. Michigan; CUPRESSACEAE.

White Cedar (Arbor Vitae)

SIZE: About 32μ

RANGE: Quebec to Saskatchewan south to New England, northern Indiana and Minnesota, south to North Carolina in mountains.

Spheroidal, widely-scattered clumps of verrucae. The pollen of the several genera and species of the Cupressaceae are probably indistinguishable.

40a Shape spherical 41

40b Shape spheroidal to angular and irregular (Fig. 126) . . . . *Taxus floridana* Nutt.

Florida Yew

SIZE: About 27μ

RANGE: Limited area in northern Florida

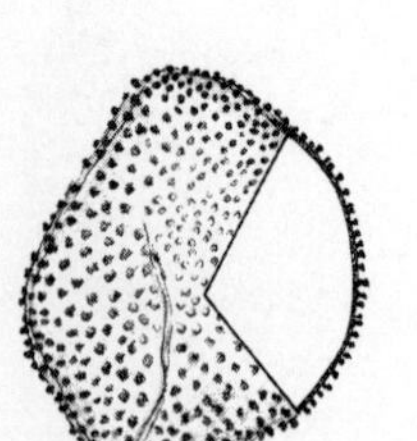

Fig. 126. *Taxus floridana* Nutt. Coll. Florida; TAXACEAE.

Shape irregular due to folds, often subtriangular, quadrate, subprolate, to spheroidal. Intectate, verrucate with a tendency to split irregularly. Virtually indistinguishable from *Taxus canadensis* Marsh.

41a Surface densely verrucate (Fig. 127, 128)............*Populus* spp.

Cottonwood

SIZE: About 45μ

RANGE: Quebec to Manitoba south to New England and widespread southward to grassland

Spheroidal; densely scabrate or fine verrucate.

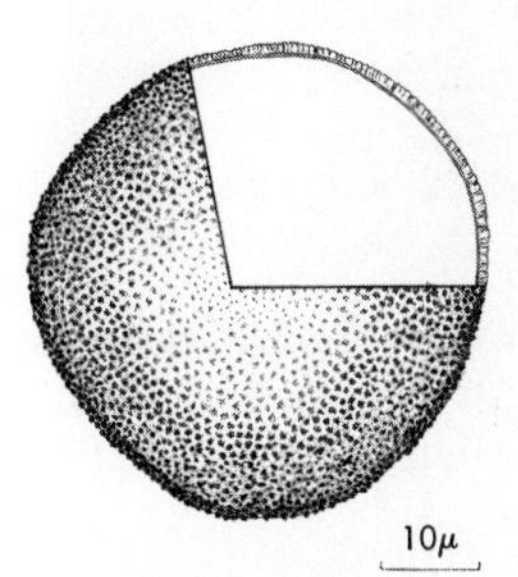

Fig. 127. *Populus deltoides* Marsh. Duke Univ. Herbarium; SALICACEAE.

Trembling Aspen

SIZE: About 45μ

RANGE: Very widespread in U.S. (except southeastern States) and Canada

Spheroidal, but sometimes folded; endexine sometimes separated from ektexine forming an internal sphere.

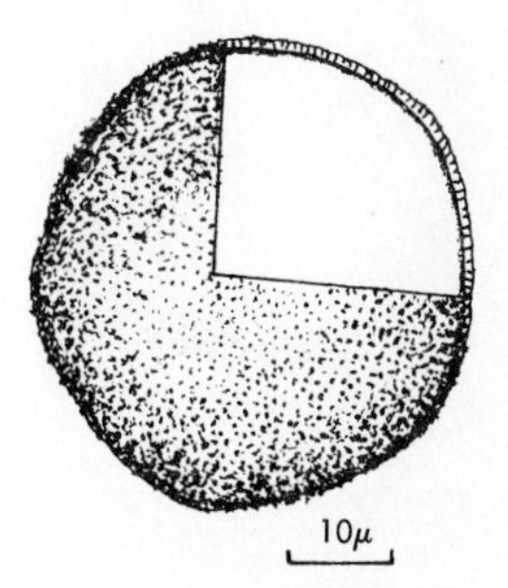

Fig. 128. *Populus tremuloides* Michx. Coll. N. Mexico; SALICACEAE.

41b Surface verrucate to clavate; columellae form a fine reticulum (Fig. 129)..............................*Potamogeton* spp.

Sheathed Pondweed

SIZE: About 32μ

RANGE: Widespread in Canada, Alaska, and northern United States

Spheroidal; surface appears perforate-tectate due to presence of partially-fused gemmae and clavae. The columellae give a somewhat reticulate appearance. Surface vaguely verrucate. The illustrated species is very similar to *P. amplifolius* and *P. pectinatus.*

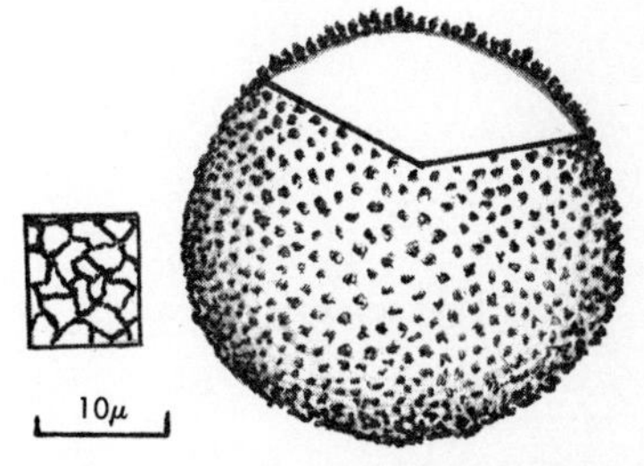

Fig. 129. *Potamogeton vaginatus* Turcz. Coll. Michigan; ZOSTERACEAE.

Potamogeton may be distinguished from *Typha* or *Sparganium* by the presence of the tectum and absence of the poroid aperture.

MONOPORATE TYPES

Monoporate pollen is characteristic of the monocotyledonous families Gramineae, Typhaceae, and Sparganaceae. In addition, pollen of the Cyperaceae may sometimes appear to have a single indistinct pore, although probably more pores are present (see Pericolpate type). Pollen of the Taxodiaceae (Gymnosperms) may have a single poroid papilla, but sometimes appears inaperturate. Some fungi have a single germination pore through which a hypha develops. Testaceous rhizopods (Protozoa, Sarcodina) may leave remains which have a single opening, thus resembling pollen or spores.

1a Diameter 100μ or more, opening without an annulus (Protozoan test) (Fig. 130) .... ***Arcella artocera***

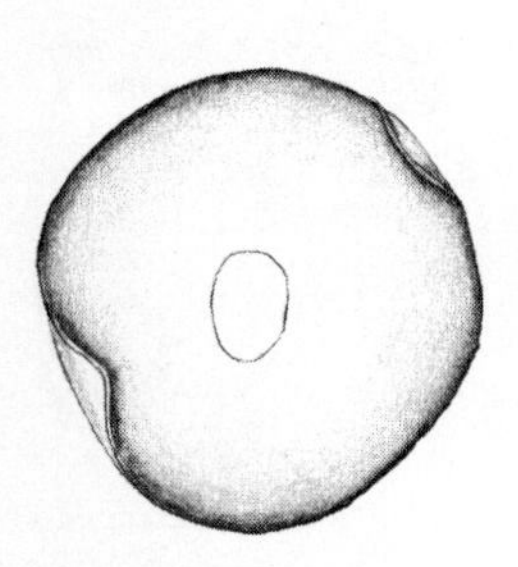

Fig. 130. *Arcella artocera;* ARCELLIDAE.

Testaceous Rhizopod

SIZE: 120μ (some species much larger)

RANGE: Widespread in fresh water

The test of this amoeboid protozoan is apparently chitinous and resistant to decay. *Gromia fluviatilus, Arcella vulgaris* and *A. nitrata* are rather similar to the species shown. *A. dentata* has a coarsely-toothed outline with 15 to 17 large marginal spines. *Difflugia* differs in having a test formed of cemented sand grains.

1b Diameter usually less than 100μ, except *Zea*, then with an annulate pore **2**

2a Wall psilate; pores, if distinct, not annulate **3**

2b Wall verrucate, reticulate or with granular appearance **6**

3a Spore unstalked **4**

3b Spore stalked **5**

4a Length 14 to 15μ (Fig. 131) .... ***Chaetomium elatum***

SIZE: About 10 x 15μ

Shape prolate, somewhat tapered at each end ("lemon-shaped"). Wall thinner at end opposite pore. Fresh spores brown.

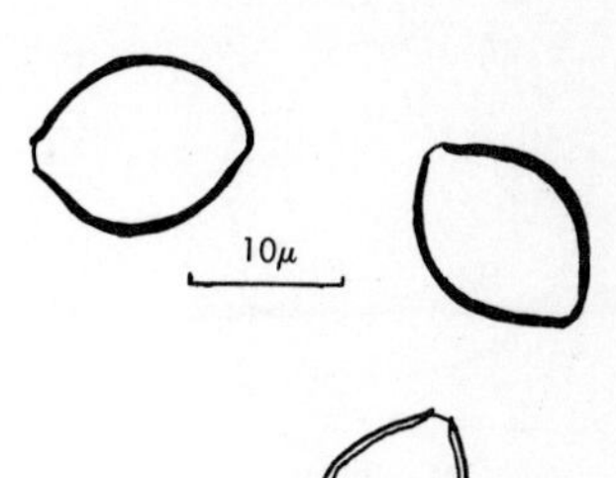

Fig. 131. *Chaetomium elatum;* CHAETOMIACEAE (ascospores).

4b Length 9 to 10μ (Fig. 132) .... ***Coprinus* sp.**

Inky-Cap Mushrooms

SIZE: 6(to 7)μ by 9(to 10)μ

Shape subspheroidal to eliptical; wall thick. Fresh spores dark.

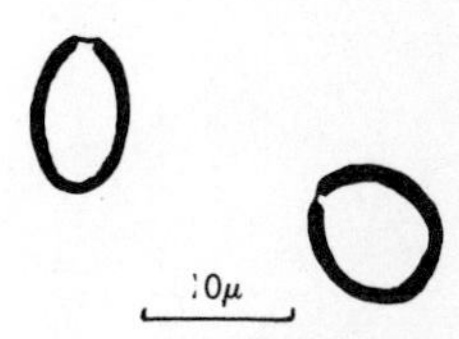

Fig. 132. *Coprinus cordisporus;* COPRINACEAE (basidiospores).

5a Shape prolate (Fig. 133) .... ***Uromyces fabae***

Broad Bean Rust

SIZE: About 24 to 35μ

Shape of spore prolate, stalked. Pore at end opposite stalk with internal thickenings of the walls near pore.

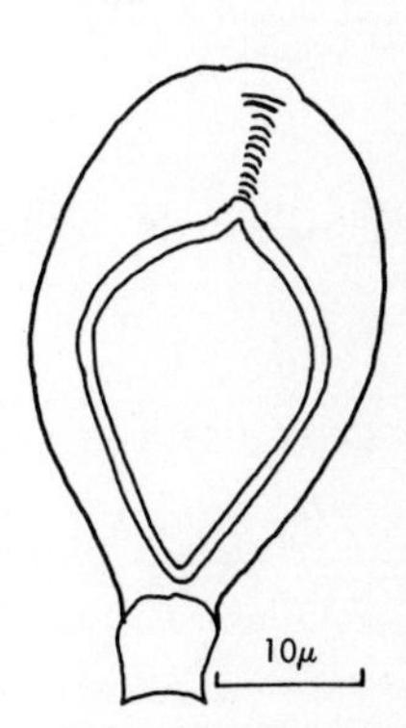

Fig. 133. *Uromyces fabae;* PUCCINIACEAE (teliospore).

5b Shape spheroidal (Fig. 134)..............*Uromyces appendiculatus*

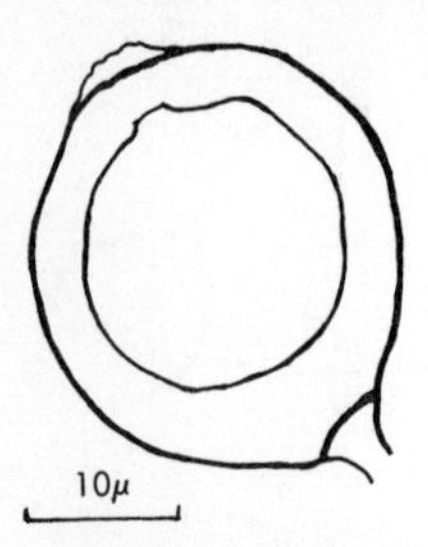

Fig. 134. *Uromyces appendiculatus;* PUCCINIACEAE (teliospore).

Bean Rust

SIZE: 25 to 28μ diameter

Subspheroidal in outline with stalk at proximal end and bulge at distal pore. Wall not thickened near the pore.

6a Pore with distinct annulus (grasses-Gramineae)..................8
6b Pore not annulate, but may be at tip of exit papilla.............7
7a Pollen grain with exit papilla; verrucate (Fig. 135, 136, 137)........ ....*Taxodiaceae*

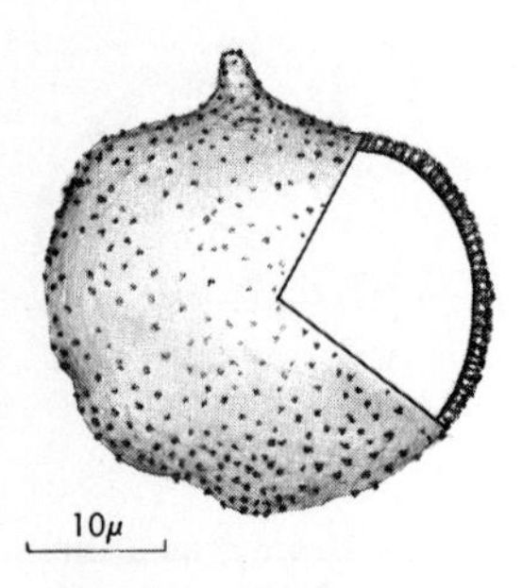

Fig. 135. *Taxodium distichum* (L.) Richard Coll. Kansas (cultivated); TAXODIACEAE.

Bald Cypress

SIZE: 30 to 32μ

RANGE: Florida to Texas north to southern New Jersey, southern Illinois and Oklahoma

Pore indistinct at tip of papilla. Shape suboblate. Exine apparently tectate with granular appearamce due to verrucae.

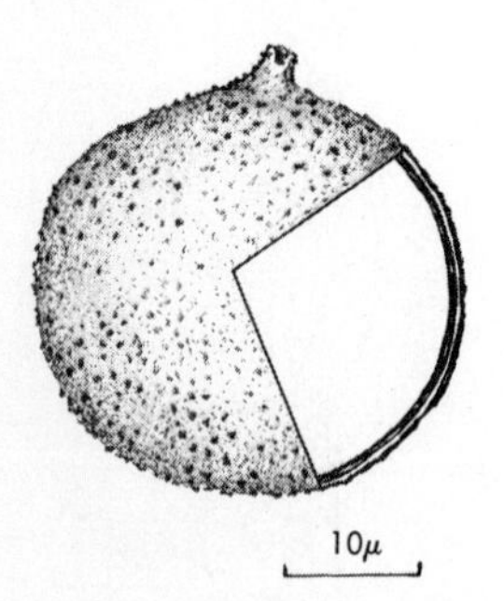

Fig. 136. *Sequoia sempervirens* (Lamb.) Endl. Coll. Arnold Arboretum; TAXODIACEAE.

Coast Redwood

SIZE: 29 to 32μ

RANGE: Coastal Northwestern United States

Aperture obscure, pore at tip of exit papilla. Shape slightly oblate. Exine apparently tectate with rather dense granular surface (verrucae). *Metasequoia* pollen is very similar to that of *Taxodium* and *Sequoia*.

Giant Redwood (Big Tree)

SIZE: About 22μ diameter

RANGE: Coastal California, west slope of Sierra Nevada Mts.

Shape spheroidal with long exit papilla (to 6 or 7μ). Very coarse verrucate wall surface.

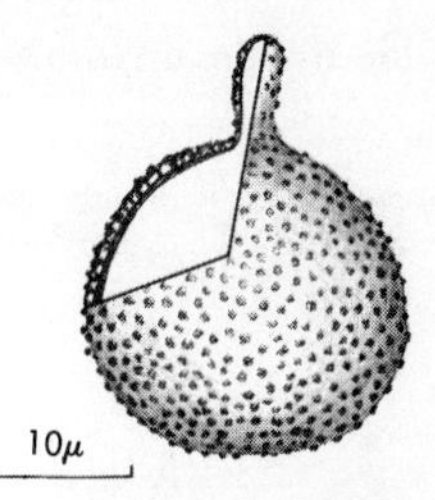

Fig. 137. *Sequoiodendron giganteum* (Lindl.) Bucholz Coll. California; TAXODIACEAE.

7b Pollen grain without exit papilla; reticulate (Fig. 138, 139)........... ....*Sparganium* (or *Typha*)

Bur-reed

SIZE: 27 to 32μ

RANGE: Throughout the United States and southern Canada

Shape spherical to more or less angular. Pore may be distinct or obscure. Probably indistinguishable from other species of *Sparganium*, or from *Typha angustifolia* (which produces monads) or single grains of *Typha latifolia* when the tetrads separate. Intectate or indistinctly perforate-tectate.

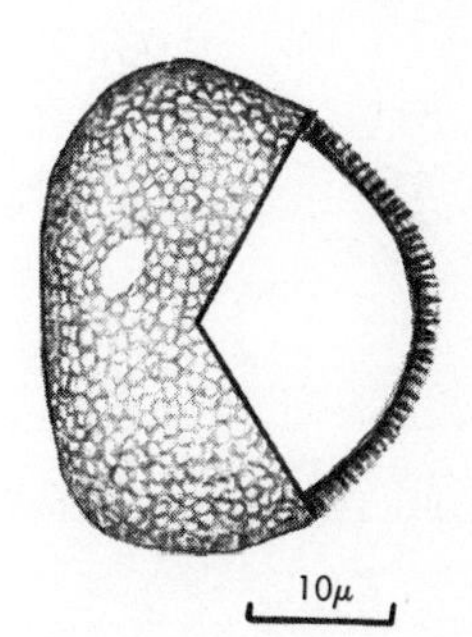

Fig. 138. *Sparganium eurycarpum* Engelm. Coll. Michigan; SPARGANACEAE.

Bur-reed

SIZE: About 32μ

RANGE: Southeastern Canada and eastern United States

Exine apparently intectate or perforate-tectate.

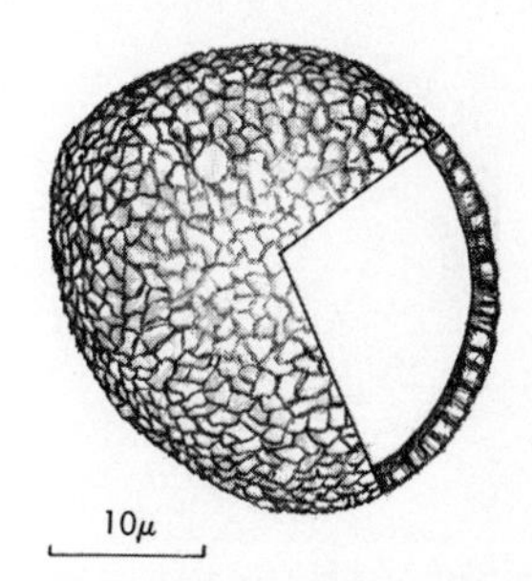

Fig. 139. *Sparganium americanum* Nutt. Coll. Texas; SPARGANACEAE.

8a Maximum diameter 70μ or greater (often greater than 100μ) (Fig. 140)..*Zea mays* L.

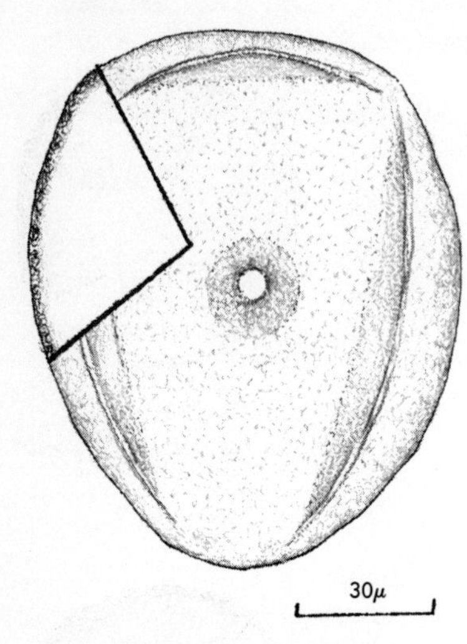

Fig. 140. *Zea mays* L. Coll. Michigan; (cultivated, hybrid); GRAMINEAE.

Maize (Corn)

SIZE: Illustrated specimen about 120μ

RANGE: Cultivated throughout

Shape spherical, often folded or wrinkled. Surface very faintly granular. Pore opening up to 7μ diameter to 15μ, including the annulus.

8b Maximum diameter less than 65μ..............................9

9a Size less than 40μ (Fig. 141)...Wild Grasses (example *Dactylis*-type)

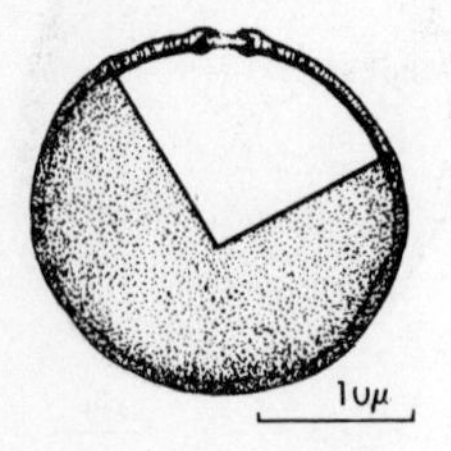

Fig. 141. *Dactylis glomerata* L. Coll. Michigan; GRAMINEAE.

Orchard-Grass

SIZE: About 25μ

RANGE: Widespread throughout United States and southern Canada

Shape spheroidal or suboblate. Exine tectate with indistinct verrucae. Columellae form a subdued microreticulate pattern. (The key to grasses in Faegri and Iversen, 1964, distinguishes four types in this size category: *Phragmites*-type; *Dactylis*-type; *Glyceria*-type; and *Festuca*-type.)

9b Size greater than 40μ (Fig. 142, 143, 144)............................ Cereals and Wild Grasses

Oats

SIZE: 40 to 60μ

RANGE: Cultivated throughout North America

Spheroidal, but densely granular appearance, wall heavy and thick. Pore up to 7μ in diameter.

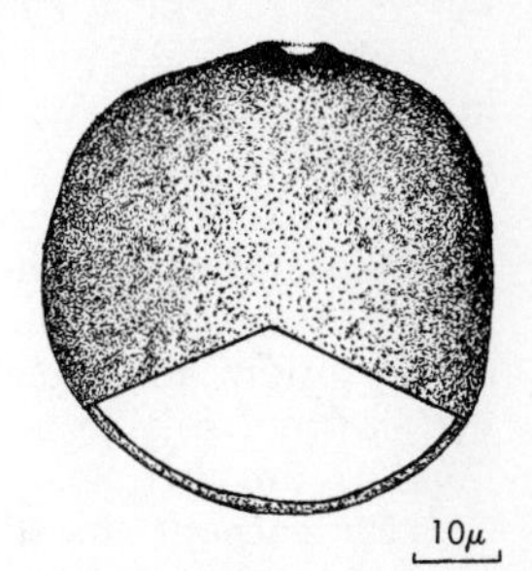

Fig. 142. *Avena sativa* L. Coll. Michigan (cultivated); GRAMINEAE.

Quackgrass

SIZE: About 50μ

RANGE: Widespread throughout

Pore 4μ in diameter. Indistinctly verrucate.

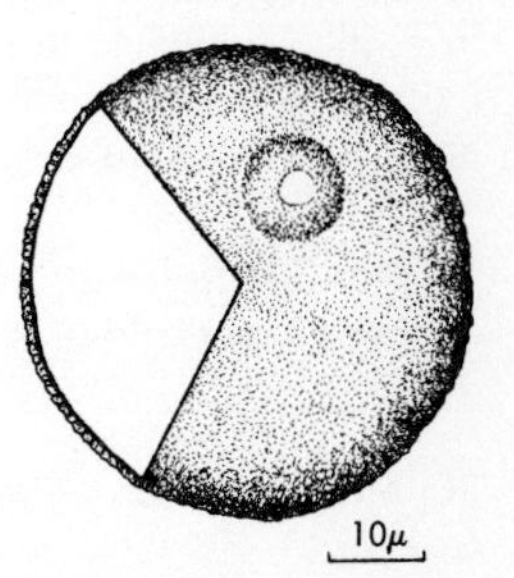

Fig. 143. *Agropyron cristatum* (L.) Beaur. Coll. Colorado; GRAMINEAE.

SIZE: About 50μ

RANGE: Wet places; Florida

Spheroidal to subprolate. Columellae form a subdued reticulate pattern.

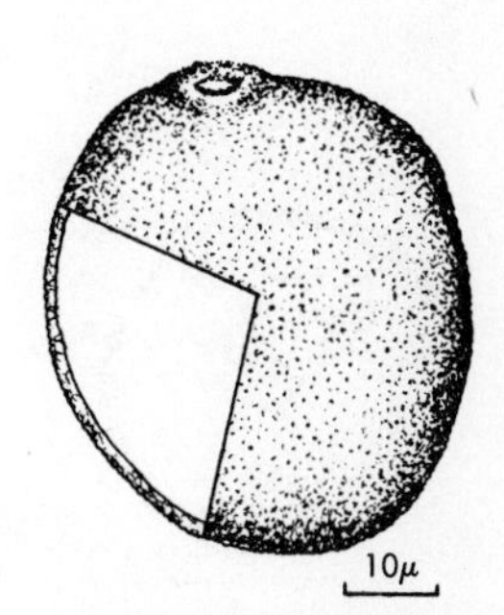

Fig. 144. *Echinochloa paludigena* Wiegand. Coll. Florida; GRAMINEAE.

MONOCOLPATE AND MONOLETE TYPES

Many genera of North American ferns, mostly in the Polypodiaceae, have spores with a single monolete scar; these spores typically have a smooth or sculptured endine and an outer perine which is frequently loosely attached or absent after processing. Some Gymnosperm pollen is monocolpate, for example, the Cycadales and Ginkgoales; pollen of the Taxaceae, Cupressaceae, and some Pinaceae (e.g. *Larix* and *Pseudotsuga*) is inaperturate, but may rupture to form an apparent furrow; vesiculate-monocolpate gymnosperms are keyed elsewhere. The single furrow at the distal pole of the pollen grain is considered to be the evolutionarily primitive condition. Accordingly, dicotyledons in the Ranales are frequently monocolpate (Nymphaeaceae; Magnoliaceae, Calycanthaceae). Most monocotyledons produce pollen with single apertures; most members of the Liliaceae, Amaryllidaceae, Iridaceae, Orchidaceae, Pontederiaceae, Araceae, Palmae and Commelinaceae are monocolpate. Since this section of the key contains only a small fraction of the many North American species with these pollen and spore types, it should be expected that types will be encountered which may key incorrectly to another species. However, this section should be helpful in determining general taxonomic affinities of unknown monocolpate or monolete specimen types.

1a Outline (lateral) of spore kidney-shaped, a single monolete scar, perine persistent or lacking (Fig. 145, 146, 147)..Monolete Fern Spores

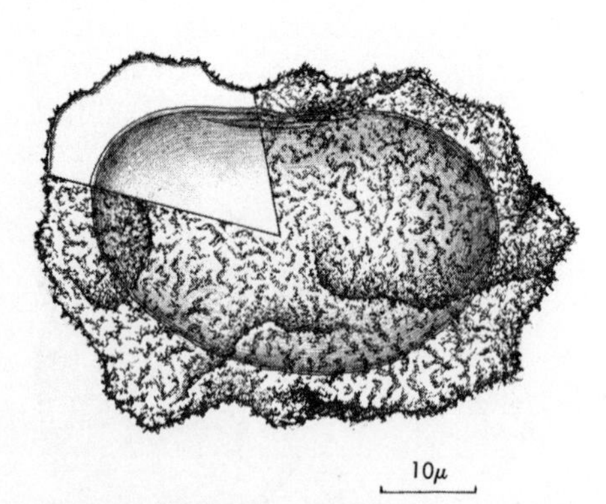

Fig. 145. *Dryopteris spinulosa* (O. F. Muell.) Watt. Coll. Indiana; POLYPODIACEAE.

Spinulose Wood-fern

SIZE: Endospore length 45μ, including perisporium 50-55μ

RANGE: The several varieties are very widespread in the eastern and northwestern United States and Canada primarily southern Labrador to Alberta, Idaho and Washington (in subalpine areas) south to North Carolina, Tennessee, Missouri and Minnesota.

Spore kidney-shaped, rugulate with a verrucate and/or subechinate loose-fitting perisporium.

Virginian Chain-fern

SIZE: Endospore about 35μ long, including perisporium about 50μ

RANGE: Florida to Texas north to southern Michigan, southern Ontario, southwestern Quebec and Nova Scotia

Endospore kidney-shaped as in many Polypodiaceae; perisporium much folded and very loosely adherent to spore. Perispore wall granular (verrucate), endospore wall psilate.

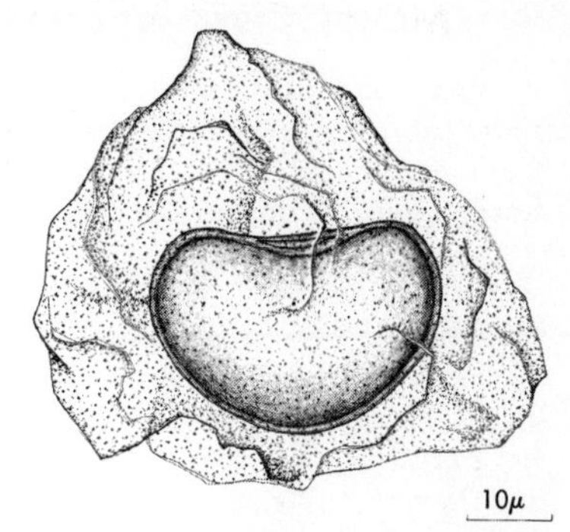

Fig. 146. *Woodwardia virginica* (L.) Sur. Coll. Michigan; POLYPODIACEAE.

SIZE: Length about 45μ

RANGE: Cretaceous, North America and Europe

Monolete microspores (isospores ?); scar about two-thirds the length of the spore. Exine with polygonal to irregular coarse-meshed reticulum. Probably fossil fern in the Polypodiaceae (?).

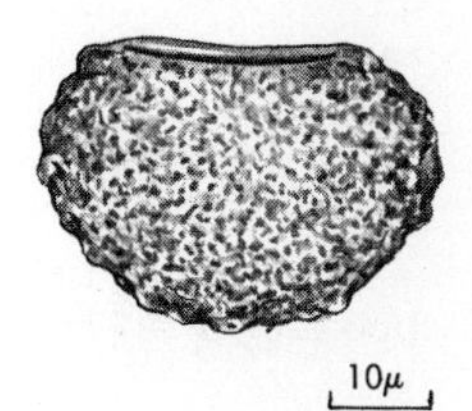

Fig. 147. *Reticuloidosporites dentatus* Pflug. Coll. Cretaceous, Montana (Oltz, 1968), Fossil.

1b Outline of pollen grain spherical, elliptical or boat-shaped; perine never present; true aperture 2

2a Surface echinate 3

2b Surface psilate, gemmate, verrucate, reticulate, etc., not echinate .. 6

3a Shape elliptical (Fig. 148) .... *Nuphar*

Yellow Pond Lily

SIZE: 35-65μ

RANGE: Florida to Texas and eastern Mexico north to Nebraska, Wisconsin, southern Michigan and New England

A single long furrow extends the entire length of grain; long, tapered, pointed spines vary from 4-7μ long. Size remarkably variable.

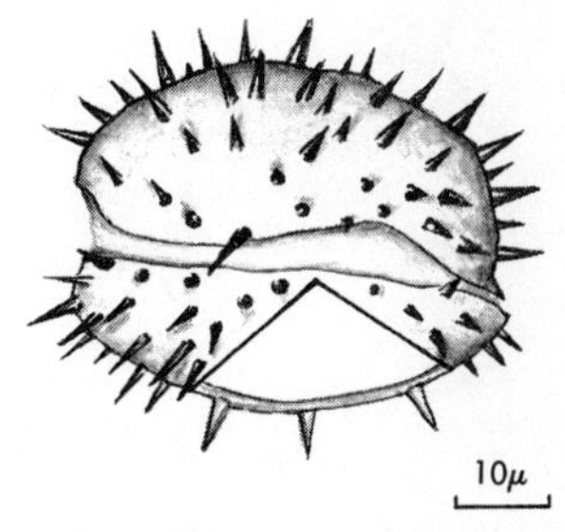

Fig. 148. *Nuphar advena* (Ait.) Ait. f. Coll. Michigan; NYMPHAEACEAE.

3b Shape approximately spherical 4

4a Aperture straight; echinae and baculae long and slender (Fig. 149) .... *Nymphaea*

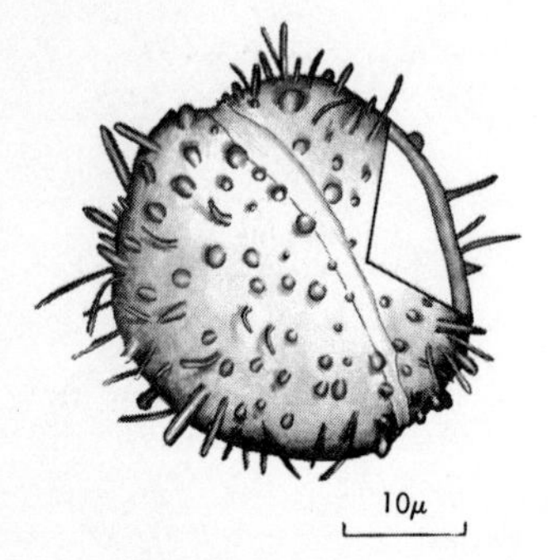

Fig. 149. *Nymphaea odorata* Ait. Coll. Michigan; NYMPHAEACEAE.

Water-lily

SIZE: Subspherical to slightly elongate, 29 x 32μ

RANGE: Florida to Louisiana north to Manitoba and Nova Scotia

Grain with numerous and variable processes (gemmae, clavae, baculae, echinae). *Nymphaea tetragona* lacks distinct projections.

4b Aperture arc-shaped; echinae short and broad-based 5

5a Size about 25μ (Fig. 150) .... *Arisaema* spp.

Jack-in-the-Pulpit

SIZE: About 25μ in diameter

RANGE: New Brunswick and Quebec to Manitoba south to South Carolina, Tennessee and eastern Kansas

Shape spheroidal; the single arc-shaped furrow somewhat poroid in character. Exine stratification obscure, apparently intectate. Widely-spaced broad-based spines are less than 1μ long. *Arisaema dracontium* (coll. Texas) is very similar.

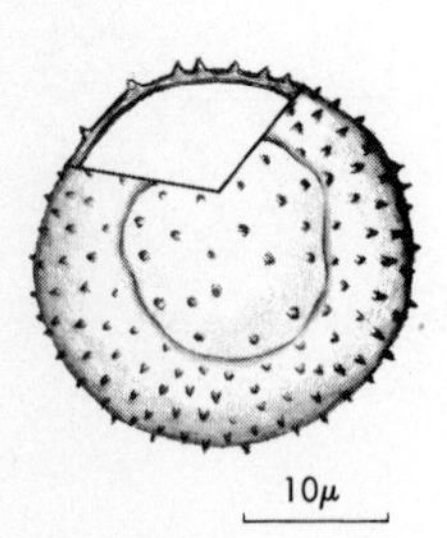

Fig. 150. *Arisaema atrorubens* (Ait.) Blume Coll. Michigan; ARACEAE.

5b Size about 18-19μ (Fig. 151) .... ***Smilax* spp.**

Carrion-Flower

SIZE: Diameter about 19μ

RANGE: Southern Quebec to Manitoba, south to Virginia and upland Georgia and Alabama, Tennessee, and Missouri

The single arc-shaped furrow is sometimes indistinct; in other cases it is obvious and appears to be nearly a circular "operculum." Surface beset with very short baculae and spines (about 0.5μ).

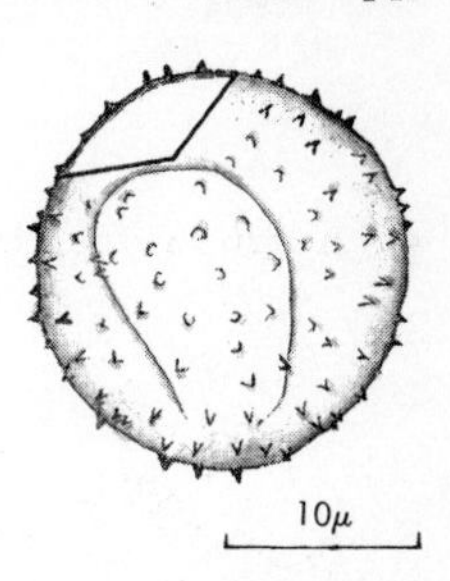

Fig. 151. *Smilax herbacea* L. Coll. Michigan; LILIACEAE.

6a Opening in wall an irregular, or often broad, gaping rupture, not a true aperture **7**

6b Opening in wall a more or less regular distal furrow, sometimes indistinct **9**

7a Surface psilate, Pinaceae in part **8**

7b Surface verrucate or gemmate (Fig. 152, 153) .... **Cupressaceae, Taxaceae, or Taxodiaceae may key here if grains are ruptured (see Inaperturate Type, p. 52)**

Alaska Cypress

SIZE: Diameter about 34μ

RANGE: Coastal Alaska to Oregon

Spherical and actually inaperturate, but almost every grain has a sulcoid tear which may be confused with a furrow. Surface densely, but irregularly, covered with verrucae, gemmae and short baculae.

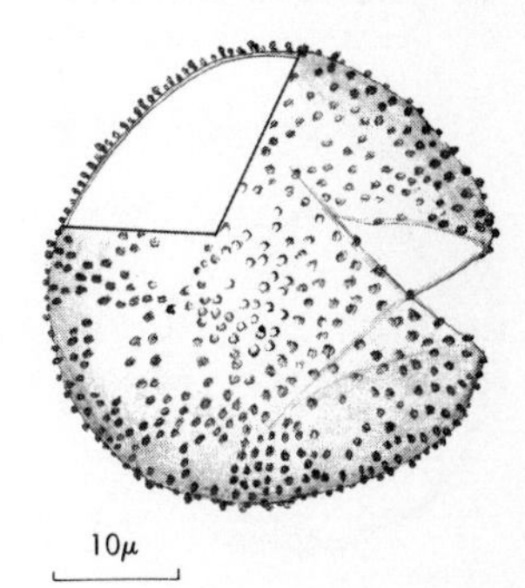

Fig. 152. *Chamaecyparis nootkanensis* (Lamb.) Spach. Coll. Washington; CUPRESSACEAE.

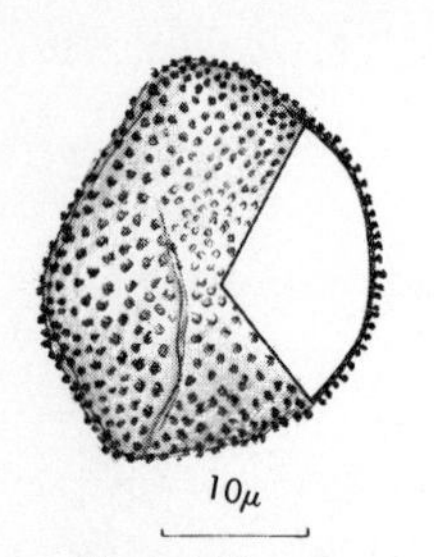

Fig. 153. *Taxus floridana* Nutt. Coll. Florida; TAXACEAE.

Florida Yew

SIZE: Maximum dimension about 27μ

RANGE: Locally in Florida

Shape variable, mostly due to folding of the wall. This grain is actually inaperturate, but has tendency to split forming an irregular furrow-like slit. Exine intectate distinctly verrucate. *Taxus canadensis* Marsh pollen is very similar (indistinquishable?).

8a Diameter usually greater than 90μ.................. *Pseudotsuga*

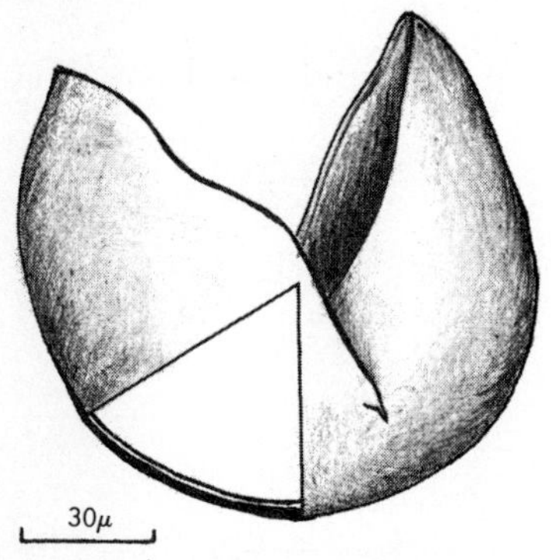

Fig. 154. *Pseudotsuga menziesii* (Mirb.) Franco Coll. Montana; PINACEAE.

Douglas Fir

SIZE: 80-110μ

RANGE: Alberta to British Columbia, south to western Texas, northern Mexico and California; uplands

Exine thin and smooth, no trace of surface sculpture or aperture; however, the grain often ruptures to form two gaping hemispheres. This split may superficially resemble a furrow.

8b Diameter usually 60-85μ (Fig. 155)...................... *Larix* spp.

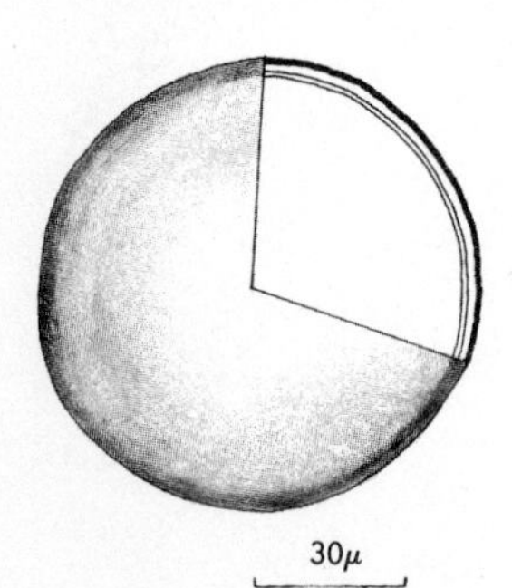

Fig. 155. *Larix laricina* (DuRoi) Koch Coll. Michigan; PINACEAE.

Larch or Tamarack

SIZE: Diameter 50-70μ

RANGE: Labrador to Alaska south to Pennsylvania, west Virginia, northern Illinois and Minnesota to British Columbia

Exine thin, psilate and inaperturate. Often ruptures to form a furrow-like area.

9a Exine reticulate..10

9b Exine psilate, verrucate, or striate..............................14

10a Size greater than 75μ (Fig. 156)............*Agave parryi* Engelm.

Century Plant

SIZE: About 100μ

RANGE: New Mexico, Arizona and northern Mexico

A single short broad furrow is evident. Exine per-reticulate (perforate tectate); reticulum coarse, shape subspherical or with distinctly flattened faces.

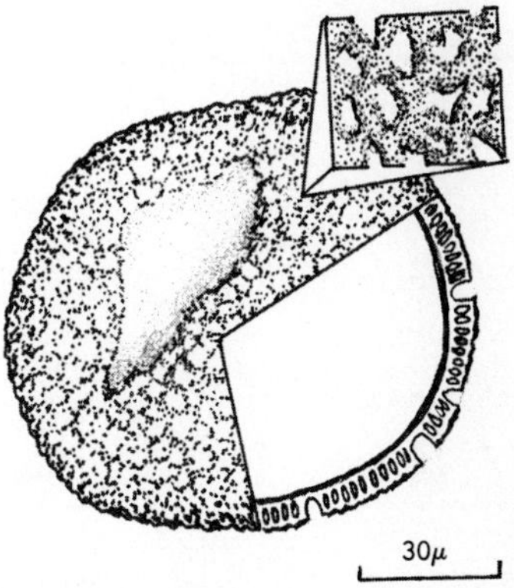

Fig. 156. *Agave parryi* Engelm. Coll. Texas; AMARYLLIDACEAE.

10b Size mostly 30-45μ, to 70μ in *Iris setosa*......................11

11a Shape spheroidal to slightly elongate (Fig. 157)................ ....*Symplocarpus foetidus* (L.) Nutt.

Skunk-cabbage

SIZE: 40μ long

RANGE: Quebec to southeastern Manitoba, south to Iowa, Indiana, Virginia and uplands south to Georgia

Shape spheroidal to somewhat elongate. Exine per-reticulate. Furrow broad, generally tapering, with persistent verrucate membrane.

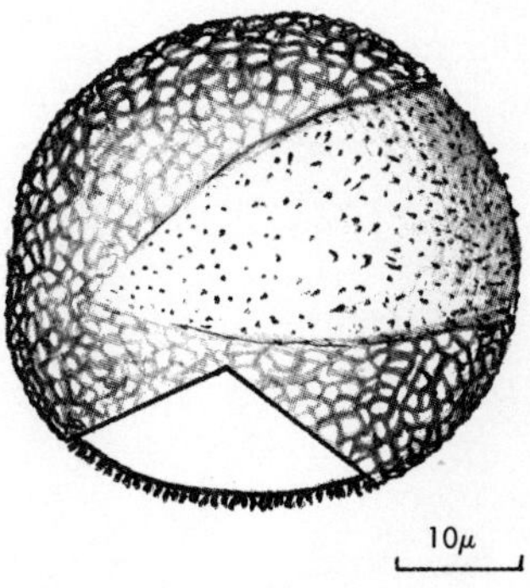

Fig. 157. *Symplocarpus foetidus* (L.) Nutt. Coll. Michigan; ARACEAE.

11b Shape elliptical..12

12a Reticulum restricted to the broad centrum of the grain, verrucate at ends (Fig. 158) .... ***Smilicina racemosa*** **(L.) Desf.**

False Spikenard

SIZE: 27 x 34μ

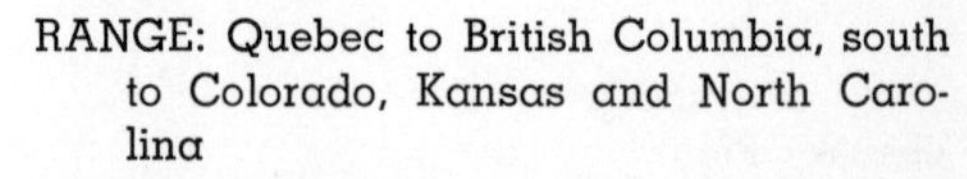

RANGE: Quebec to British Columbia, south to Colorado, Kansas and North Carolina

Shape elliptical. Exine intectate, perreticulate. Reticulum coarse at the equators grading to a verrucate surface at the ends. Furrow slender, somewhat shorter than the grain.

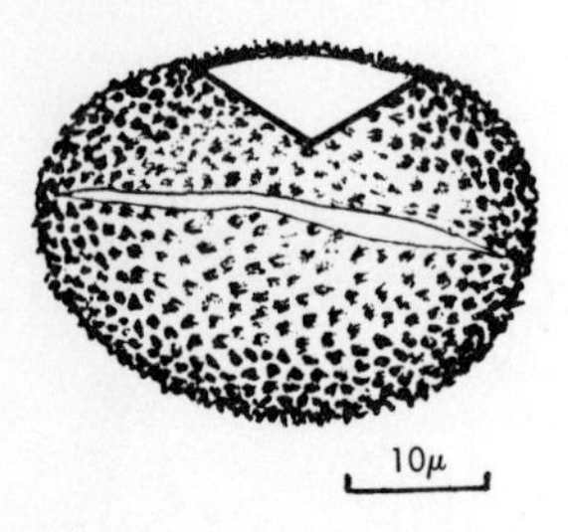

Fig. 158. *Smilicina racemosa* (L.) Desf. Coll. Michigan; LILIACEAE.

12b Reticulum throughout **13**

13a Reticulum of coalesced baculae, finer lacunae near the furrow (Fig. 159) .... ***Iris setosa*** **Pall.**

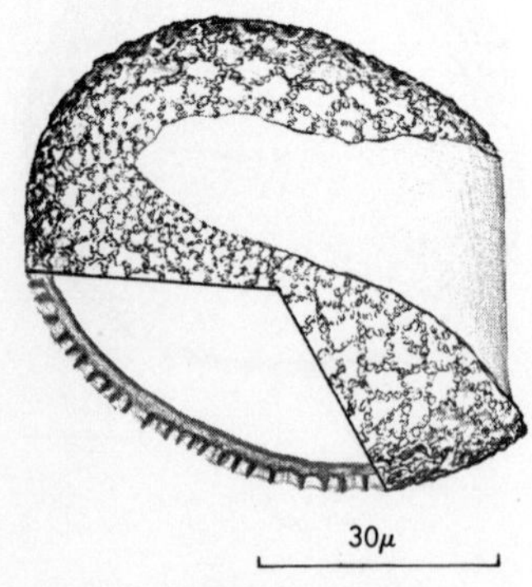

Fig. 159. *Iris setosa* Pall. Coll. Alaska; IRIDACEAE.

Wild Iris

SIZE: About 50 x 70μ

RANGE: Alaska to Labrador, Newfoundland and Maine

Reticulum coarse, except finer near furrows. Furrow very long and broad, with irregular margins. Shape eliptical.

13b Reticulum of undulating ridges, perforations in lacunae (Fig. 160) .... *Sabal minor* Becc.

Dwarf Palmetto

SIZE: 25 x 32μ

RANGE: Coastal Plain, North Carolina to Florida and Texas.

Exine vaguely reticulate due to pattern of the undulating surface. Perforations in the tectum appear to occur in the thin portions of the reticulum. Furrow very long, exceeding the length of the grain.

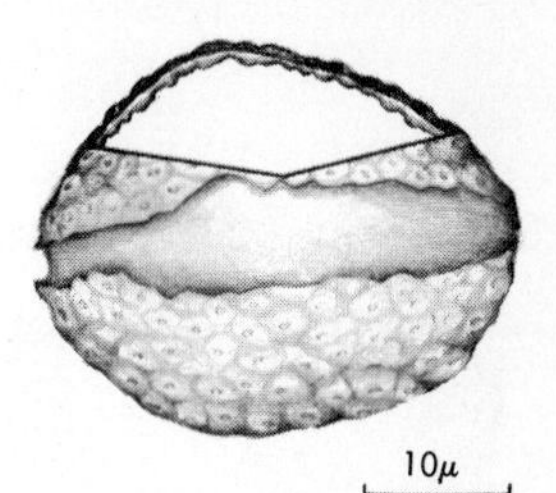

Fig. 160. *Sabal minor* Becc. Coll. Florida; PALMAE.

14a Size large, greater than 60μ 15

14b Size smaller, less than 50μ 19

15a Surface striate (Fig. 161) .... *Cabomba caroliniana* Gray

Water-shield

SIZE: 63 x 93μ

RANGE: Coastal Plain to North Carolina and Florida to Texas, north to Missouri

Very large with broad long furrow with irregular margins. Exine tectate with columellae arranged in rows to form striations.

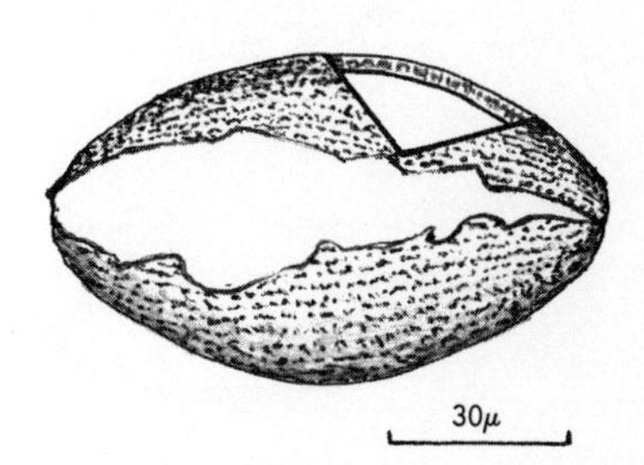

Fig. 161. *Cabomba caroliniana* Gray Coll. Florida; NYMPHAEACEAE.

15b Surface psilate, verrucate or micro-reticulate 16

16a Length 60 to 70μ (Fig. 162)............*Yucca mohavensis* Sargent

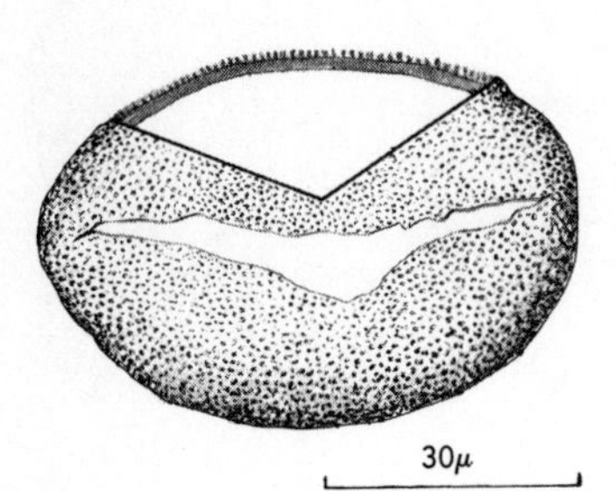

Fig. 162. *Yucca mohavensis* Sargent; LILIACEAE.

Mojave Yucca

SIZE: 44 x 66μ

RANGE: Southern Nevada, Arizona, southeastern California and Baja California

Furrow slender, but expanded at center, margin irregularly roughened. Exine tectate, but tectum very thin (oil!); columellae large and visible giving the surface a granular or fine reticulate appearance. The dicolpate pollen of *Pontederia* may occasionally be interpreted as monocolpate; it would then key to this choice.

16b Length 70μ or greater; if less than 70μ, coarsely reticulate......17

17a Exine coarsely and irregularly verrucate (Fig. 163)...*Liriodendron tulipifera* L.

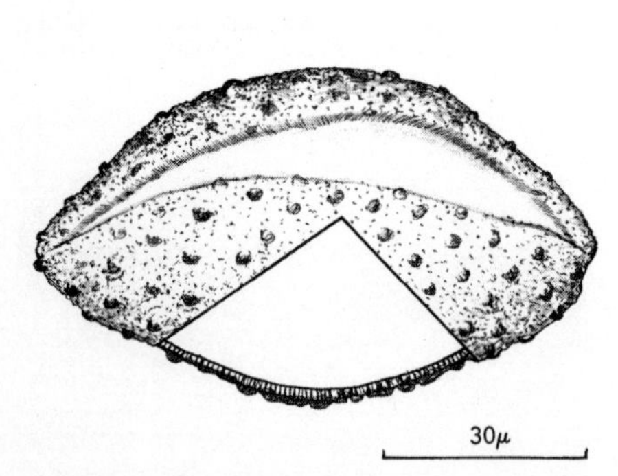

Fig. 163. *Liriodendron tulipifera* L. Coll. Arn. Arb.; MAGNOLIACEAE.

Tulip Tree

SIZE: 50 x 88μ

RANGE: Massachusetts to southern Ontario and Wisconsin south to Arkansas and northern Florida

Shape ellipsoid and narrowed at ends. Furrow long, broad at center with gradual taper. Exine tectate with large irregularly-shaped supratectate verrucae giving a distinctive surface sculpture.

17b Exine psilate or with fine verrucae..........................18

18a Furrow shorter than grain length (Fig. 164)..........*Magnolia* sp.

Southern Magnolia

SIZE: Length about 100μ; width and height about 60μ

RANGE: Coastal Plain Florida and Texas. Cultivated northward to limit of tolerance.

Shape flattened-ellipsoidal wall flat or concave on the furrow side (b). Furrow long, with rounded ends, indistinct in some specimens (a). Exine tectate.

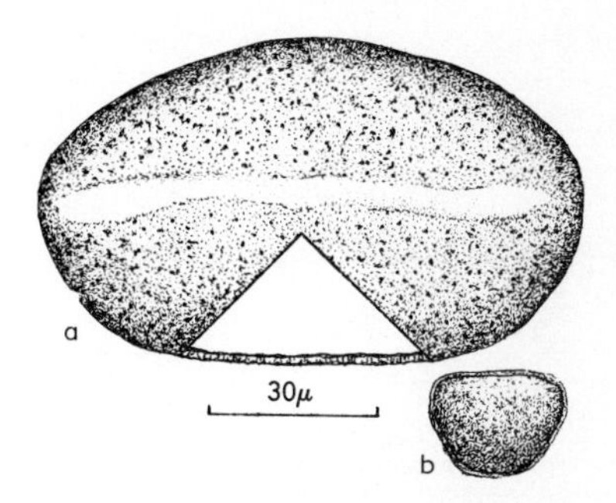

Fig. 164. *Magnolia grandiflora* L. Coll. Florida; MAGNOLIACEAE.

18b Furrow extends the full length of grain..*Hypoxis hirsuta* (L.) Coville

Stargrass

SIZE: Length about 37μ

RANGE: Florida to Texas northward to North Dakota, Manitoba, Ohio and southwest Maine

Shape elliptical; furrow very long and broad. Exine tectate with fine verrucae.

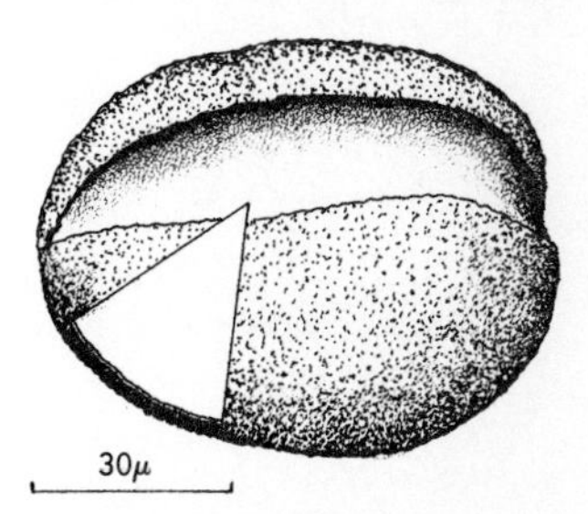

Fig. 165. *Hypoxis hirsuta* (L.) Coville Coll. Michigan; AMARYLLIDACEAE.

19a Longest dimension less than 20μ..............................20
19b Longest dimension generally greater than 30μ................25
20a Shape spherical..21
20b Shape elliptical...............*Anemopsis californica* (Nutt.) Hook.

Yerba-mansa

SIZE: About 13 x 18μ

RANGE: Western Texas to Utah, Arizona and California, southward into Mexico

Shape spherical to elliptical. Exine psilate, stratification obscure, apparently intectate. Furrow membrane persistent, verrucate to subreticulate.

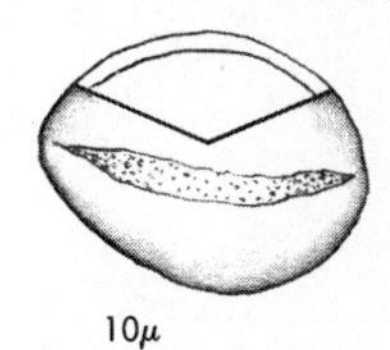

Fig. 166. *Anemopsis californica* (Nutt.) Hook. Coll. Colorado; SAURURACEAE.

21a Aperture arc-shaped, opening as a "flap" (Fig. 167)............ ....*Operculites carbonis* Newman

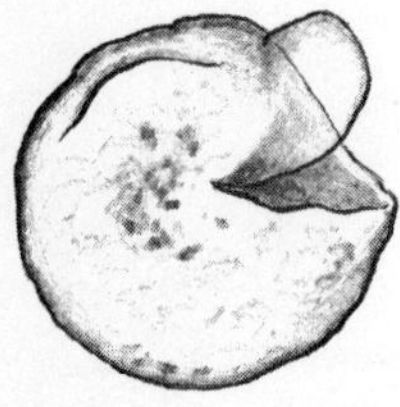

Fig. 167. *Operculites carbonis* Newman Coll. Cretaceous of Montana; FOSSIL.

SIZE: about 24μ

RANGE: Cretaceous of North America

Spherical with a curved aperture near one end from which a "flap" of exine may protrude. Surface psilate. Relationship unknown.

21b Aperture an approximately straight furrow..................22

22a Exine psilate...23

22b Exine verrucate, striate, or subreticulate......................24

23a Diameter 15-16μ (Fig. 168)...................*Acorus calamus* L.

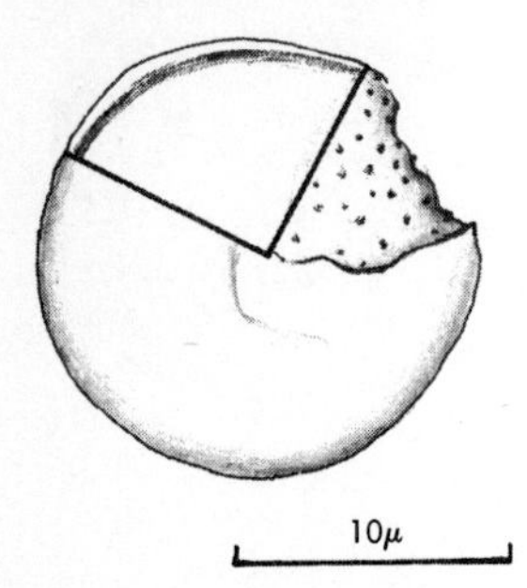

Fig. 168. *Acorus calamus* L. Coll. Indiana; ARACEAE.

Sweet Flag

SIZE: About 16μ diameter

RANGE: Prince Edward Island to Montana and Oregon south to Texas and Florida

Shape usually spherical or subelliptical; exine psilate with obscure stratification. Furrow with an irregular border, membrane flecked with scattered verrucae.

23b Diameter about 13μ (Fig. 169).............*Saururus cernuus* L.

Lizard's Tail

SIZE: About 13μ

RANGE: Florida to Texas north to southeastern Kansas; southern Michigan, southwestern Quebec and Rhode Island

Spherical and psilate grain. Furrow long, broad centered with attenuate ends and irregular margins, membrane persistent and verrucate. Exine stratification obscure or absent.

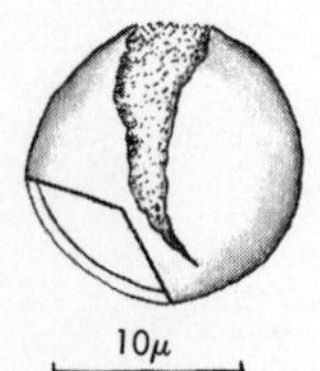

Fig. 169. *Saururus cernuus* L. Coll. Texas; SAURURACEAE.

24a Size about 11 x 13μ, slightly prolate (Fig. 170)... *Bryum caespiticium* Hedw.

SIZE: 11 x 13μ

RANGE: Widely distributed in United States and northward; also Eurasia

Spherical or discoidal, often split into two hemispheres. Wall densely verrucate. Spores of *Pohlia nutans, Leptobryum pyriforme,* and *Mnium spinulosum* resemble this species rather closely.

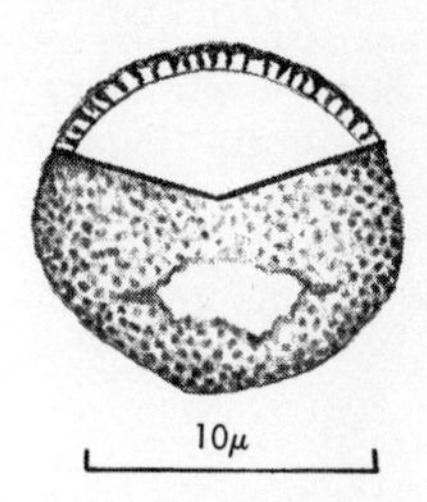

Fig. 170. *Byrum caespiticium* Hedw. Coll. Michigan; BRYACEAE (moss).

24b Size about 19μ, approximately spherical (Fig. 171)... *Funaria hygrometrica* Hedw.

SIZE: 19μ

RANGE: Cosmopolitan

Spherical or flattened on one face. Indistinct fold or "monolete" scar on one side. Heteropolar, irregularly verrucate or subreticulate on one face, more-or-less rugulate on the aperturate side.

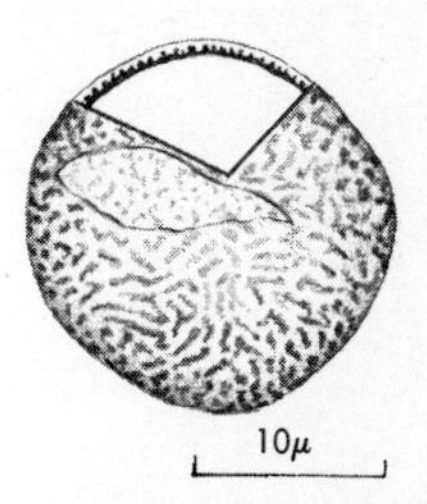

Fig. 171. *Funaria hygrometrica* Hedw. Coll. Michigan; FUNARIACEAE (moss).

25a Exine surface psilate (Fig. 172)............*Zamia floridana* A. DC.

Wild Sago

SIZE: 25 x 42μ

RANGE: Florida to lower Keys

Shape oval-boat-shaped; furrow long and with expanded ends. Furrow opens tardily and lies in a concave fold, indistinctly tectate or intectate.

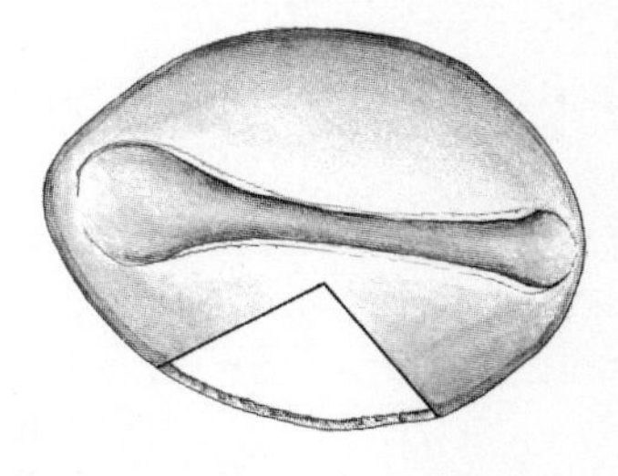

Fig. 172. *Zamia floridana* A. DC. Coll. Florida; CYCADACEAE.

25b Exine surface scabrate to verrucate **26**

26a Furrow very broad at midpoint, tapering abruptly, shorter than grain (Fig. 173) .... ***Ginkgo biloba* L.**

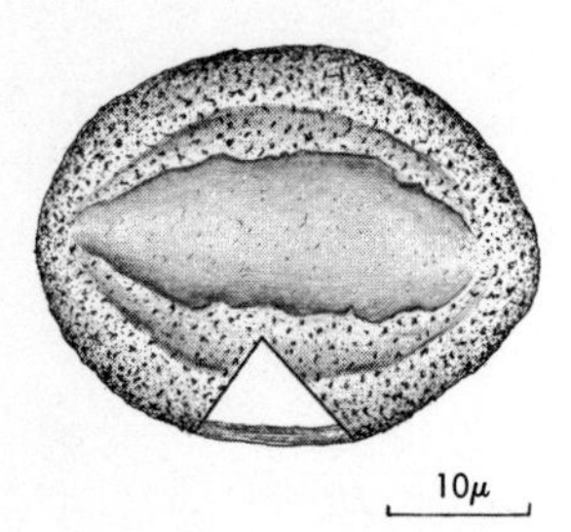

Fig. 173. *Ginkgo biloba* L. Coll. Michigan (cult.); GINKGOACEAE.

Maidenhair Tree

SIZE: About 28 x 36μ

RANGE: Introduced and widely cultivated in the United States

Boat-shaped with furrow in concave face. Furrow border very irregular. Exine tectate (somewhat indistinct) with surface verrucae.

26b Furrow otherwise **27**

27a Aperture full length of grain or longer **28**

27b Aperture shorter than grain **29**

28a Furrow slender, with irregular margin, and extending nearly around the grain (Fig. 174) .... ***Calycanthus floridus* L.**

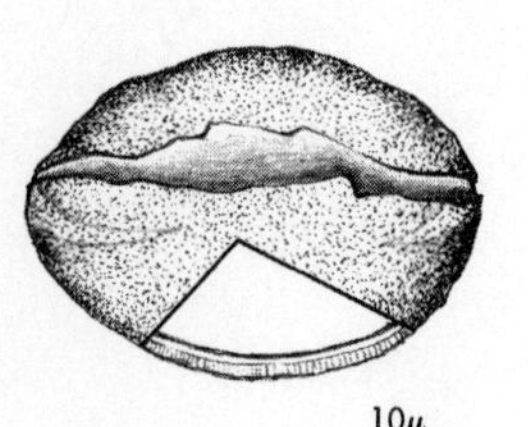

Fig. 174. *Calycanthus floridus* L. Coll. Georgia; CALYCANTHACEAE.

Carolina Allspice

SIZE: About 32 x 47μ

RANGE: Mississippi and Florida north to Virginia and West Virginia

Furrow (if single) very long extending nearly around grain. Some grains are apparently di- or tricolpate. Exine tectate, surface psilate; columellae impart granular to subreticulate appearance.

28b Furrow broad, with smooth margins, only slightly longer than grain (Fig. 175)........*Polygonatum canaliculatum* (Muhl.) Pursh.

Channelled Solomon's Seal

SIZE: About 35 x 41μ

RANGE: Connecticut to southern Manitoba south to Missouri, Oklahoma and South Carolina

Shape elliptical, furrow long and broad with parallel margins. Exine psilate, scabrate, or faintly reticulate; apparently intectate.

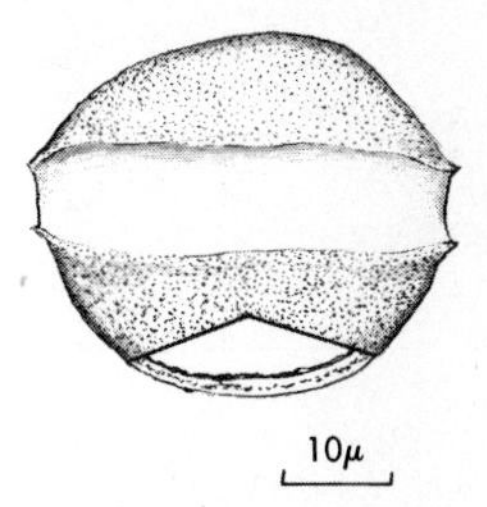

Fig. 175. *Polygonatum canaliculatum* (Muhl.) Pursh. Coll. Michigan; LILIACEAE.

29a Furrows slender, tapering to pointed ends......................30

29b Furrows broad, especially near the rounded ends (Fig. 176).......................... *Trillium flexipes* Raf. var. *Walpolei* (Farw.) Fern.

Trillium

SIZE: About 28 x 34μ

RANGE: Central New York to southern Minnesota south to Missouri, Tennessee and Maryland

Shape irregularly elliptical; exine intectate verrucate or with a faint "swirled rugulate" pattern. Furrow usually broadest at ends, with a fine-verrucate membrane.

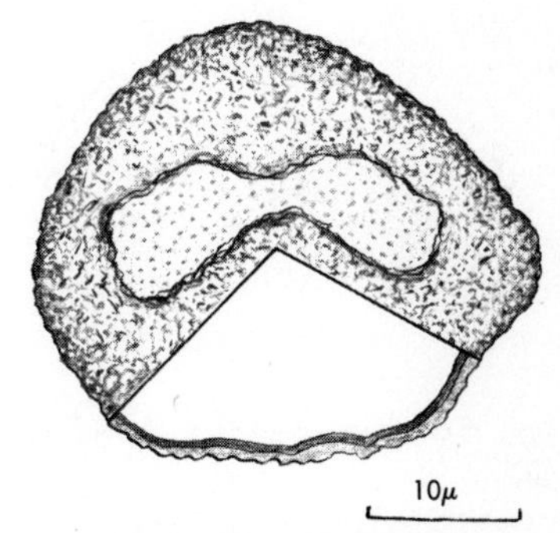

Fig. 176. *Trillium flexipes* Raf. var. *Walpolei* (Farw.) Fern. Coll. Michigan; LILIACEAE.

30a Outline boat-shaped; furrow about 3/4 as long as grain (Fig. 177) .... *Tradescantia ohioensis* Raf.

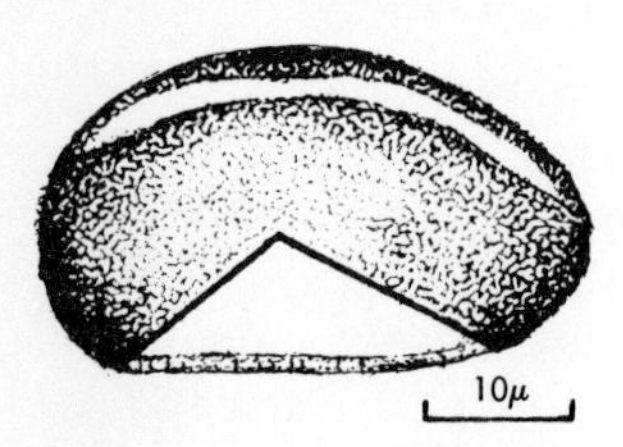

Fig. 177. *Tradescantia ohioensis* Raf. Coll. Indiana; COMMELINACEAE.

Spiderwort

SIZE: About 20 x 38μ

RANGE: Massachusetts to Minnesota south to Texas and Florida

Shape elliptical to kidney-shaped; narrow furrow on convex side. Exine apparently intectate, surface scabrate to verrucate.

30b Outline elliptical, often folded; furrow 1/3 to 1/2 as long as grain (Fig. 178)............................. *Cypripedium calceolus* L.

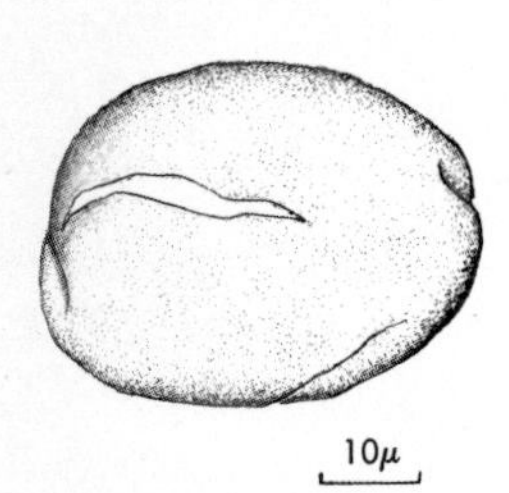

Fig. 178. *Cypripedium calceolus* L. Coll. Michigan; ORCHIDACEAE.

Yellow Lady's Slipper

SIZE: Length about 43μ

RANGE: The several varieties range from Newfoundland to British Columbia south to Washington, Utah, New Mexico and Georgia

Shape elliptical. Furrow slender and short. Exine intectate, scabrate or fine verrucate. Exine wrinkled. Either joined as pollinia and tetrads, or separate monads.

SYNCOLPATE TYPES

In addition to the illustrated genera, pollen grains with fused furrows are found in some members of the Papaveraceae, Scrophulariaceae and Ruppiaceae.

1a Furrows oriented meridionally.................................2

1b Furrows non-meridional, grain large.............................3

2a Grains small, less than 20μ, psilate (Fig. 179). . . . .*Primula*

Bird's Eye Primrose

SIZE: Polar axis 7.5-8μ, equatorial diameter 13.5-14.5μ

RANGE: Labrador to Alaska, south to Newfoundland, upper Great Lakes and British Columbia

Each of three furrows broadens and joins the others at the poles. Grain constricted along the furrows forming a strongly trilobed unit. Grains small (< 15μ) and often dimorphic (two size classes). a, polar view; b, equatorial view.

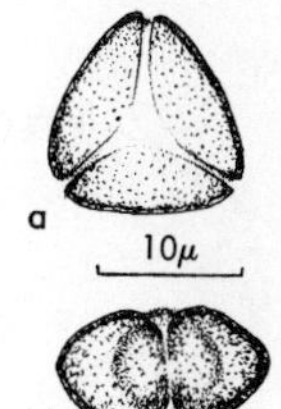

Fig. 179. *Primula mistassinica* Michx. Coll. Michigan; PRIMULACEAE

2b Grains larger, usually > 25μ, reticulate. . . .*Illicium floridanum* Ellis

Anise Tree

SIZE: 28-36μ diameter

RANGE: Southern United States

Shape spherical to suboblate, three furrows meet at both poles. Exine coarsely reticulate, intectate.

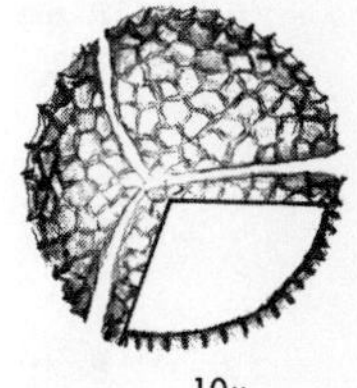

Fig. 180. *Illicium floridanum* Ellis Coll. Florida; ILLICIACEAE.

3a Furrows with a spiral orientation. 4

3b Furrows either latitudinal or longitudinal or irregular, dividing the exine into angular plates (Fig. 181). . . . . .Dinoflagellate tests

Dinoflagellate

SIZE: 50μ

RANGE: Marine, coastal Pacific

Dinoflagellate tests, such as the illustrated genus, may occasionally be found with pollen. The cysts (see Figures 41 and 42) of these protozoans are often recovered in marine or fresh water sediments. When living these protists have a transverse flagellum which lies in the equatorial girdle and a second longitudinal flagellum. For details on identification consult "How to Know the Protozoa" by Jahn or other manuals treating algae or protozoa.

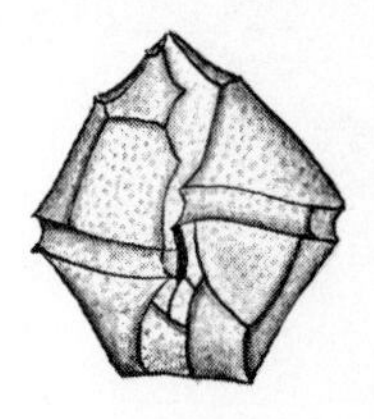

Fig. 181. *Gonyaulax polyhedra* Stein. Redrawn from: Evitt, W. R. 1964. Dinoflagellates and their use in petroleum geology. Pp. 65-72 In: *Palynology in Oil Exploration*, Soc. of Economic Paleontologists and Mineralogists (Text-figure 2).

4a Pollen grain with small spines (Fig. 182)................*Eriocaulon*

White-Buttons

SIZE: 34-35μ diameter

RANGE: Southeastern Canada, northeastern U.S. and Great Lakes region

The spiral furrows of *Eriocaulon*, and related structures in other monocots, have been interpreted as spiral endocracks. They resemble, but may not be homologous with, spiral apertures in dicotyledonous pollen grains. In *E. septangulare* the exine strips are 8-10μ wide; the spinules are 1.5-2.0μ apart.

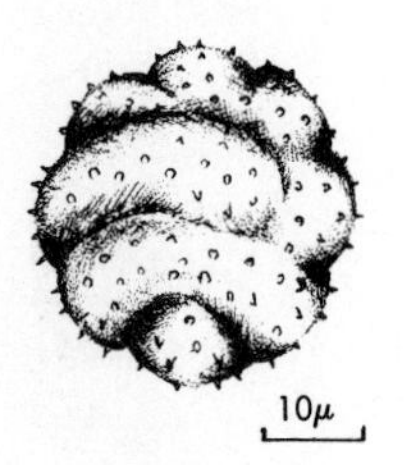

Fig. 182. *Eriocaulon septangulare* Withering Coll. Michigan; ERIOCAULACEAE.

4b Pollen grain without spines, surface irregularly pitted or verrucate (Fig. 183)..*Berberis*

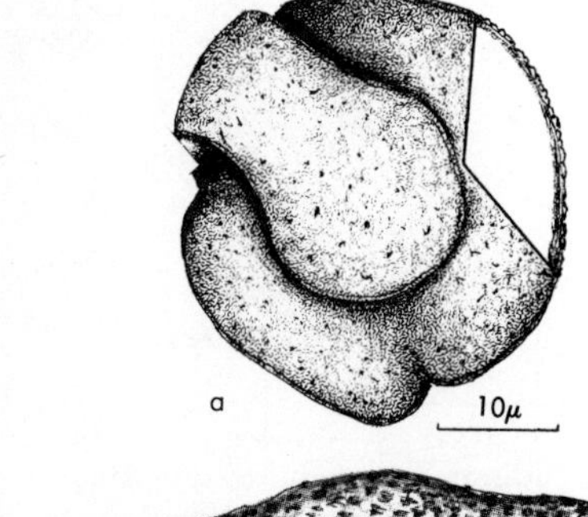

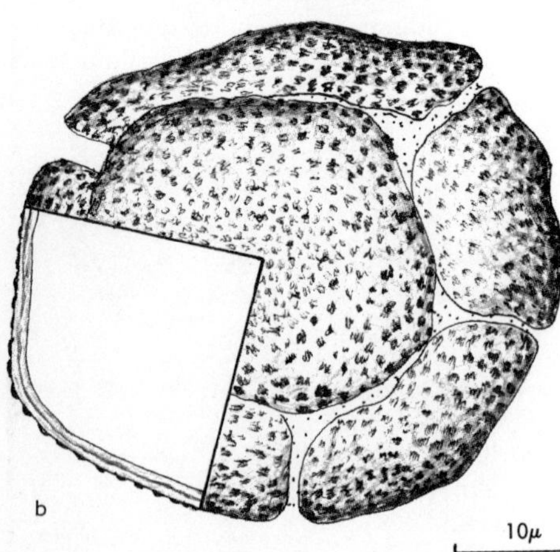

Fig. 183. *Berberis fremontii* Torr. Coll. New Mexico; BERBERIDACEAE.

Barberry

SIZE: 34μ diameter

RANGE: Colorado, Utah, New Mexico and northern Arizona

Exine surface verrucate, obscurely tectate. Apertures have appearance of four spirally-joined furrows (a). Some pollen grains of this species have an exine divided into 6 or more hexagonal plates (b); this may be a late stage of maturation or the pollen may possibly be dimorphic.

DICOLPATE TYPES

Occasionally pollen grains which normally have either one or three furrows will be dicolpate. The genus *Tofieldia*, as well as some other members of the Liliaceae, may regularly, or exceptionally, be dicolpate.

1a Colpi long; nearly dividing the pollen grains in two (Fig. 184)...... ....*Pedicularis*

Swamp Lousewort

SIZE: Polar axis about 20μ, equatorial diameter about 29μ

RANGE: Southeastern Canada and Europe

Grains may appear to be bilateral in some views or monocolpate in others. Oblate shape, P/E Index about 0.7. Figure 184 (a) is equatorial longitudinal view, polar axis is vertical; (b) equatorial end view showing two furrows.

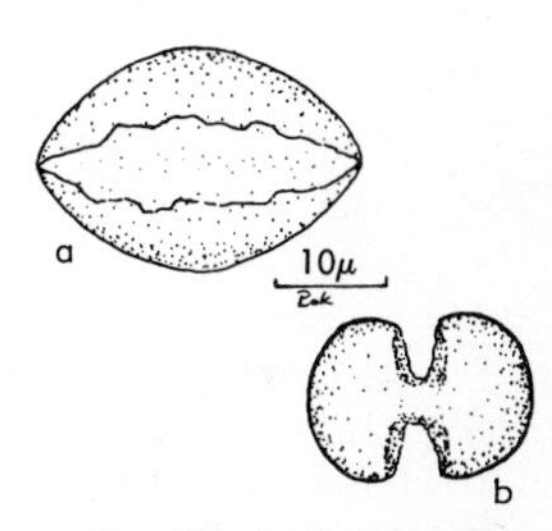

Fig. 184. *Pedicularis palustris* L.; SCROPHULARIACEAE.

1b Colpi shorter...2

2a Shape rather quadrate, furrows broad (Fig. 185).................. ....*Calla palustris* L.

Wild Calla

SIZE: 25 x 35μ

RANGE: Alaska, central and southern Canada, northeastern United States

Shape is subovate with squared ends. Exine tectate, faintly rugulate (some authors describe European specimens as finely reticulate). Furrow membranes verrucate.

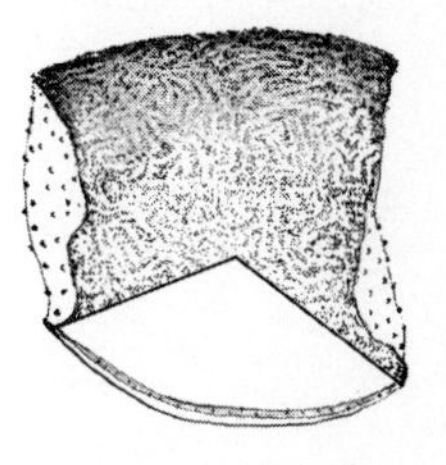

Fig. 185. *Calla palustris* L. Coll. Michigan; ARACEAE.

2b Shape elliptical to kidney-shaped, furrows tapered (Fig. 186)..*Pontederia*

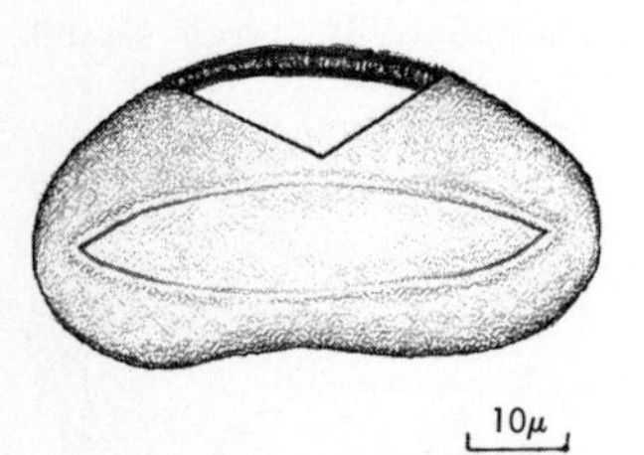

Fig. 186. *Pontederia cordata* L. Coll. Michigan; PONTEDERIACEAE.

Pickeral Weed

SIZE: About 32 x 62μ

RANGE: Prince Edward Island to southern Ontario, south to northern Florida, Missouri and Oklahoma

Shape ellipsoid or often slightly curved and kidney-shaped. Furrows broad often with a heavy persistent membrane. Exine tectate with fine verrucae. May superficially appear to be monocolpate.

TRICOLPATE TYPES

This is a very common pollen type among dicotyledons. It is common in the Ranunculaceae, Rosaceae, Aceraceae, Cruciferae, Labiatae, Geraniaceae, Saxifragaceae, Portulaceae, Primulaceae, Fagaceae (*Quercus*), Violaceae, Verbenaceae and many other families. In many cases a family may have both tricolpate and tricolporate representatives.

1a Exine reticulate..2

1b Exine psilate or scabrate, verrucate, clavate, striate, rugulate or echinate..17

2a Maximum dimension 20μ or less..3

2b Maximum dimension 20μ or greater..5

3a Reticulum very coarse; shape subprolate, P/E Index about 1.16 (Fig. 187)..............................*Menispermum canadense* L.

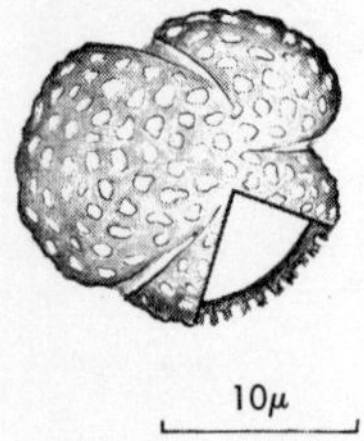

Moonseed

SIZE: Polar axis 18.5μ; equatorial diameter 16μ

RANGE: Quebec to Manitoba south to Oklahoma and Georgia

Grain subprolate (P/E Index about 1.16); exine perforate tectate, very coarsely reticulate. Erdtman (1952) describes several tricolporate members of the Menispermaceae (including *Cocculus*); either Menispermum departs from this generalization or the pore is very indistinct.

Fig. 187. Menispermum canadense L. Coll. Michigan; MENISPERMACEAE.

3b Reticulum fine 4

4a Shape spherical (P/E Index about 1.0), polar axis about 14μ (Fig. 188) .... *Arabidopsis thaliana* (L.) Heynh.

Mouse-ear-cress

SIZE: Diameter (polar and equatorial) about 14μ

RANGE: Massachusetts to Minnesota (Utah) south to Arkansas and Georgia, introduced from Europe

Exine tectate with distinct columellae; fine reticulate. Furrow broad and long, without persistent membrane; polar area very small.

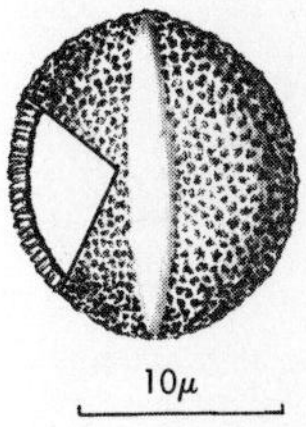

10μ

Fig. 188. *Arabidopsis thaliana* (L.) Heynh. Col. Michigan; CRUCIFERAE.

4b Shape prolate spheroidal (P/E Index about 1.1), polar axis is about 20μ (Fig. 189) .... *Tamarix gallica* L.

Tamarisk

SIZE: Polar axis 20μ, equatorial diameter 18μ

RANGE: Native and widespread in southwest along rivers, beyond in cultivation

Shape prolate spheroidal (P/E Index about 1.1); thin extension of endexine. Exine tectate, reticulate.

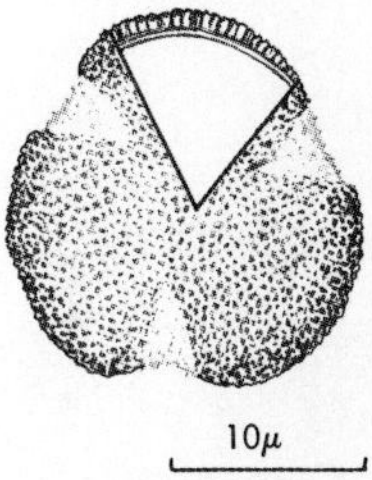

10μ

Fig. 189. *Tamarix gallica* L. Coll. Kansas, TAMARICACEAE.

5a Size 20-30μ 6

5b Size 35μ or greater 13

6a Polar area small, Polar Area Index 0.35 or less 7

6b Polar area large, Polar Area Index 0.40 or more 11

7a Shape prolate (Fig. 190)........................*Mitella diphylla* L.

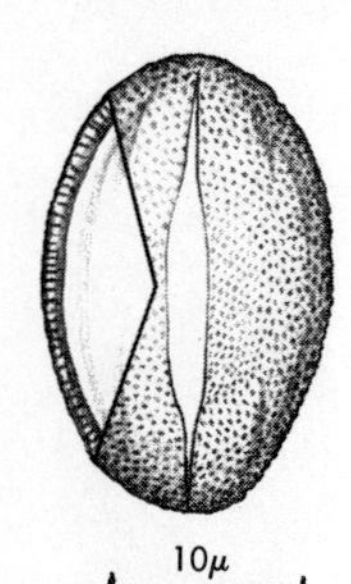

Fig. 190. *Mitella diphylla* L. Coll. Michigan; SAXIFRAGACEAE.

Miterwort

SIZE: Polar axis about 23μ; equatorial diameter 15μ

RANGE: Quebec to Minnesota south to Missouri, Mississippi, Tennessee and South Carolina in uplands

Shape prolate (P/E Index about 1.5); furrow abruptly expanded over about one-half its length in equatorial region (poroid?). Exine tectate, reticulate pattern very fine (oil!).

7b Shape spheroidal to subprolate.................................8

8a Reticulum finest near the furrows and poles (Fig. 191).......*Salix*

Willow

SIZE: Polar axis about 28μ; equatorial diameter 24μ

RANGE: Oregon to Utah and California

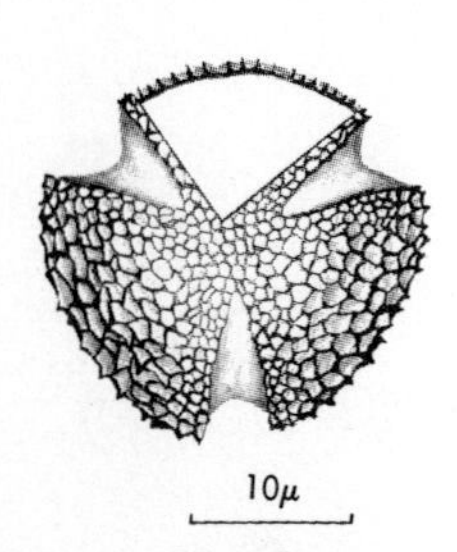

Fig. 191. *Salix laevigata* Bebb. Coll. California; SALICACEAE.

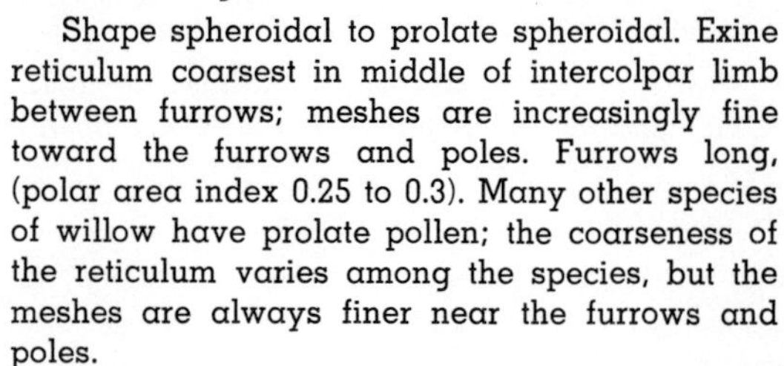

Shape spheroidal to prolate spheroidal. Exine reticulum coarsest in middle of intercolpar limb between furrows; meshes are increasingly fine toward the furrows and poles. Furrows long, (polar area index 0.25 to 0.3). Many other species of willow have prolate pollen; the coarseness of the reticulum varies among the species, but the meshes are always finer near the furrows and poles.

8b Reticulum uniform...9

9a P/E Index (shape class index) about 1.2 (subprolate)............10

9b P/E Index (shape class index) about 1.0 (spheroidal) (Fig. 192)..... ....*Catalpa speciosa* Warden

Catalpa

SIZE: Polar and equatorial diameter about 27μ

RANGE: Southeastern Iowa to southern Indiana south to Tennessee and Arkansas, naturalized beyond

Shape spherical; exine tectate coarse reticulate. Ektexine distinctly thicker in the intercolpae than next to the furrows due to longer columellae.

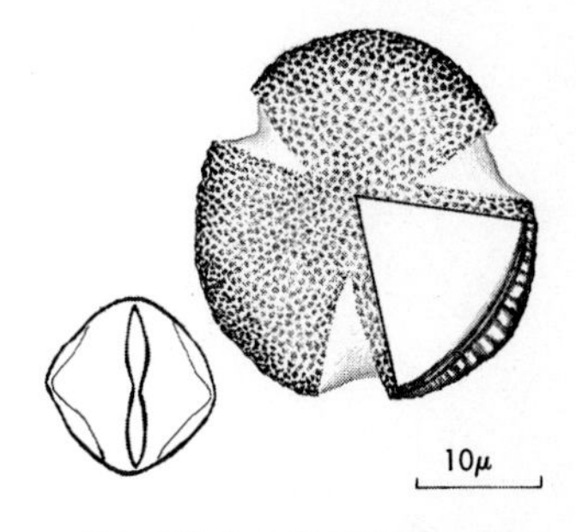

Fig. 192. *Catalpa speciosa* Warden Col. Indiana; BIGNONIACEAE.

10a Reticulum finer at poles (Fig. 193).......*Saxifraga pensylvanica* L.

Swamp Saxifrage

SIZE: Polar axis about 22μ; equatorial diameter about 19μ

RANGE: Maine to Minnesota south to Missouri and Virginia

Shape subprolate (P/E Index about 1.18). Exine tectate (indistinct); reticulum coarse near equator, finer at poles.

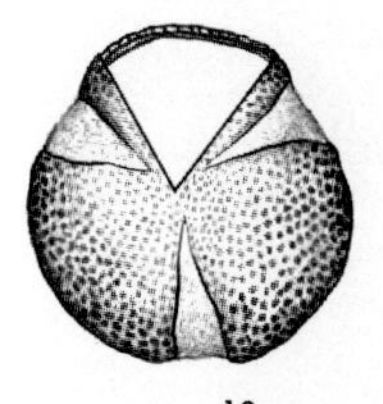

Fig. 193. *Saxifraga pensylvanica* Coll. Michigan; SAXIFRAGACEAE.

10b Reticulum uniform throughout (Fig. 194)........................ ....*Rorippa palustris* v. *hispida* (L.) Bess

Yellow Cress

SIZE: Polar axis 27μ; equatorial diameter 22.5μ

RANGE: Quebec to Michigan south to Pennsylvania and New Jersey

Shape subprolate (P/E Index about 1.2), surface coarsely and uniformly reticulate; exine intectate. Furrows long, polar area small (polar area index about 0.2).

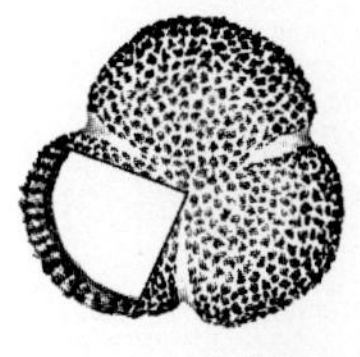

Fig. 194. *Rorippa palustris* v. *hispida* (L.) Bess Coll. Indiana; CRUCIFEREAE.

11a Reticulum finest at poles (Fig. 195)......... *Marrubium vulgare* L.

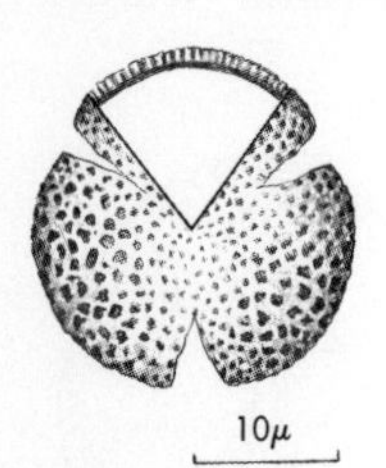

Fig. 195. *Marrubium vulgare* L. Coll. New Mexico; LABIATAE.

Horehound

SIZE: Polar axis about 25μ; equatorial diameter 22μ

RANGE: Widespread (naturalized from Europe) throughout the continental United States and parts of adjacent Canada

Shape spheroidal (P/E Index about 1.0); exine tectate or perforate tectate. Reticulum coarse (1 to 1.5μ), except finer at the poles.

11b Reticulum approximately uniform throughout.................12

12a Furrows very short (polar area index about 0.7), furrow membranes absent (Fig. 196).................*Fraxinus quadrangulata* Michx.

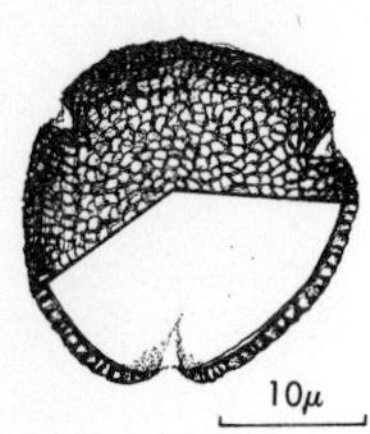

Fig. 196. *Fraxinus quadrangulata* Michx. Coll. Missouri; OLEACEAE.

Blue Ash

SIZE: Polar axis 19μ; equatorial diameter about 22μ

RANGE: Southern Ontario to Wisconsin south to Oklahoma and Alabama

Shape suboblate to oblate spherical (P/E Index about 0.86). Furrows short, polar area large (polar area index 0.6-0.7).

12b Furrows longer (polar area index about 0.45), furrow membrane distinctly and coarsely verrucate (Fig. 197)..*Platanus occidentalis* L.

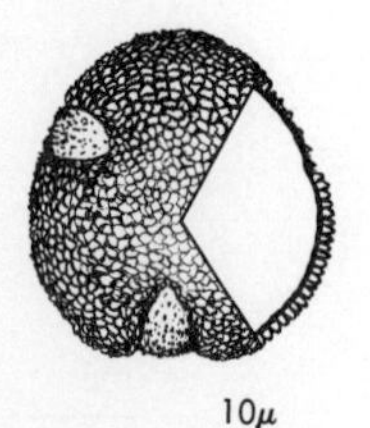

Fig. 197. *Platanus occidentalis* L. Coll. Michigan; PLATANACEAE.

Sycamore

SIZE: Polar and equatorial diameter 19-22μ

RANGE: Maine to Ontario and Nebraska south to Texas and Florida

Spherical (P/E Index about 1.0); exine tectate, fine reticulate. Furrow membrane persistent, distinctly flecked with verrucae.

13a Longest dimension about 35-45μ 14

13b Longest dimension usually greater than 75μ (some species resembling *Hoffmanseggia repens* in the Caesalpinoideae may be smaller, see Tsukada, 1963) 16

14a Shape prolate, P/E Index about 1.4 (obscurely tricolporate) (Fig. 198) . . . . *Trifolium pratense* L.

Red Clover

SIZE: Polar axis about 45μ; equatorial diameter about 32μ

RANGE: Introduced from Europe, widespread throughout the United States and southern Canada

Shape prolate (P/E Index about 1.4); pore indistinct, furrow often with merely a roughened constriction at the equatorial limb. Surface coarsely reticulate. Probably indistinguishable from pollen of other species of clover or other Legumes in the Papilionoideae.

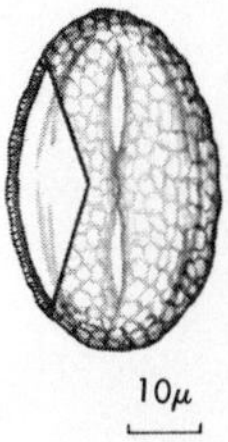

Fig. 198. *Trifolium pratense* L. Coll. Michigan; LEGUMINOSAE.

14b Shape subprolate to prolate spheroidal 15

15a Reticulate throughout (Fig. 199) . . . . *Dentaria laciniata* Muhl.

Toothwort

SIZE: Polar axis about 40μ; equatorial diameter about 36μ

RANGE: Quebec to Minnesota and Nebraska south to Louisiana and Florida

Shape prolate spheroidal (P/E Index about 1.1). Exine thickest in the intercolpi, reticulate. Polar area index about 0.3.

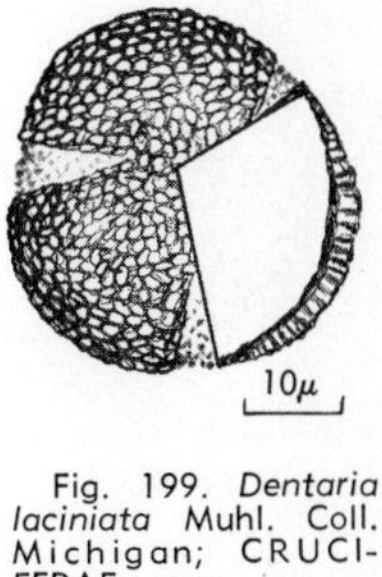

Fig. 199. *Dentaria laciniata* Muhl. Coll. Michigan; CRUCIFERAE.

15b Reticulate in intercolpate regions, verrucate-granular at the poles and adjacent to the furrows (Fig. 200)..*Podophyllum peltatum* L.

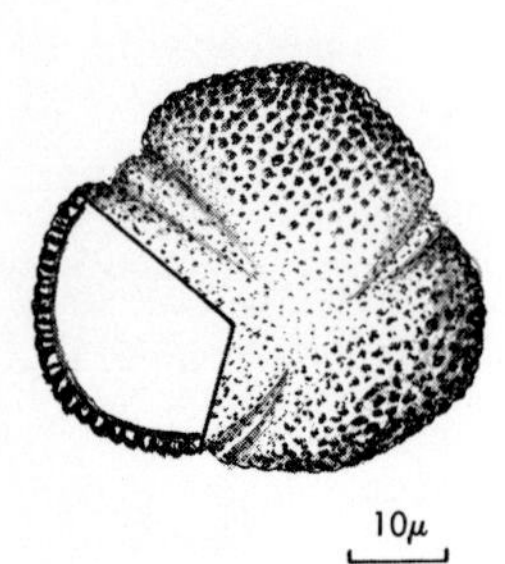

Fig. 200. *Podophyllum peltatum* L. Coll. Michigan; BERBERIDACEAE.

May-apple

SIZE: Polar axis about 45μ; equatorial diameter about 38μ

RANGE: Florida to Texas north to Minnesota and western Quebec

Shape subprolate (P/E Index about 1.18), furrows very long, polar area small (polar area index about 0.16). Exine tectate in the intercolpi, verrucate at the furrow margins. Furrow membranes persistent and finely verrucate.

16a Exine with very coarse baculate and clavate excresences throughout, these forming a reticulum (Fig. 201)................*Geranium*

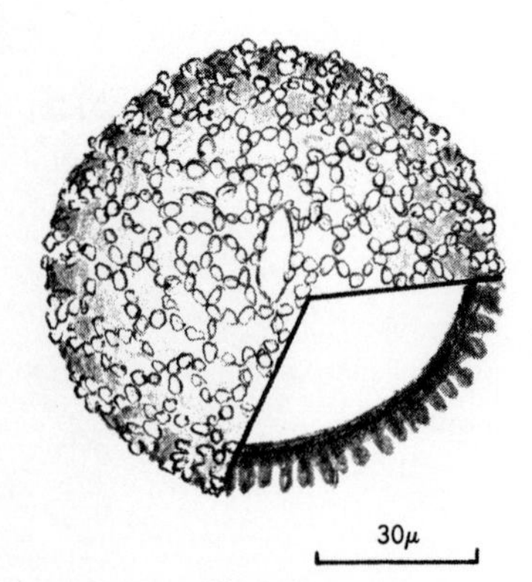

Fig. 201. *Geranium maculatum* L. Coll. Michigan; GERANIACEAE.

Spotted Cranesbill

SIZE: Polar axis about 80μ; equatorial diameter about 86μ

RANGE: Maine to Manitoba south to Kansas, Tennessee and Georgia

Shape oblate spherical (P/E Index about 0.92). Exine intectate with very coarse baculae and clavae, these forming a coarse imperfect reticulum. Furrows short, usually less than 20μ long.

16b Exine with regular fine reticulum, each furrow short, but with long, broad, striate to rugulate margo (Fig. 202). .*Hoffmanseggia* spp.

Hoffmanseggia

SIZE: Polar axis about 125μ; equatorial diameter about 115μ

RANGE: This species restricted to Utah; the genus widespread in the southwestern United States and southern plains

Shape prolate spherical; large. Exine reticulate in the intercolpi, this sculpture sharply differentiated from the striate or rugulate furrow margins (a). Furrows short, about 28μ long (b). The exine morphology illustrated by *Hoffmanseggia* is rather widespread in the tribe Eucaesalpinieae of the Legume subfamily Caesalpinoideae. It is rather like *Poincianella, Nicarago, Biancaea, Caesalpinia, Guimasia, Mesoneureum, Erythrostemon, Stahlia* and other central American genera.

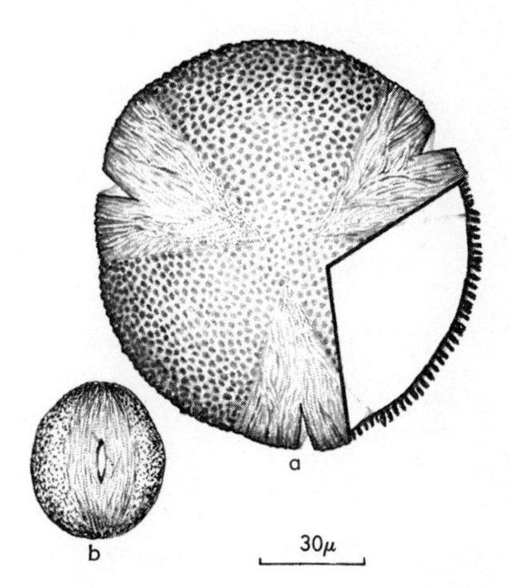

Fig. 202. *Hoffmanseggia repens* (Gastw.) Cock. Coll. Utah; LEGUMINOSAE.

17a Surface psilate or essentially so (sometimes scabrate or with structural elements which give granular appearance)..........18

17b Surface with a sculpturing pattern............................26

18a Size usually 20μ or greater..................................20

18b Size extremely small, less than 14μ..........................19

19a Shaped in equatorial view like a "dumbell," 3 x 7μ (Fig. 203)... ....*Myosotis*

Forget-me-not

SIZE: Polar axis 7μ; equatorial diameter 3μ or less

RANGE: Alberta to Colorado and Alaska

Shape perprolate (P/E Index about 2.3), very small. Triangular in polar view (b), with three faint colpi; hour glass shaped in equatorial view (a); furrows often indistinct. Exine thin psilate.

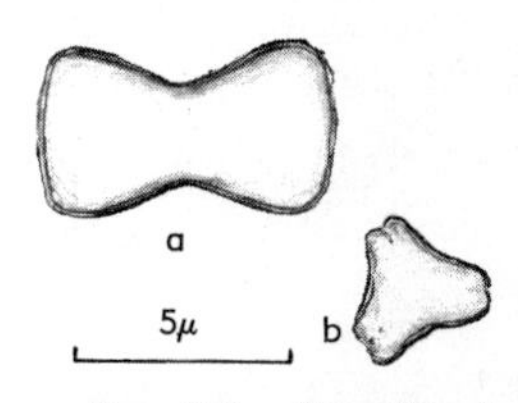

Fig. 203. *Myosotis alpestris Schmidt*. Coll. Montana; BORAGINACEAE.

19b Shape in equatorial view subovate; polar axis about 14μ (Fig. 204) .... *Dodocatheon* sp.

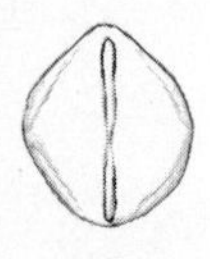

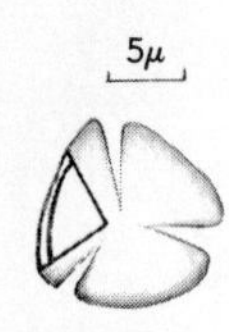

Fig. 204. *Dodocatheon cylindrocarpum* Rydb. Coll. Montana; PRIMULACEAE.

Shooting Star

SIZE: Polar and equatorial diameter about 12μ

RANGE: Eastern Washington to Montana

Shape spherical (P/E Index about 1.0). Furrows very long, constricted at the equator, polar area very small (polar area index 0.12 to 0.16). Surface psilate; exine stratification obscure.

20a Maximum dimension 20-30μ.......... 21

20b Maximum dimension 30-40μ.......... 23

21a Furrow membranes persistent and bearing distinct baculae (Fig. 205) .... *Oxalis grandis* Small

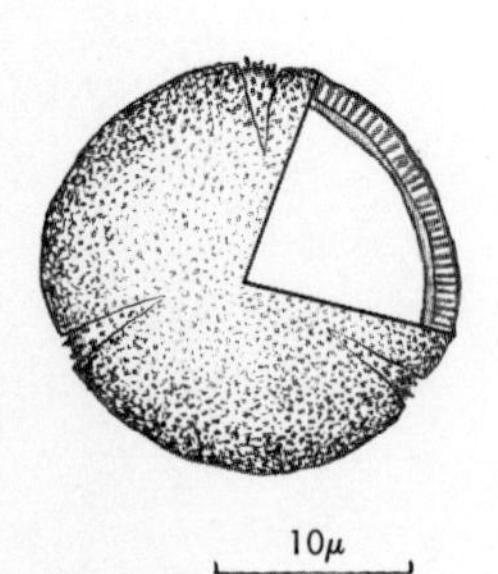

Fig. 205. *Oxalis grandis Small* Earlham College Herbarium; OXALIDACEAE.

Large Wood Sorrel

SIZE: Polar and equatorial axes approximately 20μ

RANGE: Pennsylvania to southern Illinois southward to Georgia and Alabama

Shape spherical (P/E Index about 1.0); furrows of intermediate length (Polar area index about 0.4). Exine tectate with distinct columellae, surface psilate or scabrate; a faint reticulate pattern in surface view is due to the pattern of the columellae. Furrow membranes persistent, ornamented with baculae. Some species of *Oxalis* are subprolate in shape and nearly twice as large.

21b Furrow membrane lacking or not persistent..................22

22a Exine thin adjacent to furrows forming margo (Fig. 206)........... ....*Physocarpus opulifolius* (L.) Maxim.

Ninebark

SIZE: Polar axis about 23μ; equatorial diameter about 19μ

RANGE: Quebec to Hudson Bay, west to Minnesota and Colorado, southward to Illinois, Arkansas and South Carolina

Shape subprolate (P/E Index about 1.2) exine psilate, tectate with distinct columellae which give a pebbled appearance under high-low focus (L-O analysis). Furrows distinctly marginate due to thinning of the ektexine.

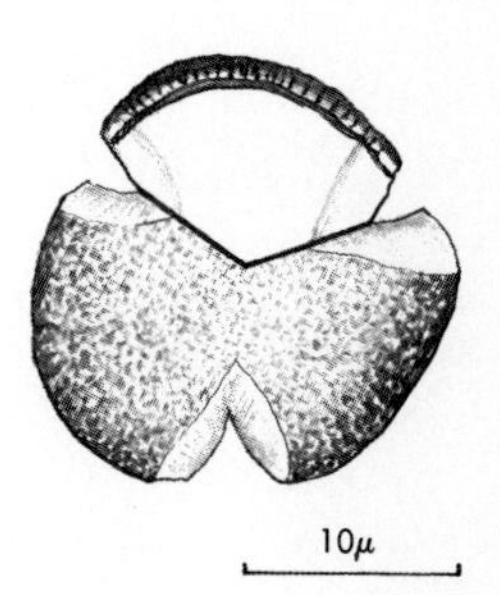

Fig. 206. *Physocarpus opulifolius* (L.) Maxim. Coll. Michigan; ROSACEAE.

22b Exine not thinned near aperture, furrow essentially without a margo (Fig. 207)...........................*Dryas octapetala* L.

Mountain Avens

SIZE: Polar axis about 29μ; equatorial diameter about 24μ

RANGE: Greenland to Colorado, Washington and Alaska

Shape subprolate (P/E Index about 1.18), exine tectate, surface psilate. Columellae distinct and coarse, imparting a granular character under L-O analysis.

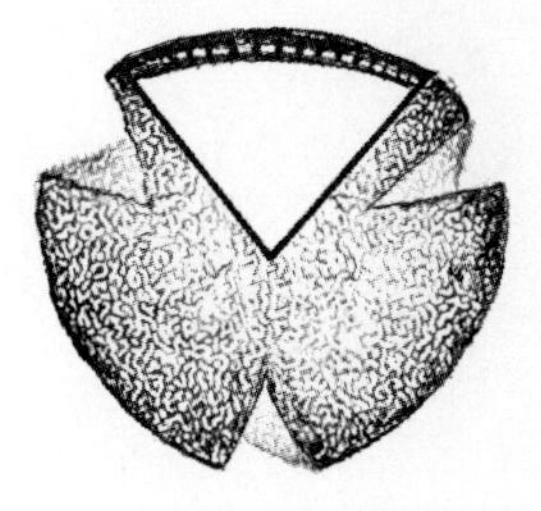

Fig. 207. *Dryas octapetala* L. Coll. Montana; ROSACEAE.

23a Furrow complex (Fig. 208) .... ***Verbena* (in part)**

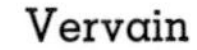

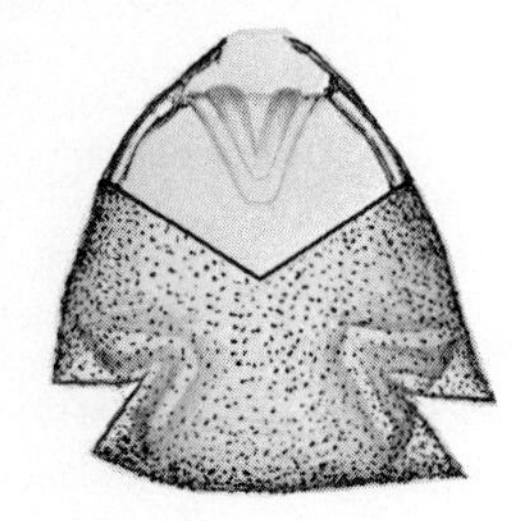

Vervain

SIZE: Polar axis about 35μ; equatorial diameter about 38μ

RANGE: The species extends from western Texas to Arizona and southward to southern Mexico

Shape oblate spheroidal (P/E Index about 0.9). Furrow very complex, with two pseudocolpi flanking each furrow; pore apparently elongated as a transverse furrow.

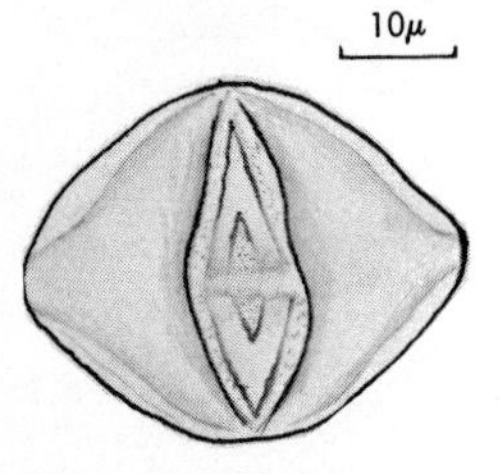

Fig. 208. *Verbena ciliata* Benth. v. *longidentata* Perry Coll. Texas; VERBENACEAE.

23b Furrow simple **24**

24a Exine thick with distinct stratification and stout columellae (Fig. 209) .... ***Hepatica* spp.**

Liverleaf (Hepatica)

SIZE: Equatorial diameter 33-35μ; spherical to slightly oblate

RANGE: Nova Scotia to Manitoba south to Missouri, Alabama and northern Florida

Furrows long (polar area index about 0.29), membrane persistent and coarsely verrucate. Exine tectate and surface psilate, columellae very stout and widely spaced appearing distinctly in L-O analysis. Endexine slightly thickened at the furrow margin, especially in the equatorial limb.

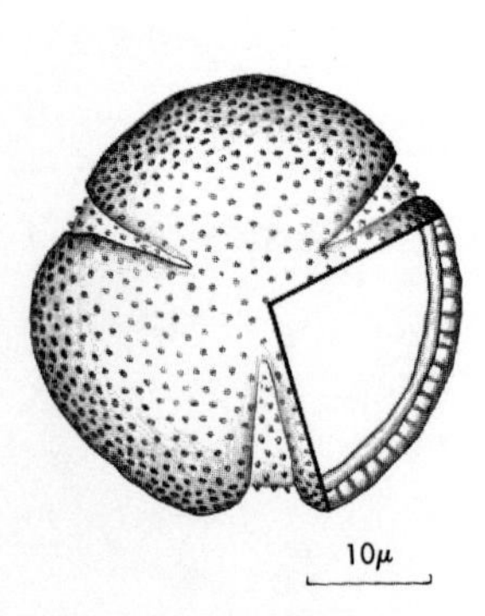

Fig. 209. *Hepatica americana* (DC.) Ker. Coll. Michigan; RANUNCULACEAE.

24b Exine thinner, lacking distinct columellae **25**

25a Shape suboblate (P/E about 0.9); furrows interrupted or "bent" at the equator (Fig. 210)................................... *Viola*

Arrow-Leaved Violet

SIZE: Polar axis about 30μ; equatorial diameter about 36μ

RANGE: Massachusetts to Minnesota south to eastern Texas and Georgia

Shape oblate spheroidal to suboblate (P/E Index about 0.88). Furrows long and broadly-tapered to the poles, irregularly folded or arched at the equatorial limb, pore indistinct. Exine surface psilate.

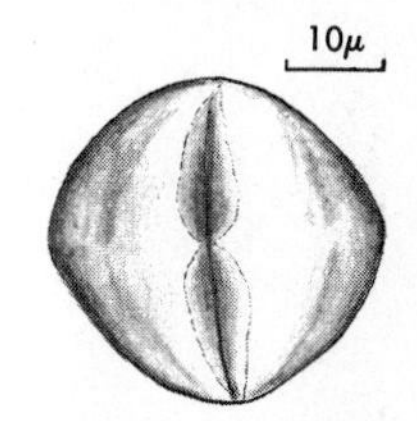

Fig. 210. *Viola sagittata* Ait. Coll. Michigan; VIOLACEAE.

25b Shape spherical or prolate spheroidal (P/E Index about 1.06); furrows continuous at equator (Fig. 211)... *Crataegus phaenopyrum* (L.f.) Medic.

Washington Thorn

SIZE: Polar axis about 36μ; equatorial diameter about 34μ

RANGE: Pennsylvania to Florida westward to Missouri and Arkansas; escaped from cultivation beyond

Shape spherical or prolate spheroidal (P/E Index about 1.06); exine apparently intectate (columellae absent or obscure), surface psilate. Furrows of intermediate length, polar area index about 0.5.

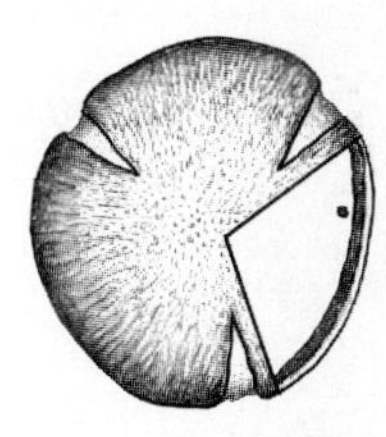

Fig. 211. *Crataegus phaenopyrum* (L.f.) Medic. Coll. Arnold Arboretum; ROSACEAE.

26a Exine rugulate, at least in the intercolpate areas..............26

26b Exine sculpture neither rugulate nor reticulate.................27

27a Rugulate sculpturing uniformly distributed on exine; pore distinct or obscure (Fig. 212).........*Helianthemum canadense* (L.) Michx.

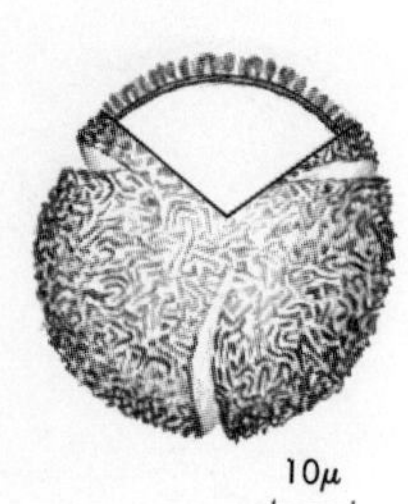

Fig. 212. *Helianthemum canadense* (L.) Michx. Coll. Michigan; CISTACEAE.

Frostweed

SIZE: Polar axis about 41μ; equatorial diameter about 45μ

RANGE: Nova Scotia to Wisconsin south to Missouri and North Carolina

Shape oblate spheroidal (P/E Index about 0.9); pollen of some species of *Helianthemum* is spherical to prolate spheroidal. Furrows with an indistinct pore ("roughening") at the equator. Exine baculate and either intectate or perforate tectate. Surface pattern faintly reticulate to substriate. Some members of the Cistaceae are reticulate, others, including some species of *Helianthemum*, are striate.

27b Rugulate sculpturing restricted primarily to intercolpate regions (Fig. 213)................................*Claytonia virginica* L.

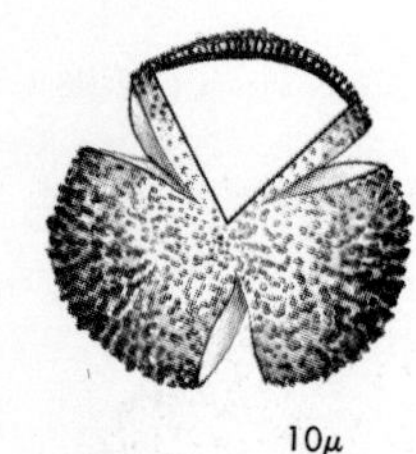

Fig. 213. *Claytonia virginica* L. Coll. Michigan; PORTULACACEAE.

Spring Beauty

SIZE: Polar and equatorial axes about 45-48μ

RANGE: Quebec and western New England west to Minnesota, southward to Texas and Georgia

Shape approximately spherical; furrows long and polar area small (polar area index about 0.23). Exine tectate or perforate tectate, surface subrugulate or subreticulate; sculpturing much coarser in the intercolpate areas than at the poles or furrow margins.

28a Sculpturing either gemmate, baculate, clavate, or subechinate..29

28b Sculpturing either striate or verrucate......................36

29a Surface with spines; these may be small broad-based spines or blunt (in *Centaurea*)....................................30

29b Surface gemmate, clavate, or baculate......................33

30a Exine perforate tectate or frustillate; pore absent (Fig. 214)..*Lewisia rediviva* Pursh.

Bitter-root

SIZE: Equatorial diameter about 38μ

RANGE: Montana to Colorado, Arizona and California

Shape spherical to slightly-oblate spheroidal. Furrows with persistent coarsely verrucate membranes which are frequently ruptured giving appearance of furrows with broad, thin margins. Exine semitectate, frustillate with several large columellae often coalesced at ends forming a partial tectum. These frustillae bear short supratectate spines.

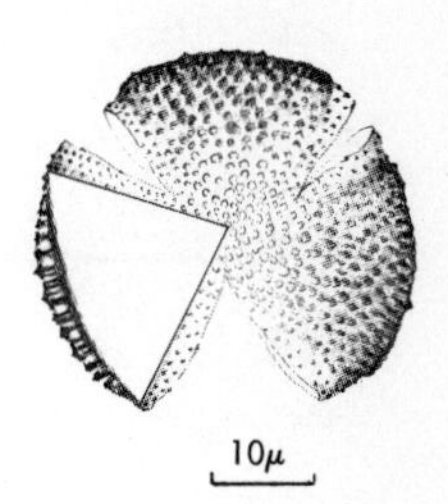

Fig. 214. *Lewisia rediviva* Pursh. Coll. Montana; PORTULACACEAE.

30b Exine tectate; if tricolporate, pore obscure....................31

31a Furrow very short (Fig. 215)............*Dipsacus sylvestris* Huds.

Wild Teasel

SIZE: About 100μ diameter

RANGE: Southeastern Canada, northeastern United States south to North Carolina and Missouri

Apertures elongate, either short furrows or elongate pores; apertures with a strong margo (annulus). Exine very thick (endexine 2μ, ektexine 8.5μ, including 2μ echinae). Pore membranes bear large clavae and baculae.

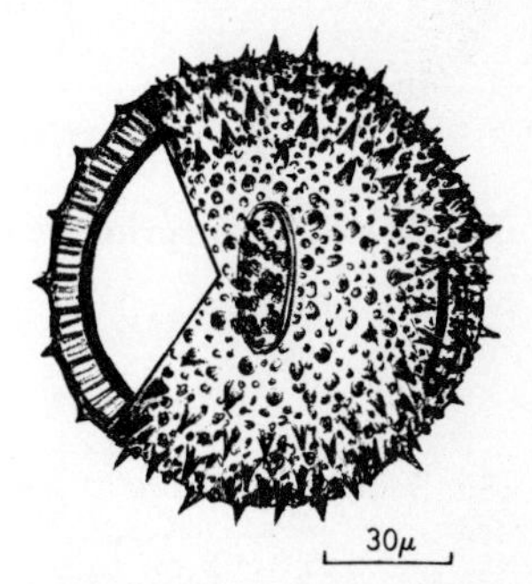

Fig. 215. *Dipsacus sylvestris* Huds. Coll. Michigan; DIPSACACEAE.

31b Furrows long and broad..................................32

32a Supratectate spines broad-based and low but pointed (Fig. 216)..
...*Artemisia* spp.

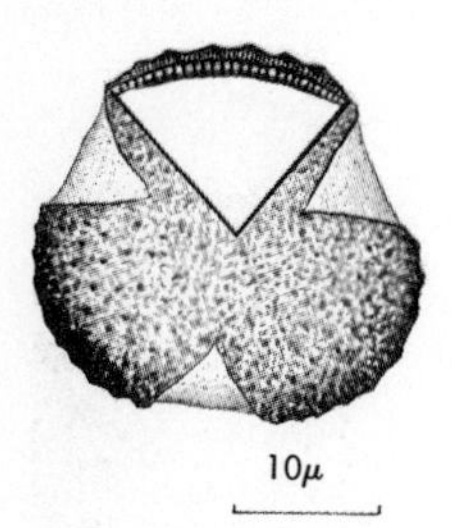

Fig. 216. *Artemisia caudata* Michx. Coll. Indiana; COMPOSITAE.

Wormwood

SIZE: Polar axis about 27μ; equatorial diameter about 25μ

RANGE: Florida to Texas and Arizona, north to Washington, Manitoba and southern Maine

Shape prolate spheroidal (P/E Index about 1.1); furrow broad and expanded, pore distinct or obscure. Exine distinctly stratified; the radial baculae outside the tectum are more or less fused at the distal ends, especially at the bases of the spines. Surface subechinate or spines obscure; surface sometimes appears sub-reticulate. Exine thinner next to the furrows forming a margo.

32b Supratectate spines subdued, blunt, surface undulate (Fig. 217)...
.......................................*Centaurea picris* Pall.

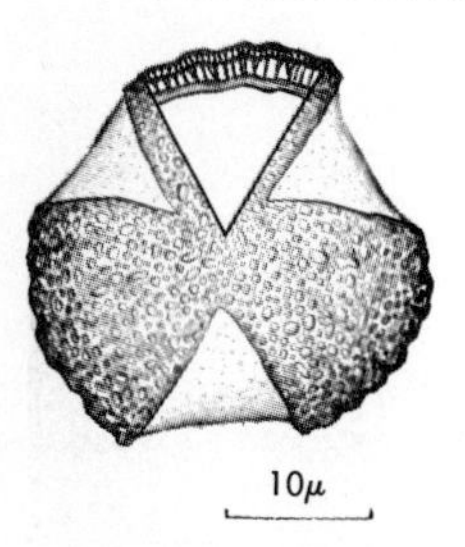

Fig. 217. *Centaurea picris* Pall. Coll. New Mexico; COMPOSITAE.

Star Thistle

SIZE: Equatorial diameter about 27μ

RANGE: Michigan to Washington south to Missouri, Texas and southern California

The *Centaurea*-type of Compositae pollen is subprolate, furrow membrane smooth to slightly verrucate. Exine about $3\text{-}7.5\mu$ thick, thinner at the poles. Coarse subtectate baculae about 1.5-2.5μ long; fine, supratectate radial columellae fused distally forming a surface membrane. Surface undulate, apparently vestigial spines (Stix, 1960).

33a Size $35\text{-}40\mu$, surface with large clavae and baculae (Fig. 218)...*Ilex*

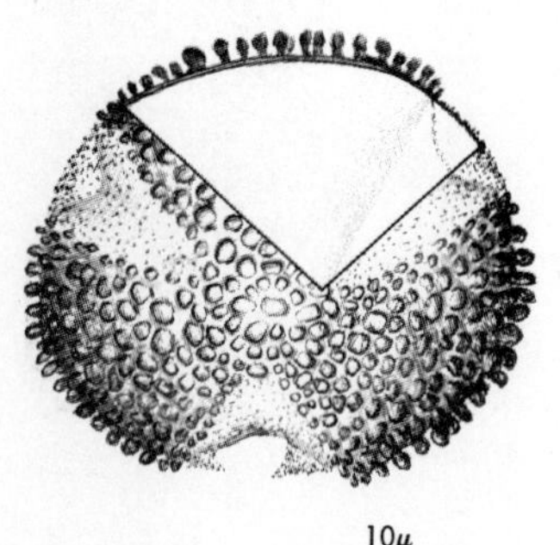

Fig. 218. *Ilex vomitaria* Ait. Coll. Texas; AQUIFOLIACEAE.

Cassena

SIZE: Polar axis about 43μ; equatorial diameter about 34μ (-38μ)

RANGE: Florida to Texas northward to Arkansas and Virginia, primarily on coastal plain

Shape subprolate (P/E Index about 1.25), furrow broad and long, polar area small (polar area index about 0.24). Furrow membrane persistent, fine verrucate; pore often indistinct or obscure. Exine intectate, with very coarse clavate elements.

33b Size 60μ or greater 34

34a Furrows short, less than 1/2 length of polar axis (Fig. 219)... *Linum* (*Geranium* may key to this point, though considered reticulate, see Figure 201.)

Flax

SIZE: Polar axis about 58μ; equatorial diameter about 64μ

RANGE: Saskatchewan to Wyoming south to Texas and Arizona

Shape oblate spheroidal (P/E Index approximately 0.9). Exine intectate with a thick endexine and a coarsely gemmate verrucate surface. Furrows very short (about 18μ) but bordered by a margin which extends meridionally about three-fourths the length of the polar axis. Other species of *Linum* vary in shape from spherical to prolate; some have longer furrows; *L. usitatissimum* pollen is occasionally hexacolpate.

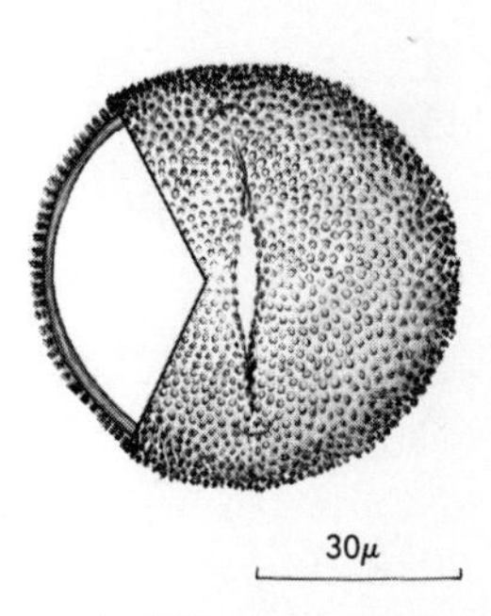

Fig. 219. *Linum pratense* (Norton) Small Coll. Montana; LINACEAE.

34b Furrows longer, more than 1/2 the length of the polar axis 35

35a Shape spherical, furrow sometimes obscure .... (Fig. 220) *Croton*

Hogwort

SIZE: Polar and equatorial axes about 80μ

RANGE: Georgia to Texas northward to Iowa and southern New York

Shape spheroidal; furrows indistinct or obscure. Erdtman (1952) states that they are frequently inaperturate. This species, and many others in the Crotonoideae have coarse, regularly-arranged triangular excrescences (clavae or gemmae); many species in the subfamily have similar morphology, but may vary in size. For example, *Cnidoscolus texanus* (from Texas) is similar in structure, but has more obscure furrows and is up to 105μ in size.

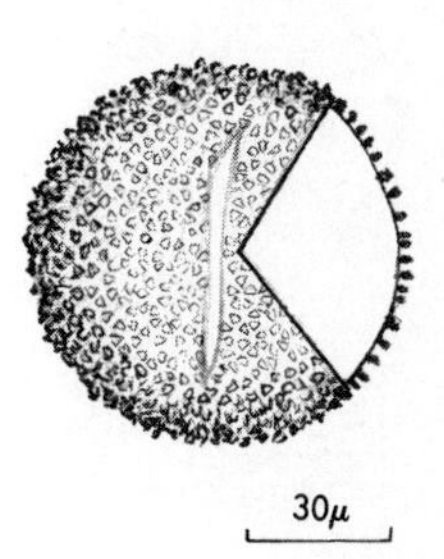

Fig. 220. *Croton capitatus* Michx. v. *albinoides* (Ferg.) Small Coll. Texas; EUPHORBIACEAE.

35b Shape subprolate, furrows obvious (Fig. 221) .... ***Plumbago capensis* Thunb.**

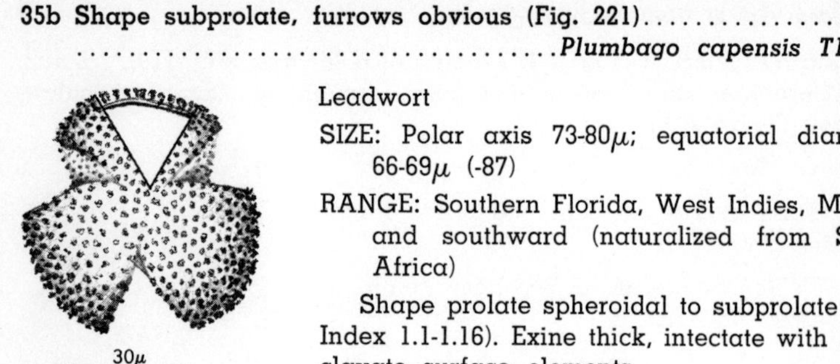

Fig. 221. *Plumbago capensis* Thunb. Coll. Massachusetts; PLUMBAGINACEAE (cultivated).

Leadwort

SIZE: Polar axis 73-80μ; equatorial diameter 66-69μ (-87)

RANGE: Southern Florida, West Indies, Mexico and southward (naturalized from South Africa)

Shape prolate spheroidal to subprolate (P/E Index 1.1-1.16). Exine thick, intectate with large clavate surface elements.

36a Exine surface striate **37**

36b Exine surface verrucate **41**

37a Furrow border wavy or ragged, an indistinct pore often present (Fig. 222) .... ***Prunus virginiana* L.**

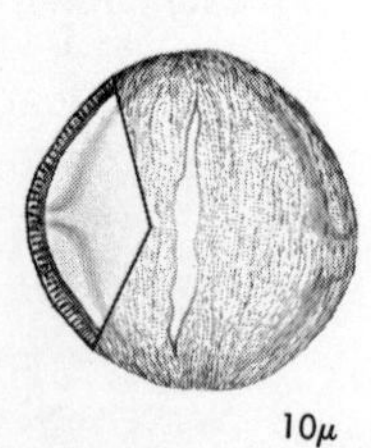

Fig. 222. *Prunus virginiana* L. Coll. Michigan; ROSACEAE.

Choke Cherry

SIZE: Polar and equatorial axes about 24μ

RANGE: Newfoundland to Saskatchewan south to Kansas, Tennessee and North Carolina

Shape spheroidal (P/E Index about 1.0). Furrow margins irregular, pores indistinct or obscure, especially in face view. Exine tectate and under oil immersion is strongly striate.

37b Furrow border straight, pore absent **38**

38a Grain large (>40μ) and prolate (Fig. 223)..*Jeffersonia diphylla* (L.) Pers.

Twin Leaf

SIZE: Polar axis about 42μ; equatorial diameter about 33μ

RANGE: New York and southern Ontario to Wisconsin and Iowa south to Alabama

Shape subprolate (P/E Index about 1.25). Furrows generally broadly expanded in acetolyzed pollen grains. Exine indistinctly tectate, fine reticulate or faintly striate pattern due to columellae.

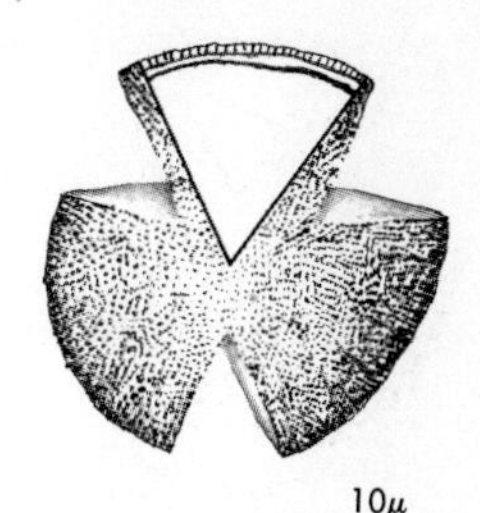

Fig. 223. *Jeffersonia diphylla* (L.) Pers. Coll. Michigan; BERBERIDACEAE.

38b Grain smaller, 15-35μ..39

39a Striations heavy and conspicuous, arranged in blocks of differing orientation (Fig. 224)......................*Menyanthes trifoliata* L.

Buck Bean

SIZE: Equatorial diameter about 30-32μ

RANGE: Labrador to Alaska south to Wyoming, Nebraska, Ohio and Maryland

Shape subspherical, furrows expanded often with pores indistinct. Wall surface distinctly striate, with blocks of striae intersecting each other in an angular pattern.

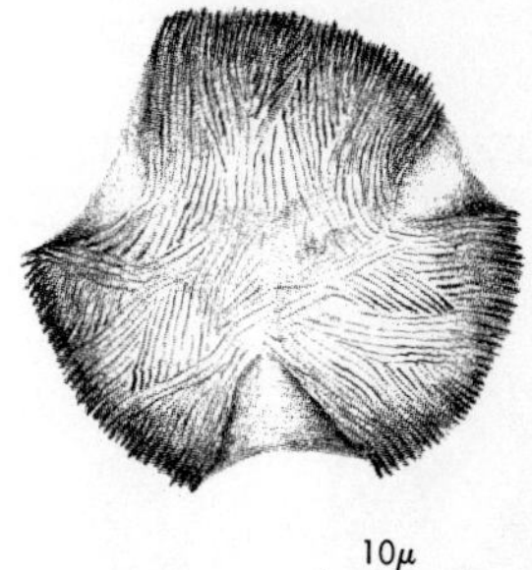

Fig. 224. *Menyanthes trifoliata* L. Coll. Michigan; GENTIANACEAE.

39b Striations conspicuous or faint, arranged longitudinally, short and oriented in varying angles to one another; this characteristic not always obvious under low or intermediate magnification........40

40a Size less than 20μ (Fig. 225)....... *Fragaria virginiana* Duchense

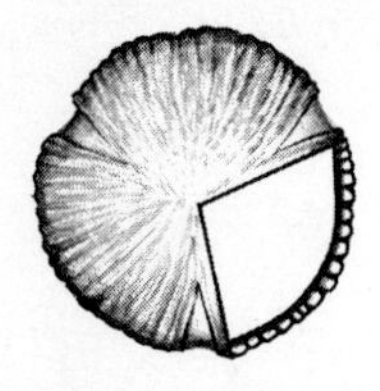

Fig. 225. *Fragaria virginiana* Duchense. Coll. Michigan; ROSACEAE.

Strawberry

SIZE: Polar axis about 18μ; equatorial diameter about 16μ

RANGE: Newfoundland to Alberta south to Oklahoma and Georgia

Shape subprolate (P/E Index about 1.2); furrows of intermediate length (polar area index 0.45-0.50). Exine apparently intectate, distinctly striate.

40b Size mostly 22-35μ (Fig. 226, 227, 228, 229).............. *Acer* spp.

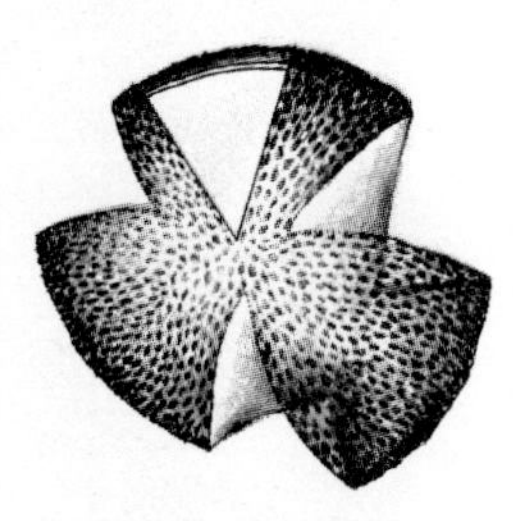

Fig. 226. *Acer pensylvanicum* L. Coll. Michigan; ACERACEAE.

Striped Maple

SIZE: Polar and equatorial axes 22-25μ

RANGE: Quebec to Manitoba south to Michigan, Tennessee and Georgia

Shape approximately spherical (P/E Index about 1.0). Columellae arranged in striations, often clustered in patches of various orientations; intectate. Furrows long (extend to about 3μ from poles). This species is most similar to *A. macrophyllum,* which is, however, significantly larger.

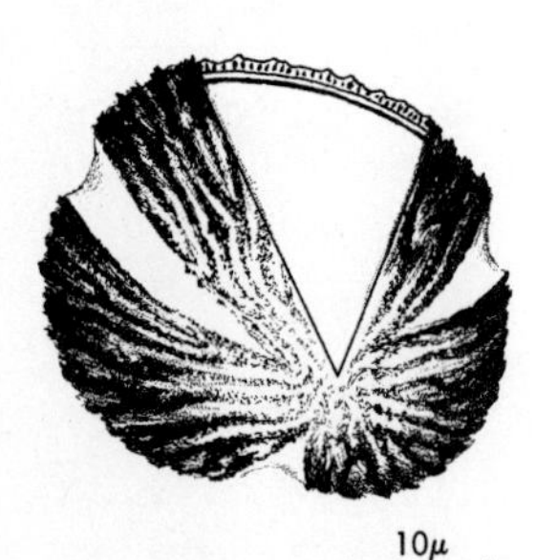

Fig. 227. *Acer glabrum* Torr. Coll. Colorado; ACERACEAE.

Smooth Maple

SIZE: Polar axis 25-31μ; equatorial diameter 23-28μ

RANGE: South Dakota and Wyoming to New Mexico and Utah

Shape approximately spherical (P/E Index about 1.0). Exine strongly striate due to arrangement of columellae, some of which are distally fused to form longitudinal tectate ridges. Furrows are wide with a finely granular membrane. Rather similar to *A. rubrum* in having tectate striae, but the pollen of A. rubrum does not have such strongly projecting ridges as *A. glabrum.* Pollen of *A. tripartitum* and *A. torreyi* is said to be indistinguishable from that of *A. glabrum.*

Chalk Maple

SIZE: Polar axis 25-34μ; equatorial diameter 24-32μ

RANGE: North Carolina to Georgia, Louisiana and Arkansas (coastal plain)

Shape prolate spheroidal (P/E Index about 1.1). Tectum forms longitudinal ridges which give the striate appearance. The striations are interconnected to form a reticulum which is visible at high focus. Furrows wide, taper to about 3μ from poles (polar area small); furrow membranes have scattered conspicuous flecks (verrucae). Very similar to *A. floridanum*. Differs from *A. saccharum* and *A. nigrum* in its more pronounced striations.

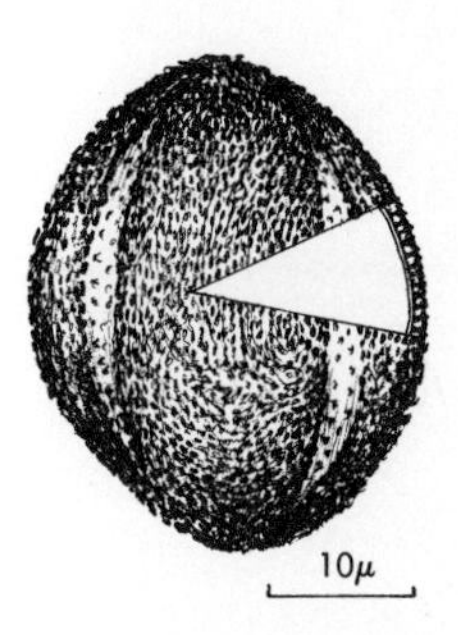

Fig. 228. *Acer leucoderme* Small. Coll. Alabama; ACERACEAE.

Sugar Maple

SIZE: Polar axis about 27-37μ; equatorial diameter 23-32μ

RANGE: Gaspe Peninsula to eastern Manitoba south to northeastern Texas and Georgia

Shape prolate spheroidal (P/E Index about 1.08). Surface appears to be finely reticulate with striations most prominent at the poles and near the furrow margins. Furrow membranes with a fine granular appearance. Very similar to *A. nigrum*, but the reticulum of *A. nigrum* is coarser.

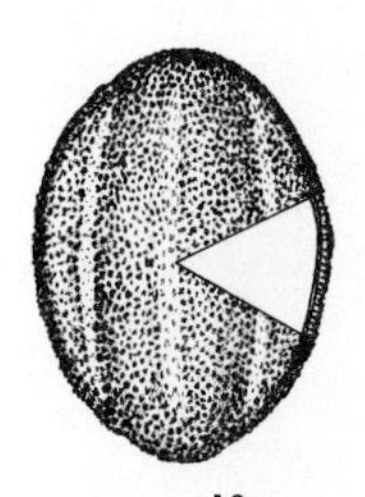

Fig. 229. *Acer saccharum* Marsh. Coll. Arnold Arboretum; ACERACEAE.

41a Furrows typically open, without a persistent membrane.......42

41b Furrow typically with a persistent, verrucate membrane........43

42a Exine apparently with small uniform verrucae, or sometimes exine pattern appears to be reticulate or to have short irregular furrows (Fig. 230) . . . . *Acer negundo* L.

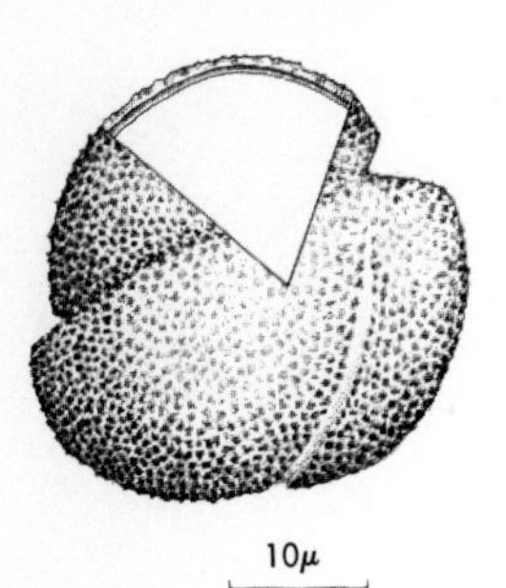

Fig. 230. *Acer negundo* L. Coll. Michigan; ACERACEAE.

Box-Elder

SIZE: Polar axis 24-30μ; equatorial diameter 23-27.5μ

RANGE: Florida to Texas north to Minnesota and New England

Shape prolate spheroidal (P/E Index about 1.1). Exine indistinctly tectate with short irregular ("fingerprint") swirls, not striate. Very similar to *A. saccharinum* except that in Silver Maple pollen the exine pattern becomes finer along the furrow margin, whereas in *A. negundo*, the exine pattern is uniform throughout.

42b Exine verrucate; surface appears to have a fine granular structure with larger scattered verrucae (Fig. 479); furrows often irregular or folded at the equatorial limb. Tricolpate *Actaea rubra* (usually tetracolpate) and some cactus pollen (e.g. *Epiphyllum*) will also key to this point; they are not illustrated (Fig. 231, 232, 233, 234) . . . . *Quercus* spp.

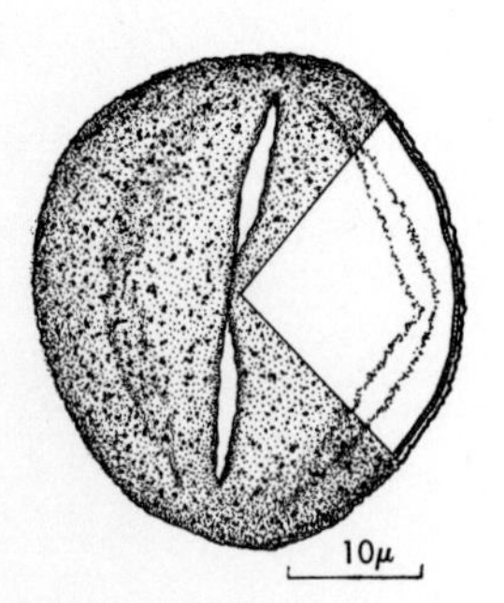

Fig. 231. *Quercus bicolor* Willd. Coll. Michigan; FAGACEAE.

Swamp White Oak

SIZE: Polar axis about 40μ; equatorial diameter 35-36μ

RANGE: Southern Maine to southern Minnesota south to Nebraska and Oklahoma and upland Georgia

Shape subprolate (P/E Index about 1.15). Surface with the verrucate surface typical of oaks. Furrows constricted at the equator.

Black Oak

SIZE: Equatorial diameter about 40-45μ

RANGE: Florida to eastern Texas, north to Nebraska, Minnesota, southern Ontario and Maine

Shape subprolate. Like many species of oaks, Q. *velutina* has a distinctive exine pattern; a fine granulated background is apparently due to structural elements of the ektexine, probably obscure columellae. The surface is, in addition, sculptured with irregularly spaced, and shaped, warty verrucae.

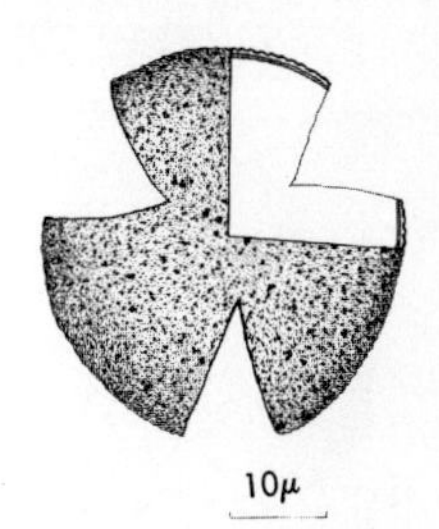

Fig. 232. *Quercus velutina* Lam. Coll. *Michigan;* FAGACEAE.

Shrub Live Oak

SIZE: Polar axis about 36μ; equatorial diameter about 29μ

RANGE: Colorado and western Texas to California and northern Mexico

Shape subprolate (P/E Index about 1.25). The exine surface of Q. *turbinella* is not as coarsely verrucate as Q. *gambelii*.

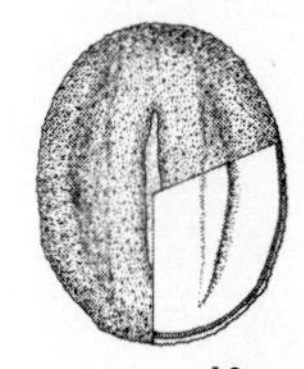

Fig. 233. *Quercus turbinella* Greene Coll. New Mexico (?); FAGACEAE.

Gambel Oak

SIZE: Equatorial diameter about 35μ

RANGE: Colorado to Nevada south to northern Mexico

The exine of this species of oak has the roughest surface of all of the oak species examined, due to the very large, coarse verrucae.

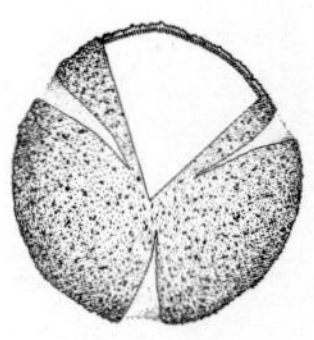

Fig. 234. *Quercus gambelii* Nutt. Coll. Colorado; FAGACEAE.

43a Lateral edge of furrow rather irregular, tapers abruptly to blunt ends (Fig. 235).....................*Cercocarpus parvifolius* Nutt.

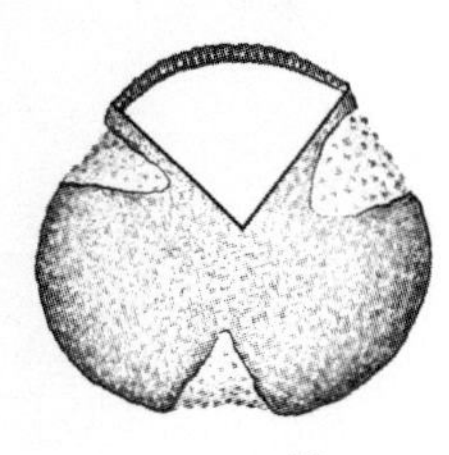

Fig. 235. *Cercocarpus parvifolius* Nutt.; Coll. Colorado; ROSACEAE.

Mountain Mahogany

SIZE: Polar axis about 29μ; equatorial diameter about 27μ

RANGE: South Dakota to Montana south to Utah, New Mexico and western Kansas

Shape prolate spheroidal (P/E Index about 1.1); size variable, equatorial diameter ranging from 27-38μ. Furrows broad and with irregular borders tapering gradually to blunt ends; membranes persistent and sparsely verrucate. Exine tectate, surface verrucate.

43b Lateral edge of furrow straight, tapers gradually to acute ends (Fig. 236, 237, 238).......................Ranunculaceae (in part)

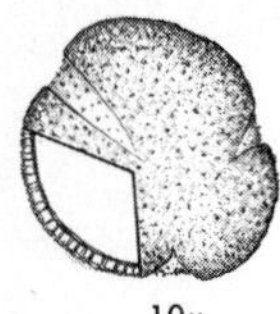

Fig. 236. *Aquilegia canadensis* L. Coll. Massachusetts; RANUNCULACEAE.

Wild Columbine

SIZE: Polar axis about 19μ; equatorial diameter about 18μ

RANGE: Quebec to Manitoba south to Iowa, Oklahoma, Texas and Georgia (several varieties)

Shape prolate spheroidal (P/E Index about 1.05). Exine tectate with verrucate surface. Furrows moderately long (polar area index about 0.4); furrow membrane persistent and verrucate.

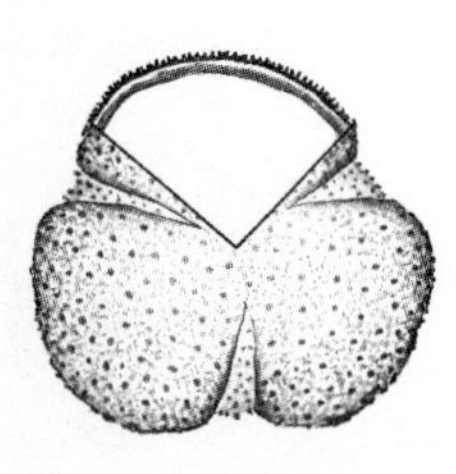

Fig. 237. *Caltha palustris* L. Coll. Massachusetts; RANUNCULACEAE.

Marsh-marigold

SIZE: Polar axis about 27.5μ; equatorial diameter about 26μ

RANGE: Labrador to Alaska south to South Carolina, Tennessee, Nebraska and Iowa

Shape prolate spheroidal (P/E Index about 1.05). Exine indistinctly tectate; surface verrucate to subreticulate. Furrows moderately long (polar area index about 0.3); furrow membrane persistent and coarsely verrucate.

Kidneyleaf Buttercup

SIZE: Polar axis about 22μ; equatorial diameter 23μ

RANGE: Florida to Oklahoma north to Saskatchewan and New England

Shape spherical to oblate spheroidal (P/E Index about 0.95). Exine tectate with a few stout columellae; surface verrucate. Furrow long (polar area index about 0.2); furrow membrane persistent with distinct verrucae.

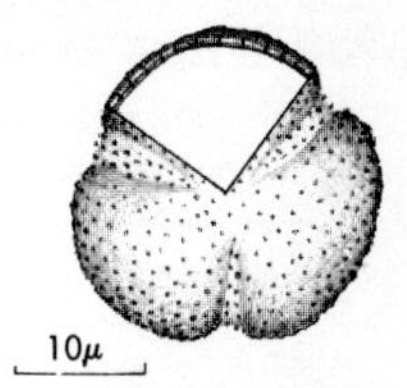

Fig. 238. *Ranunculus abortivus* L. Coll. Michigan; RANUNCULACEAE.

STEPHANOCOLPATE TYPES

1a Furrows four 2

1b Furrows five or more in number 3

2a Exine verrucate (Fig. 239) .... *Actaea rubra* (Ait.) Willd.

Red Baneberry

SIZE: 32μ diameter

RANGE: Widespread across southern Canada and northern United States, south in Rocky Mts. to Colorado and Utah.

Three or four colpate, shape spherical. Exine coarsely verrucate (*Quercus*-like) and tectate.

Fig. 239. *Actaea rubra* (Ait.) Willd. Coll. Massachusetts; RANUNCULACEAE.

2b Exine reticulate 4

3a Furrows five (4-6), broad (Fig. 240) .... *Hippuris* spp.

Mare's Tail

SIZE: Polar axis 38μ; equatorial diameter 32μ

RANGE: Greenland to Alaska south to new England, the Great Lakes States, Nebraska and New Mexico

Five (4-6) colpate, with broad, verrucate furrow membranes and narrow or obscure colpoid apertures. Intercolpate region is psilate and when seen in polar view forms a star-pattern.

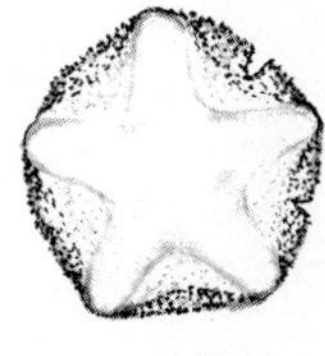

Fig. 240. *Hippuris vulgaris*. Coll. Michigan; HIPPURIDACEAE.

3b Furrows six or more in number **5**

4a Outline of grain rounded-rectangular (Fig. 241) .... ***Impatiens***

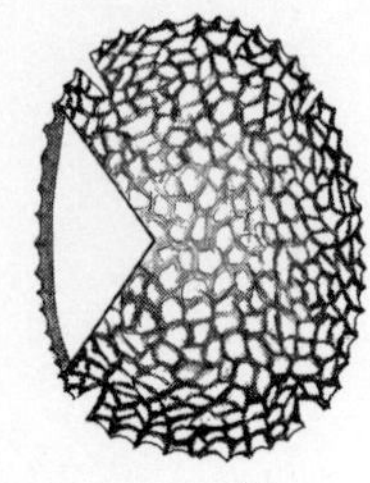

Fig. 241. *Impatiens capensis* Meerb. Coll. Michigan; BALSAMINACEAE.

Spotted Touch-Me-Not

SIZE: 23μ x 29μ

RANGE: Widespread in eastern United States, southern Canada to Alaska

Four short furrows are located at the "corners" of the rectangular grain. Exine intectate, reticulation coarse (openings about 1.5μ).

4b Outline of grain squarish (Fig. 242) .... ***Fraxinus americana* L.**

Fig. 242. *Fraxinus americana* L. Coll. Michigan; OLEACEAE.

White Ash

SIZE: 38-41μ diameter

RANGE: Newfoundland to Minnesota south to eastern Texas and northern Florida

The four short furrows are located at "corners" of the subquadrate grain. Exine tectate and reticulate; reticulum fine (openings about 0.5μ). The four-furrowed species of ash which may be confused with White Ash include *F. pennsylvanica* and *F. texensis*.

5a Colpi six or seven, transverse furrows lacking **6**

5b Colpi 9 or more, transverse furrows present, but occasionally obscure 7

6a Furrows six, intercolpae flat (Fig. 243).......... Labiatae (in part)

Horsemint

SIZE: Polar axis 54μ, equatorial diameter 47μ

RANGE: The variety extends from New England to Minnesota, south to Tennessee, Missouri, and Oklahoma

Hexacolpate, with long broad furrows. Furrow membrane sometimes remains partially intact. Intercolpate exine finely reticulate (openings 0.5-0.7μ). Exine perforate tectate. Many members of the Labiatae are tricolpate, others are hexacolpate; among the hexacolpate reticulate forms are species of *Salvia, Collinsonia, Thymus, Lycopus, Hyssopus, Agastacke, Blephilia, Pycnanthemum, Nepeta, Acinos, Clinopodium, Calamintha, Mentha, Prunella,* and *Dracocephalum.*

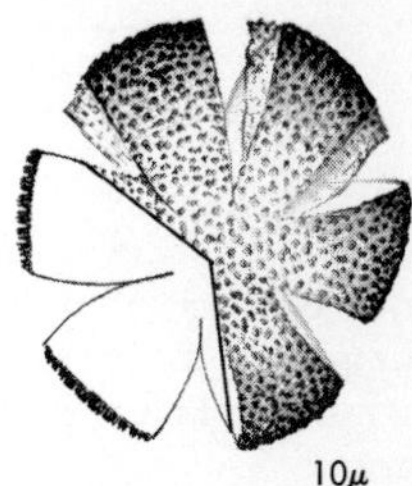

Fig. 243. *Monarda punctata* L. v. *villicaulis* Pennell. Coll. Michigan; LABIATAE.

6b Furrows seven (6-10 in some species), narrow; intercolpae rounded (Fig. 244)................................ Rubiaceae (in part)

Bedstraw

SIZE: Polar axis 22.5μ; equatorial diameter 19μ

RANGE: Eastern United States, southern Canada to Alaska

Subprolate, typically seven furrowed. Indistinctly tectate (?), surface appears slightly verrucate or sub-reticulate.

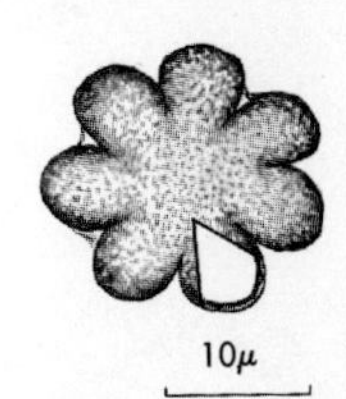

Fig. 244. *Galium aparine* L. Coll. Michigan; RUBIACEAE.

7a Furrows 9 or more; grain circular in equatorial view (Fig. 245)..... .... *Polygala*

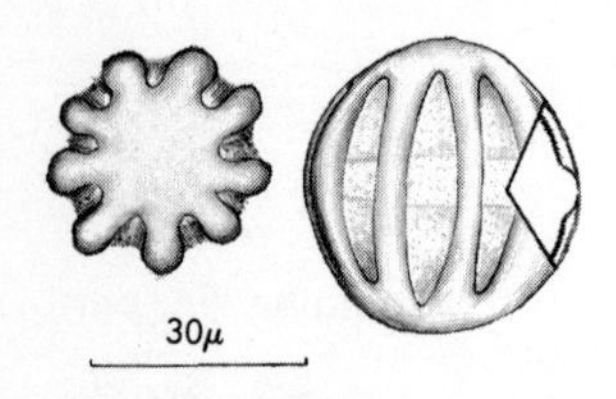

Fig. 245. *Polygala alba* Nutt. Coll. Texas; POLYGALACEAE.

White Milkwort

SIZE: Polar axis 38μ; equatorial diameter 32μ

RANGE: Washington to Mexico, east to Minnesota, Nebraska, Kansas, Oklahoma and Texas

Nine furrows of moderate length; polar area large; some species have thin poroid areas in the polar cap. The transverse furrows associated with each pore are joined to form a girdle around the equator. The transverse furrow is not always obvious, especially in polar view; therefore, this type is keyed both in this category and with the stephanocolporate types, where it is properly placed. Some species of *Polygala* have 21-24-28 colpi.

7b Furrows usually 12 or more; grain with a distinctly swollen ridge at the equator (seen in equatorial view) (Fig. 246)...... *Utricularia*

Bladderwort

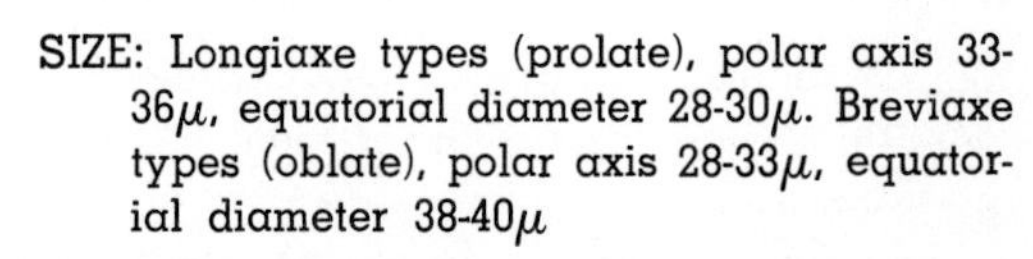

SIZE: Longiaxe types (prolate), polar axis 33-36μ, equatorial diameter 28-30μ. Breviaxe types (oblate), polar axis 28-33μ, equatorial diameter 38-40μ

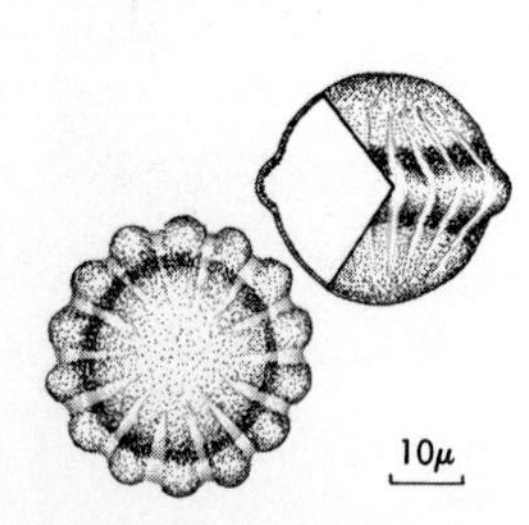

Fig. 246. *Utricularia minor* L. Redrawn from: Thanikaimoni, G. 1966. Pollen morphology of the genus *Utricularia*. *Pollen et Spores*, Vol. VIII(2):265-284 (Plate I, Figs. 3-4).

RANGE: Widespread in northern United States, Canada and Eurasia

Pollen of aquatic species of *Utricularia* may have 11-28 colporate apertures. Since the equatorial girdle may not always be properly interpreted as fused transverse furrows (pores), this type is keyed with both stephanocolpate and stephanocolporate pollen types. This species may produce pollen of two shapes, "longiaxe" (prolate) and "breviaxe" types.

PERICOLPATE TYPES

1a Grains with minute spines; furrows small and numerous (about 30) (Fig. 247)....................................*Portulaca oleracea* L.

Common Purslane

SIZE: 60μ diameter

RANGE: Cosmopolitan weed introduced in North America

Furrows slender and about 10μ long, arranged in an approximate pentagonal pattern. Exine stratification distinct; collumellae present; spines (about 1.5μ long) are supratectate. *Polygonum amphibium* has about 30 short furrows, but may be distinguished from *Portulaca* by the absence of spines and presence of a heavy reticulate exine.

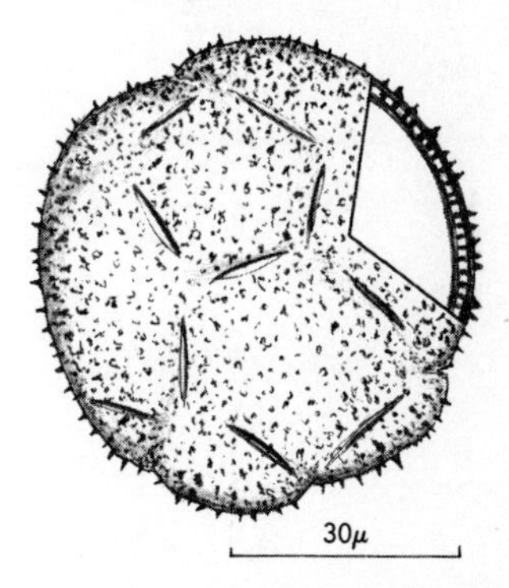

Fig. 247. *Portulaca oleracea* L. Coll. Michigan; PORTULACACEAE.

1b Grains psilate, baculate or clavate; without spines..............2

2a Apertures six large, open, furrows; exine psilate (Fig. 248)........ ....*Dicentra canadensis* (Goldie) Walp.

Squirrel-Corn

SIZE: 48-55μ diameter

RANGE: Northeastern United States and southeastern Canada

Surface of grain more or less lobed in the intercolpate areas. The six furrows are arranged symmetrically at 120° angles, providing 4 positions in which three furrows nearly touch; grain may in brief observation appear tricolpate. Exine faintly reticulate.

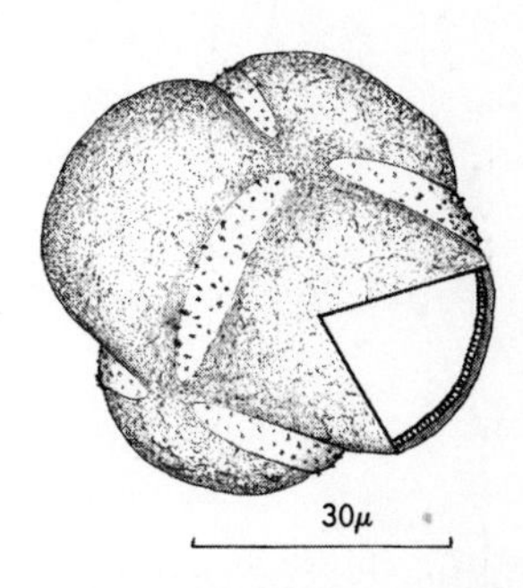

Fig. 248. *Dicentra canadensis* (Goldie) Walp. Coll. Michigan; PAPAVERACEAE.

2b Apertures 18-20, short furrows; grain with high clavate or baculate muri, fenestrate (Fig. 249) .... ***Opuntia* spp.**

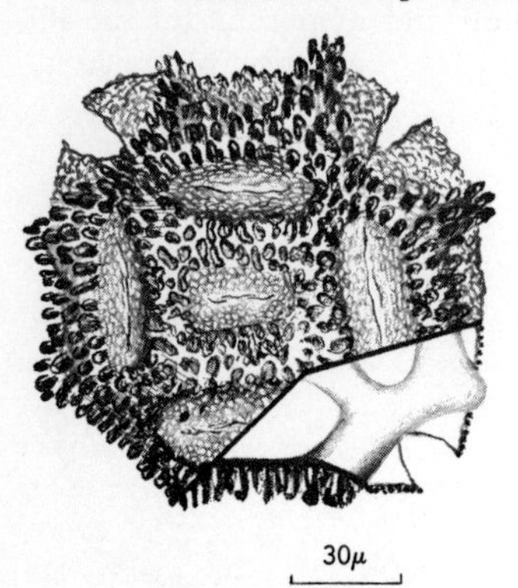

Fig. 249. *Opuntia polycantha* Haw. Coll. Montana; CACTACEAE.

Prickly Pear Cactus

SIZE: 110-120μ diameter

RANGE: Widespread in plains and Rocky Mt. states

Exine of muri very thick (8μ), composed of clavae or baculae; exine of "windows" thin (1μ) reticulate. Shape of grain more or less octagonal. Pollen of *O. humifusa* Raf. is virtually identical with *O. polycantha*. Other genera of Cactaceae may be pericolpate (e.g. *Rhizophyllum, Neobesseya, Pereskia, Rhipsalis, Echinocactus*), tricolpate (e.g. *Epiphyllum, Echinocercus, Carnegia, Ferocactus*), or periporate (*Opuntia fulgida, O. pulchella, O. acanthocarpa, O. echinocarpa, O. exaltata*); all are moderately to very large (40-150μ).

HETEROCOLPATE TYPE

In addition to *Lythrum*, some species of *Verbena* and certain members of the Boraginaceae (e.g. *Mertensia* and *Myosotis*, and *Amsinckia*, see Fig. 475) have pseudocolpi and thus are in this category.

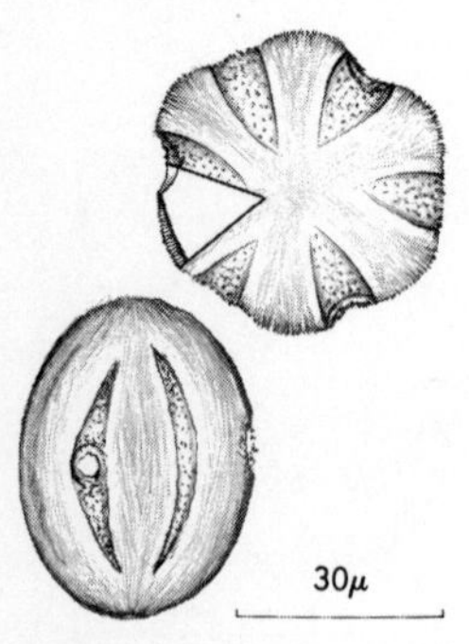

Fig. 250. *Lythrum salicaria* L. Coll. Michigan; LYTHRACEAE.

Striped Loosestrife

SIZE: Polar axis 45μ; equatorial diameter 34μ

RANGE: Naturalized from Europe; southeastern Canada and northern United States

Pollen has three true apertures (tricolporate) with three additional furrows (pseudocolpi) which lack germinal pores. Exine striate between the furrows, verrucate on the furrow membranes. Shape subprolate to prolate; P/E Index about 1.3.

TRICOLPORATE TYPES

This is the most common pollen type among dicotyledons. The pollen grains typically have tectate exines with the pore formed by thinning of the inner exine layer in the equatorial portion of the furrow. Frequently the pore is broader than the furrow itself. If the pore is elongated (at least twice as long as wide) along the equatorial arc, the pore is called a transverse furrow.

The pollen of the following families is typically tricolporate: Anacardiaceae, Aquifoliaceae, Caprifoliaceae, Cistaceae, Compositae, Cornaceae, Eleagnaceae, Gentianaceae, Nyssaceae, Rhamnaceae, Rutaceae, Saxifragaceae, Scrophulariaceae, Solanaceae, Umbelliferae, Vitaceae. In addition, some genera or species in the following families have tricolporate pollen: Euphorbiaceae, Fagaceae, Leguminosae, Polygonaceae, Rosaceae, Rubiaceae, and Violaceae. The tricolporate grains which characterize most members of the Empetraceae, Ericaceae, and Pyrolaceae are usually joined as tetrads.

In a few instances the pores may be indistinct or are frequently poorly displayed. Insofar as possible, these types are also keyed with the tricolpate types.

In some instances, grains with a psilate surface may appear to be reticulate, striate, or verrucate due to arrangement of the columellae within the ektexine. In several such cases, the grains have been keyed with those which are sculpturally reticulate, striate, or verrucate. Because of difficulties in interpretation of exine structure and sculpture, the reader is advised to seek identification in more than one section of this key.

1a Surface of exine psilate (may be finely verrucate or reticulate on the intercolpi in *Nyssa*)....................................2

1b Surface of exine reticulate, echinate, striate-rugulate, or verrucate 29

2a Longest dimension 20μ or less....................................3

2b Longest dimension greater than 22μ....................................9

3a Pore round or indistinct....................................4

3b Pore transversely elongated....................................6

4a Polar area very small (polar area index less than 0.25)..........5

4b Polar area larger (polar area index greater than 0.35); pore visible (Fig. 251) .... ***Larrea tridentata*** **(DC.) Coville.**

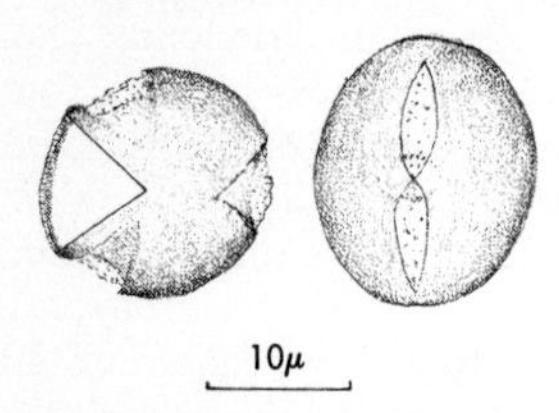

Fig. 251. *Larrea tridentata* (DC.) Coville. Coll. Arizona; ZYGOPHYLLACEAE.

Creosote-Bush

SIZE: Polar axis about 18μ; equatorial diameter about 15μ

RANGE: Western Texas to southern Utah, California, and northern Mexico

Shape subprolate (P/E Index about 1.2). Furrows sharply contracted at the equator, pore round to slightly elongate, exceeding the furrow borders. Furrow membrane persistent. Exine thin, apparently tectate, but stratification indistinct.

5a Pore indistinct (Fig. 252) .... ***Dodocatheon cylindrocarpum*** **Rydb.**

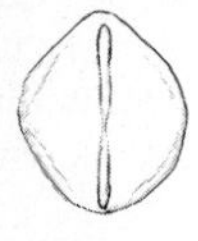

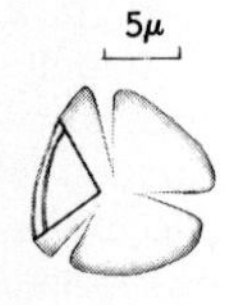

Fig. 252. *Dodocatheon cylindrocarpum* Rydb. Coll. Montana; PRIMULACEAE.

Shooting Star

SIZE: Polar and equatorial diameter about 12μ

RANGE: Eastern Washington to Montana

Shape spherical (P/E Index about 1.0). Furrows very long, constricted at the equator, polar area very small (polar area index 0.12 to 0.16). Surface psilate; exine stratification obscure.

5b Pore distinct (Fig. 253) .... ***Rhamnus frangula*** **L.**

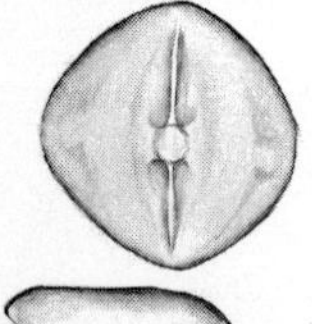

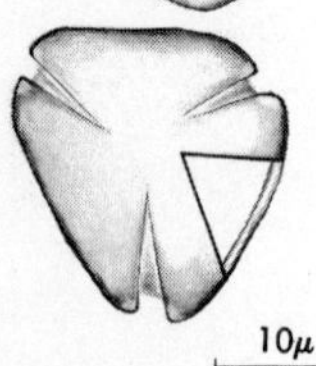

Fig. 253. *Rhamnus frangula* L. Coll. Michigan; RHAMNACEAE.

Alder Buckthorn

SIZE: Polar axis about 17μ; equatorial diameter about 19μ

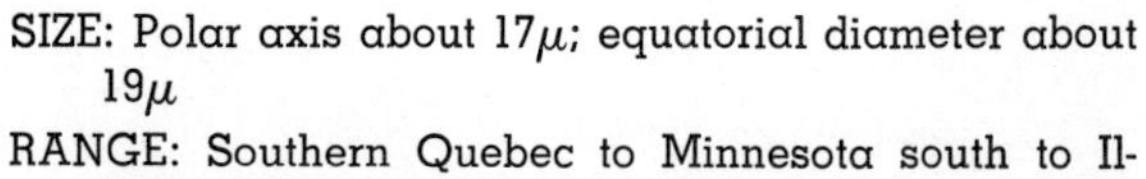

RANGE: Southern Quebec to Minnesota south to Illinois, Ohio, and New Jersey

Shape oblate spheroidal (P/E Index about 0.9); triangular in polar view. Furrows long and slender, polar area small (polar area index about 0.35). Furrow margins thickened (margo); pore round and exceeding the borders of the furrows. Exine smooth. *Rhamnus purshiana* DC. nearly identical but it has blunter furrow ends, shape subprolate (27 x 22μ).

6a Shape suboblate (P/E Index 0.8 to 0.9)........................7

6b Shape prolate (P/E Index 1.35 to 1.7)............................8

7a Transverse furrows connected to form a continuous equatorial girdle (Fig. 254).............................. *Solanum dulcamara* L.

Nightshade

SIZE: Polar axis about 11μ; equatorial diameter about 13μ

RANGE: Introduced from Europe, throughout the eastern United States

Shape subspheroidal (P/E Index about 0.85). Furrows long, polar area very small (polar area index < 0.2). Transverse furrows joined by a narrow equatorial girdle. Exine thin, psilate, without apparent stratification.

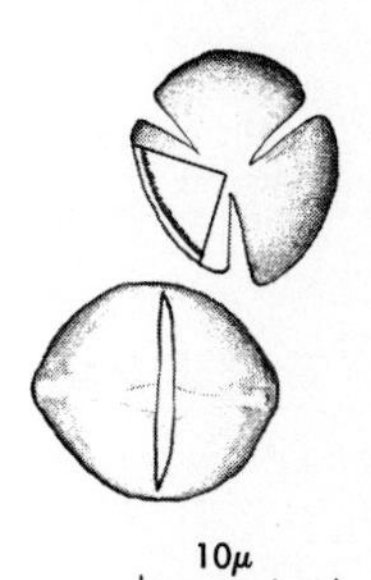

Fig. 254. *Solanum dulcamara* L. Coll. Michigan; SOLANACEAE.

7b Transverse furrows short and not interconnected (Fig. 255)........ ....*Ceanothus americanus* L.

New Jersey Tea

SIZE: Polar axis about 18μ; equatorial diameter about 20μ

RANGE: Florida to Alabama north to southern Manitoba, southern Quebec and Maine

Shape oblate spheroidal (P/E Index about 0.9); diamond-shaped in equatorial view, triangular in polar view. Furrows long and slender, polar area very small. Furrows with margo, especially near the equator; transverse furrows with rounded ends. Exine stratification indistinct, apparently tectate.

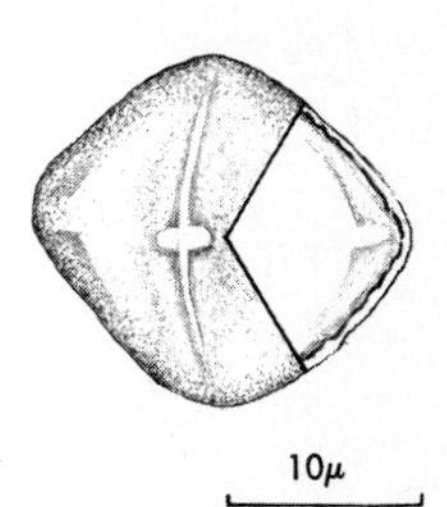

Fig. 255. *Ceanothus americanus* L. Coll. North Carolina; RHAMNACEAE.

8a P/E Index about 1.4 (Fig. 256)..... *Castanea dentata* (Marsh.) Borkh.

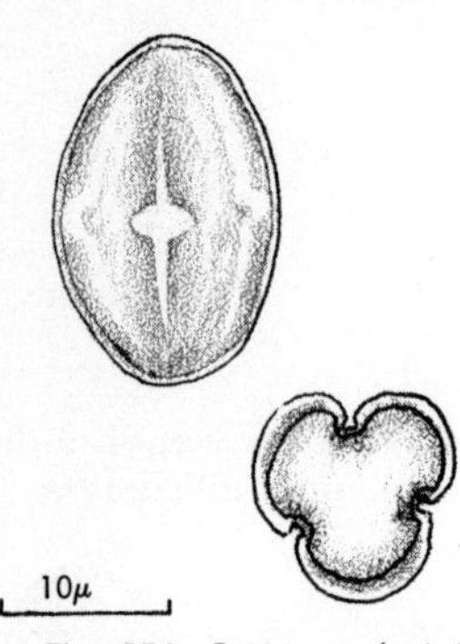

Fig. 256. *Castanea dentata* (Marsh.) Borkh. Coll. New York; FAGACEAE.

Chestnut

SIZE: Polar axis about 18μ; equatorial diameter about 13μ

RANGE: Georgia to Mississippi north to southern Minnesota, southern Ontario and southern Maine

Shape prolate (P/E Index 1.35 to 1.4), trilobate in polar view. Furrows narrow and slender, transverse furrows elliptical with pointed ends.

8b P/E Index about 1.66 (Fig. 257)...................................
.... *Lithocarpus densiflora* (H. & A.) Rehd.

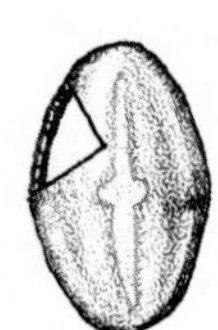

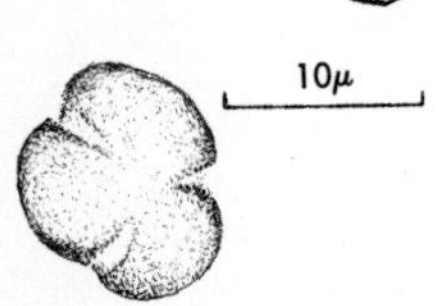

Fig. 257. *Lithocarpus densiflora* (H. & A.) Rehd. Coll. Arn. Arb.; FAGACEAE.

Tanoak

SIZE: Polar axis about 15μ; equatorial diameter about 9μ

RANGE: Southwestern Oregon and California

Shape prolate (P/E Index about 1.66). Furrow long, pore transversely elongate with broadly-rounded ends. Exine tectate, but columellae indistinct. Surface psilate, but with indistinct granular appearance due to columellae. Similar to *Castanea dentata* but smaller and with more rounded ends on the transverse furrow.

9a Size 22-30μ (to 35μ in *Acer circinatum*)........................10

9b Size greater than 35μ...20

10a Transverse furrow present..11

10b Transverse furrow absent...14

11a Shape prolate, P/E Index about 1.4 (Fig. 258)..*Cryptotaenia canadensis* (L.) DC.

Honewart

SIZE: Polar axis about 26μ; equatorial diameter about 18μ

RANGE: New Brunswick to Manitoba south to Texas and Georgia

Shape prolate (P/E Index about 1.45); rounded at poles. Furrow slender and of intermediate length, transverse furrow elliptical, with pointed ends. Exine tectate, with distinct columellae. Surface psilate, but with pebbled appearance due to columellae.*

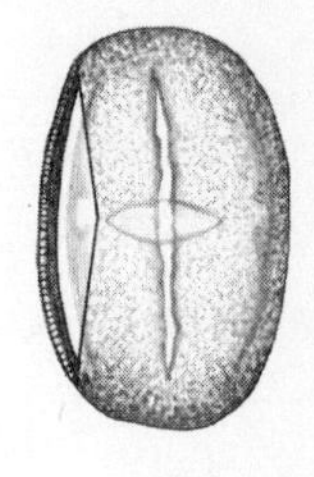

Fig. 258. *Cryptotaenia canadensis* (L.) DC. Coll. Michigan; UMBELLIFERAE.

11b Shape subprolate or prolate spheroidal (P/E Index about 1.15 to 1.3)..12

12a Exine with distinct stratification, columellae visible, ends of transverse furrows "squarish" (Fig. 259)..................................*Bupleurium americanum* Coutt. and Rose.

Throughwax

SIZE: Polar axis about 24μ; equatorial diameter about 19μ

RANGE: Alaska, and Mackenzie, south to Idaho and Wyoming

Shape subprolate (P/E Index about 1.25), diamond-shaped in equatorial view, but with exine bulging at the transverse furrows. Transverse furrows rectangular or sometimes auriculate-expanded at the ends. Endexine thickest at the apertures, ektexine tectate, with distinct columellae.

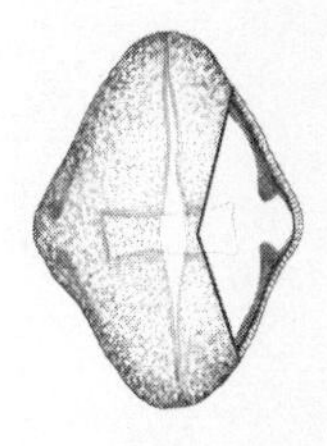

Fig. 259. *Bupleurium americanum* Coutt. and Rose. Coll. Montana; UMBELLIFERAE.*

12b Exine stratification indistinct, columellae absent or obscure......13

*Other Umbelliferae are also psilate, but many exhibit a reticulate structure in L/O analysis. These are keyed with reticulate types since they might be considered reticulate by some microscopists.

13a Ektexine (sexine and nexine) separated at the transverse furrow to form a pore cavity (Fig. 260)....... *Lycopersicum esculentum* Mill.

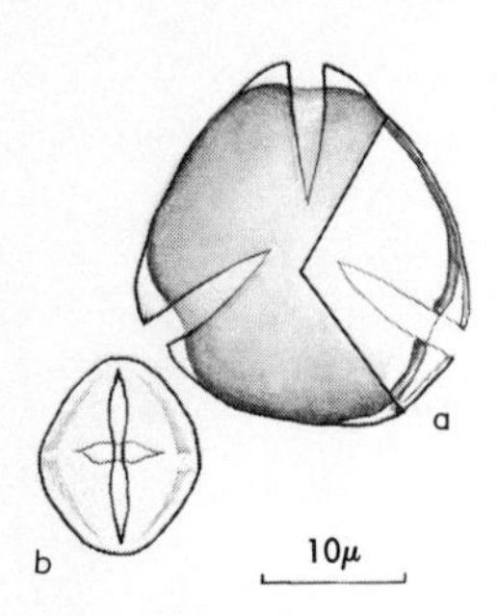

Fig. 260. *Lycopersicum esculentum* Mill. Coll. Indiana; SOLANACEAE.

Tomato

SIZE: Polar axis about 25μ; equatorial diameter about 22μ

RANGE: Widespread under cultivation

Shape prolate spheroidal or subprolate (P/E Index about 1.15). Furrows very long, polar area small (polar area index about 0.2). Pore transversely elongate, pore cavity very distinct in polar view. a. polar view; b. equatorial view.

13b Ektexine does not form a pore cavity (Fig. 261).. *Physalis virginiana* Mill.

10μ

Fig. 261. *Physalis virginiana* Mill. Coll. Indiana; SOLANACEAE.

Ground Cherry

SIZE: Polar axis about 23μ; equatorial diameter about 19μ

RANGE: Florida to Texas, northward to Manitoba and Connecticut

Shape subprolate (P/E Index about 1.2); furrows long, polar area small. Exine stratification obscure, surface psilate. Pore elongated as transverse furrow which is tapered-pointed at the ends.

14a Exine thick with a complex undulating supratectate layer......... *Centaurea* (see description of *Centaurea picris*, page 153, Figure 320)

14b Exine thinner without supratectate undulation.................15

15a Shape spherical to suboblate (P/E Index 0.88 to 1.0)............16

15b Shape subprolate (P/E Index 1.2 to 1.3)......................18

16a Ektexine with distinct columellae (Fig. 262)..*Pentstemon digitalis* Nutt.

Beard-Tongue

SIZE: Polar and equatorial axes about 30μ

RANGE: Maine and southern Quebec to South Dakota, southward to Texas, Alabama, and Virginia

Shape approximately spherical (P/E Index about 1.0). Furrows very long, polar area small; pores more or less round, confined within the furrow margins. Exine tectate with psilate or scabrate surface.

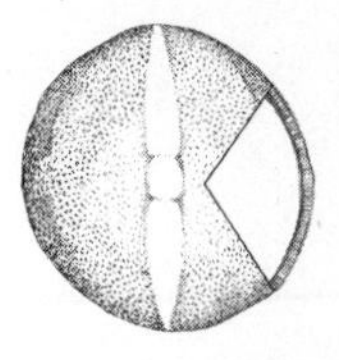

Fig. 262. *Pentstemon digitalis* Nutt. Coll. Michigan; SCROPHULARIACEAE.

16b Ektexine structure appears simpler, columellae lacking or indistinct 17

17a Furrow with an equatorial constriction or fold, pore indistinct (Fig. 263)...................................*Viola sagittata* Ait.

Arrow-Leaved Violet

SIZE: Polar axis about 30μ, equatorial diameter about 36μ

RANGE: Massachusetts to Minnesota south to eastern Texas and Georgia

Shape oblate spheroidal to suboblate (P/E Index about 0.88). Furrows long and broadly-tapered to the poles, irregularly folded or arched at the equatorial limb, pore indistinct. Exine surface psilate.

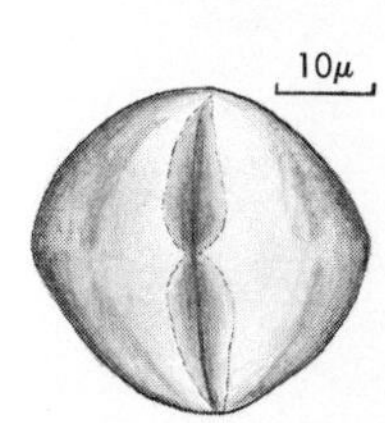

Fig. 263. *Viola sagittata* Ait. Coll. Michigan; VIOLACEAE.

17b Furrow with a well developed pore (Fig. 264)....................
......................................*Viola novae-angliae* House.

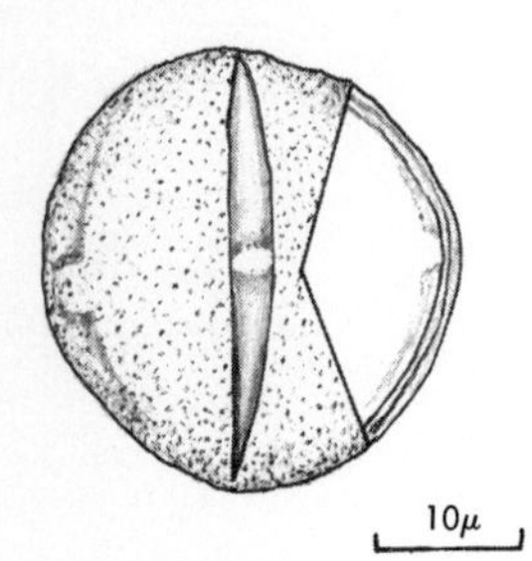

Fig. 264. *Viola novae-angliae* House. Coll. Michigan; VIOLACEAE.

New England Violet

SIZE: Polar axis about 28μ; equatorial diameter 30.5μ

RANGE: New Brunswick to Minnesota

Shape oblate spheroidal (P/E Index about 0.9). Furrows long and polar area small, pore round to oblate included within the furrow margins. Exine surface psilate, scabrate, to subverrucate.

18a Furrow with an equatorial constriction; pore indistinct (Fig. 265)..
......................................*Robinia pseudoacacia* L.

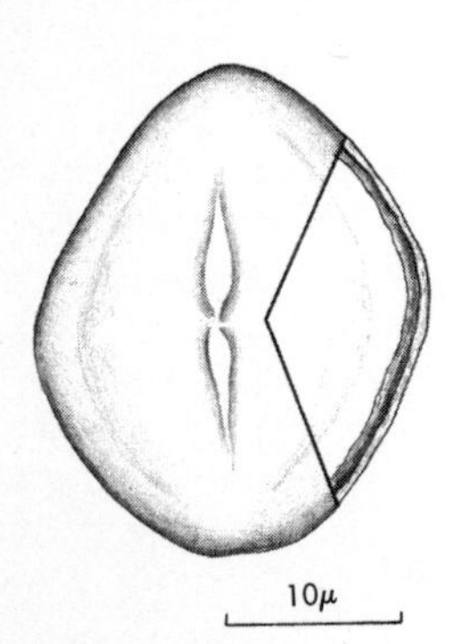

Fig. 265. *Robinia pseudoacacia* L. Coll. Michigan; LEGUMINOSAE.

Black Locust

SIZE: Polar axis about 27μ; equatorial diameter about 23μ

RANGE: Georgia to Louisiana and Oklahoma northward to Iowa and Pennsylvania; naturalized farther north.

Shape subprolate (P/E Index about 1.2). Furrow rather short, contracted at the equator, and with a distinct margo; pore may be obscure in face view. Exine stratification obscure; surface psilate.

18b Furrow not constricted; pore distinct..........................19

19a Pore with a marginal thickening (annulus?) (Fig. 266)..*Acer circinatum* Pursh.

Vine Maple

SIZE: Polar axis 22 to 35μ; equatorial diameter 20.5 to 27μ

RANGE: British Columbia to California

Shape subprolate (P/E Index about 1.2). Furrows long and polar area small (polar area index about 0.2), pore round, slightly exceeding the width of the furrows. Very similar to *Acer spicatum* Lam. in that the exine is faintly swirled-striate near the equator and more distinctly striate near the poles, due to the pattern of the columellae. *A. douglasii* Hook. also has tricolporate pollen (other species are tricolpate), but it is uniformly striate over the entire intercolpal area. Surface unsculptured.

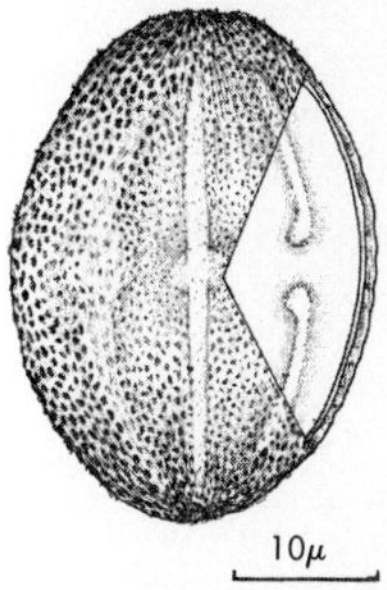

10μ

Fig. 266. *Acer circinatum* Pursh. Coll. Washington; ACERACEAE.

19b Pore lacking marginal thickening (Fig. 267)...*Aesculus hippocastanum* L.

Horse-Chestnut

SIZE: Polar axis about 29μ; equatorial diameter varies from 18 to 23μ

RANGE: Widely cultivated in eastern North America, introduced from southeastern Europe

Shape subprolate to prolate (P/E Index varies from 1.25 to 1.6). Furrows with persistent membrane which is ornamented with very coarse granules (verrucae and baculae); pore distinct, round, about the same width as the furrow. *Aesculus flava* pollen is larger, the furrow membranes are less coarsely granulate, and the pore extends beyond the furrow margins. *A. parviflora* has a faintly reticulate ektexine.

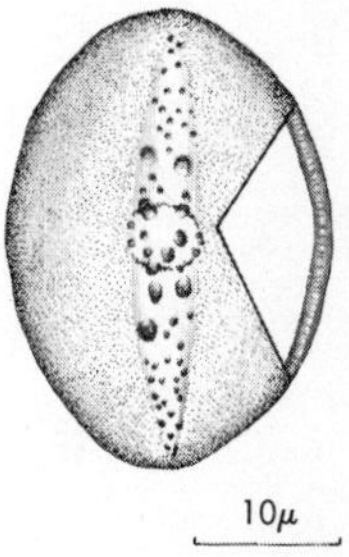

10μ

Fig. 267. *Aesculus hippocastanum* L. Coll. Michigan; HIPPOCASTANACEAE.

20a Shape subprolate or prolate..21

20b Shape approximately spheroidal, suboblate, or oblate...........24

21a Transverse furrow present..22

21b Transverse furrow lacking, pore round or indistinct 23

22a Furrow narrowest near the equator; polar axis about 40μ (Fig. 268) ....*Cornus racemosa* Lam.

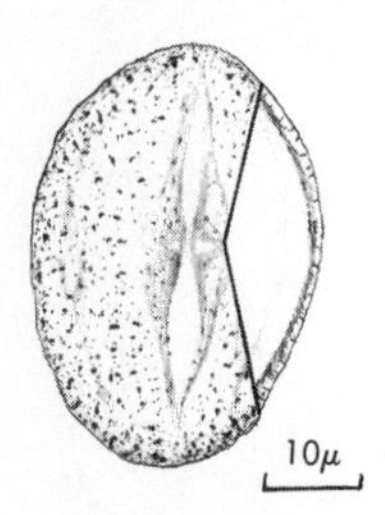

Fig. 268. *Cornus racemosa* Lam. Coll. Michigan; CORNACEAE.

Gray Dogwood

SIZE: Polar axis about 41μ; equatorial diameter about 30μ

RANGE: Central Maine to Minnesota south to Missouri, Alabama and Georgia

Shape subprolate (P/E Index about 1.35), subtriangular in polar view. Furrows strongly marginate, contracted at the equator; pores elongated as narrow transverse furrows. Exine tectate, but columellae thin and weak; surface psilate.

22b Furrow of uniform width or broadest near the equator; polar axis about 55μ (Fig. 269)......................*Diospyros virginiana* L.

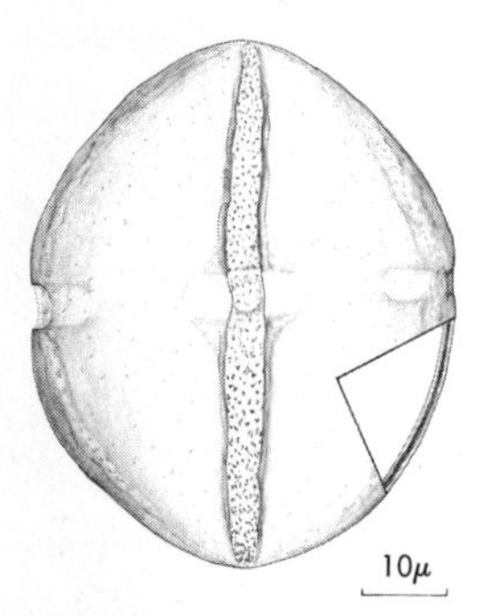

Fig. 269. *Diospyros virginiana* L. Col. Texas; EBENACEAE.

Common Persimmon

SIZE: Polar axis about 57μ; equatorial diameter about 48μ

RANGE: Florida to Texas northward to eastern Kansas, southern Iowa, Ohio, and southeastern New York

Shape subprolate (P/E Index about 1.2). Furrow long with a persistent, coarsely verrucate membrane, furrow with a margo which is thickest near the equator. Transverse furrow with broadly rounded ends. Exine stratification obscure, columellae apparently lacking; wall surface psilate.

23a Furrow lacks an equatorial constriction; pore apparent (Fig. 270).. ....*Prosopsis juliflora* (Swartz) DC.

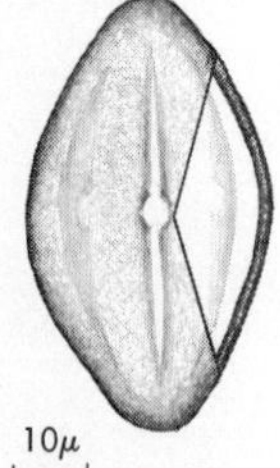

Fig. 270. *Prosopsis juliflora* (Swartz) DC. Coll. Texas; LEGUMINOSAE

Mesquite

SIZE: Polar axis about 60μ; equatorial diameter about 34μ

RANGE: Southern Kansas to southeastern California, south into Mexico

Shape prolate (P/E Index about 1.75); somewhat diamond-shaped in equatorial view. Furrow of intermediate length, more or less marginate; pore may be indistinct. Exine tectate, columellae short; surface psilate.

23b Furrow with an equatorial constriction; pore somewhat obscure in face view (Fig. 271, 272)...............numerous Leguminosae

Two-leaved Senna

SIZE: Polar axis about 38μ; equatorial diameter about 25μ

RANGE: Western Texas and southern New Mexico to Mexico

Shape prolate (P/E Index about 1.5). Furrow contracted at the equator; pore somewhat indistinct. Exine tectate with short columellae; surface psilate.

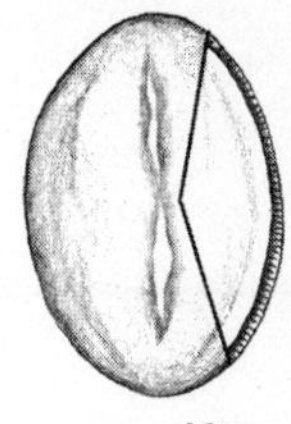

Fig. 271. *Cassia roemeriana* Scheele. Coll. Texas; LEGUMINOSAE.

Alfalfa

SIZE: Polar axis about 38μ; equatorial diameter about 33μ

RANGE: Introduced and cultivated throughout the United States and southern Canada

Shape prolate spheroidal (P/E Index about 1.15). Furrows long, contracted at the equator; pore somewhat indistinct in face view. Exine tectate, with very short columellae (oil); surface psilate.

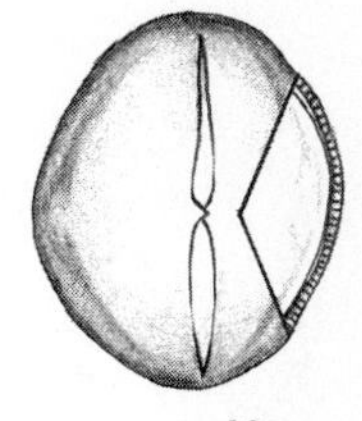

Fig. 272. *Medicago sativa* L. Coll. Michigan; LEGUMINOSAE.

24a Furrows short, bordered by dark mesexinous thickenings (Fig. 273)...*Tilia* spp.

Basswood

SIZE: About 45 to 47μ equatorial diameter

RANGE: Quebec to Manitoba, south to Alabama and Texas

The three apertures, (properly short colpi bearing pores) may be confused with simple pores. They are surrounded by mesexinous thickenings which are dark after acetolysis. Exine surface is very finely reticulate.

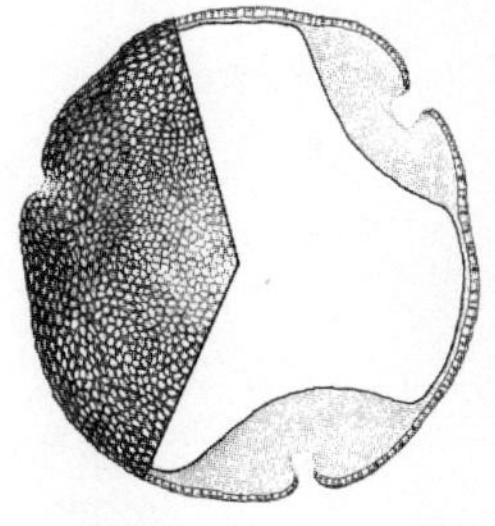

Fig. 273. *Tilia americana* L. Coll. New York; TILIACEAE.

24b Furrows longer, without mesexinous border....................25

25a Transverse furrow present (Fig. 274)..........*Cornus florida* L.

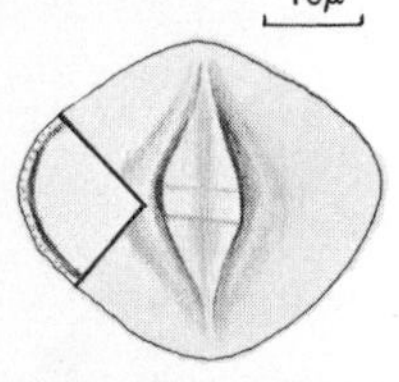

Fig. 274. *Cornus florida* L. Coll. Michigan; CORNACEAE.

Flowering Dogwood

SIZE: Polar axis about 32μ; equatorial diameter about 40μ

RANGE: Florida to Texas and Mexico, northward to Kansas, Illinois, southern Michigan and southwestern Maine

Shape suboblate (P/E Index about 0.8); diamond-shaped in equatorial view, subtriangular in polar view. Furrows of intermediate length, with prominent margo; transverse furrow rectangular in shape.

25b Transverse furrow lacking..............................26

26a Equatorial limb subtriangular in polar view; pores often obscure 27

26b Equatorial limb round in polar view; pores distinct and round (Fig. 275)...............................*Fagus grandifolia* Ehrh.

Beech

SIZE: Polar axis about 46μ; equatorial diameter 49μ

RANGE: Prince Edward Island to eastern Wisconsin southward in Great Lakes region to West Virginia and Virginia

Shape oblate spheroidal (P/E Index 0.9 to 0.95. Furrow of intermediate length and very slender, pore round, considerably exceeding the furrow. Exine indistinctly tectate; fine granular appearance apparently due to columellae.

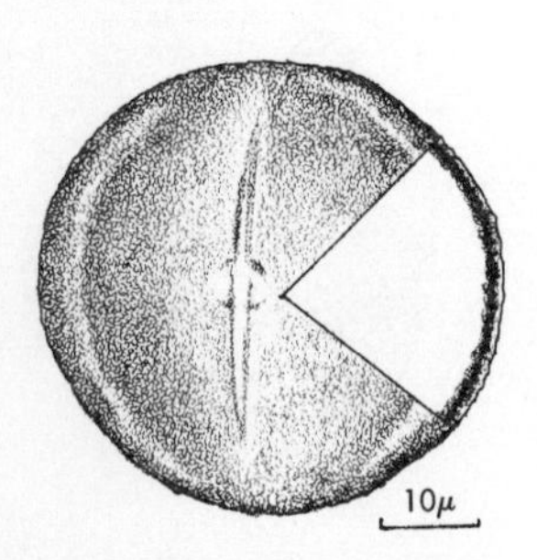

Fig. 275. *Fagus grandifolia* Ehrh. Coll. Michigan; FAGACEAE.

27a Furrow complex with two flanking supplementary furrows adjacent to the primary colpus (Fig. 276).. ....*Verbena ciliata* Benth. v. *longidentata* Perry

Vervain

SIZE: Polar axis about 35μ; equatorial diameter about 38μ

RANGE: The species extends from western Texas to Arizona and southward to southern Mexico

Shape oblate spheroidal (P/E Index about 0.9). Furrow very complex, with two pseudocolpi flanking each furrow; pore apparently elongated as a transverse furrow.

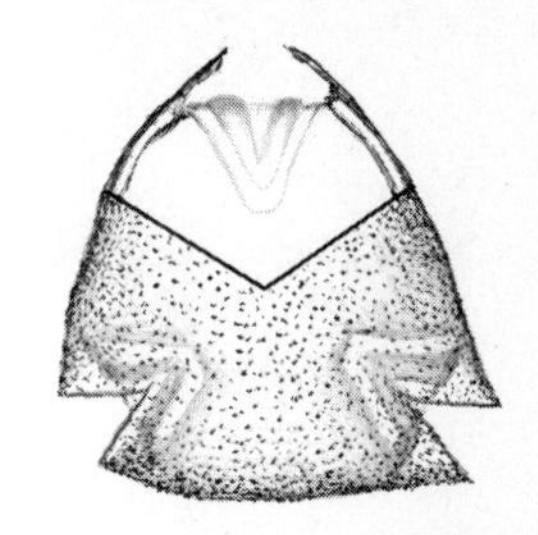

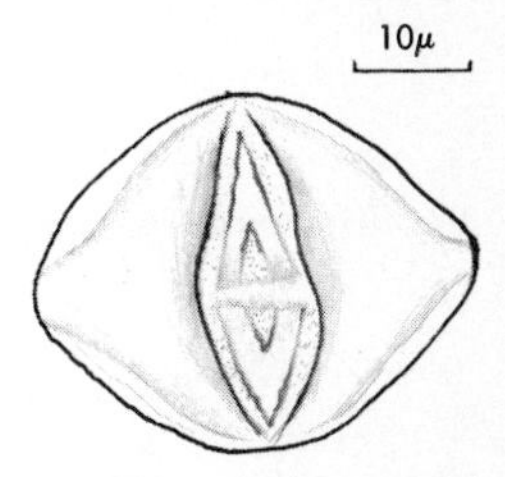

Fig. 276. *Verbena ciliata* Benth. v. *longidentata* Perry Coll. Texas; VERBENACEAE.

27b Furrows simple...28

28a Apertures not protruding, furrows with thickened margo (Fig. 277, 278)..*Nyssa* spp.

Black Gum

SIZE: Polar and equatorial axes about 49μ

RANGE: Maine to Missouri south to Florida, Texas and Mexico

Shape approximately spheroidal (P/E Index about 1.0) in equatorial view; subtriangular in polar view. Furrows relatively long (polar area index about 0.35), with a distinct margo. Pores somewhat indistinct. Surface texture subverrucate in intercolpae to psilate near furrows.

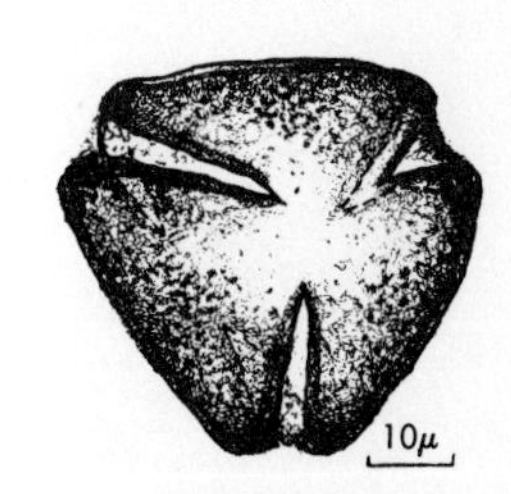

Fig. 277. *Nyssa sylvatica* Marsh. Coll. Arn. Arb.; NYSSACEAE.

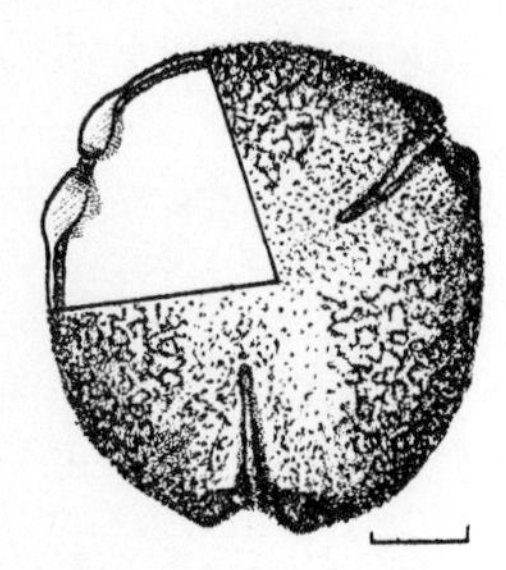

Fig. 278. *Nyssa aquatica* L. Coll. Arn. Arb.; NYSSACEAE.

Tupelo-Gum

SIZE: Variable; equatorial diameter varies from 45 to 54μ

RANGE: Coastal Plain and from Virginia to Florida and Texas north to Illinois

Shape varies from oblate spheroidal to spherical and prolate spheroidal (P/E Index 0.95 to 1.05); shape in polar view varies from nearly round to subtriangular. Furrows with distinct margos, particularly thickened at the pores making them appear annulate; furrow membranes persistent. Furrows without margos, and less distinct near the poles. Exine may be fine reticulate on the intercolpate areas, but psilate at the furrow margins.

28b Apertures protruding (aspidate) (Fig. 279, 280)..........*Elaeagnus*

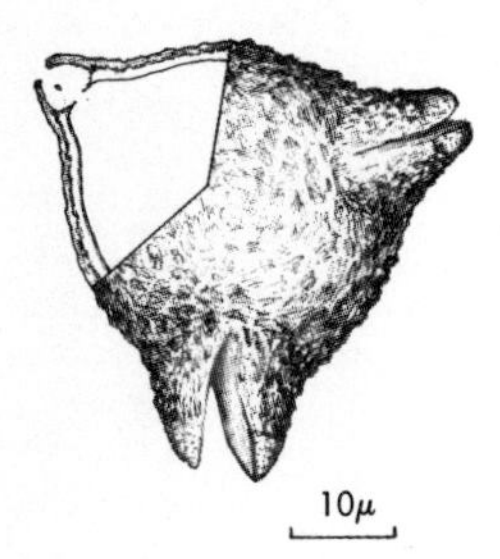

Fig. 279. *Elaeagnus argentea* Pursh. Coll. Michigan; ELAEAGNACEAE.

Silver Berry

SIZE: Polar axis about 24 to 26μ; equatorial diameter about 36 to 40μ

RANGE: Gaspe Peninsula to Alaska, southward to Utah, South Dakota, and southern Quebec

Shape oblate (P/E Index about 0.66), and heteropolar due to one pole of grain having a convex contour, the other pole being flat or slightly concave. Triangular in polar view. Furrows rather short (polar area index about 0.5). Endexine thicker than ektexine forming a pore cavity at the equator.

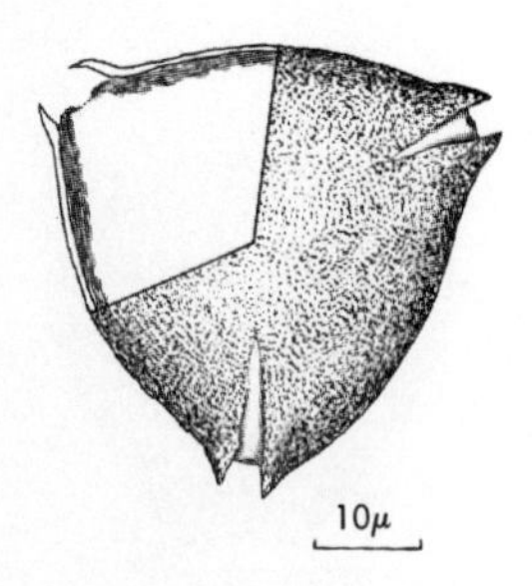

Fig. 280. *Elaeagnus umbellata* Thunb. Coll. Michigan; ELAEAGNACEAE.

Oleaster

SIZE: Equatorial diameter about 40μ

RANGE: Maine to New Jersey and Pennsylvania and beyond (introduced from Asia)

Shape oblate spheroidal to oblate; in polar view rounded with protruding (aspidate) apertures, and a distinct pore cavity. Endexine slightly thickened and roughened near apertures. Exine indistinctly tectate lacking columellae. Surface granular.

29a Exine reticulate, sometimes because of supratectate reticulum, otherwise due to a distinctly reticulate structure................30

29b Exine striate-rugulate, baculate-clavate, echinate, or verrucate...64

30a Transverse furrow present....................................31

30b Transverse furrow lacking....................................46

31a Shape oblate spheroidal to subprolate (P/E Index 0.9 to 1.3).....32

31b Shape prolate or perprolate (P/E Index 1.33 to 2.2 or greater)....38

32a Transverse furrow constricted at its midpoint..................36

32b Transverse furrow unconstricted, borders generally parallel....33

33a Exine reticulate throughout.................................34

33b Exine reticulate in the intercolpal areas, striate-verrucate along the furrow margins (Fig. 281)...*Ailanthus altissima* (Mill.) Swingle

Tree-of-Heaven

SIZE: Polar and equatorial axes about 39μ

RANGE: Widespread in United States and southern Canada. Naturalized from Asia

Shape more or less spheroidal (P/E Index about 1.0). Erdtman (1952, p. 407) indicates that this pollen is smaller (30 X 27μ) and subprolate. Furrows very long and slender, transverse furrow long (4 times as long as wide), the ends round-pointed. Exine striate-verrucate near the furrows, coarsely reticulate in the intercolpal regions.

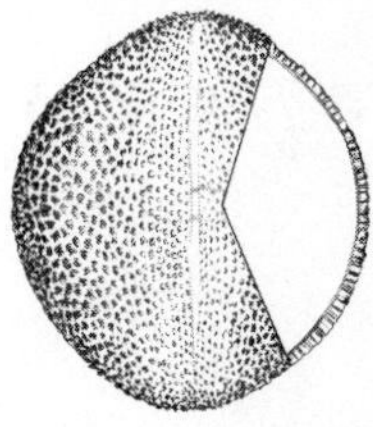

Fig. 281. *Ailanthus altissima* (Mill.) Swingle. Coll. Massachusetts; SIMARUBACEAE.

34a Shape subprolate (P/E Index about 1.2) (Fig. 282)................ ....*Philadelphus inodorum* L.

Mock Orange

SIZE: Polar axis about 22μ; equatorial diameter about 18μ

RANGE: Florida to Alabama, northwest to Tennessee and Virginia

Shape subprolate (P/E Index about 1.2). Furrow long, somewhat contracted at the equator. Transverse furrow small elliptical. Exine tectate, columellae form a faint reticulum.

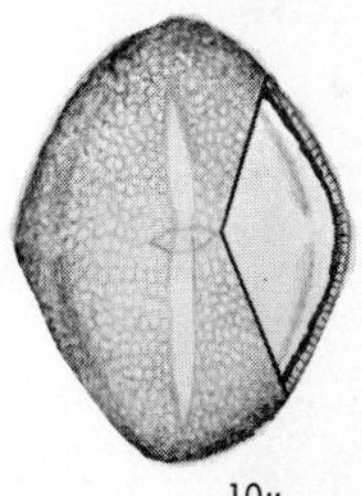

Fig. 282. *Philadelphus inodorum* L. v. *grandiflorus* (Wild.) Gray. Coll. Arn. Arb.; SAXIFRAGACEAE.

34b Shape oblate spheroidal to prolate spheroidal................35

35a Size greater than 40μ (Fig. 283) . . . . ***Mitchella repens* L.**

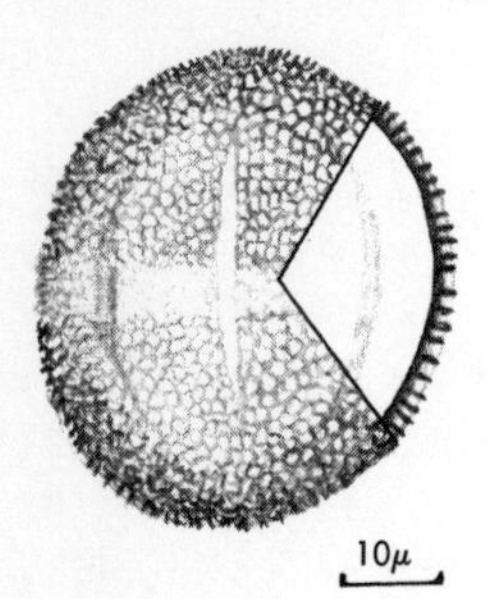

Fig. 283. *Mitchella repens* L. Coll. Michigan; RUBIACEAE.

Partridgeberry

SIZE: Polar axis about 48μ; equatorial diameter about 45μ

RANGE: Florida to Texas north to Newfoundland and Minnesota

Shape prolate spheroidal (P/E Index about 1.06). Furrows rather short; transverse furrows elongated and joined forming a band completely around the equator. Exine intectate reticulate.

35b Size approximately 18 to 20μ (Fig. 284, 285) . . . . **Caprifoliaceae (in part)**

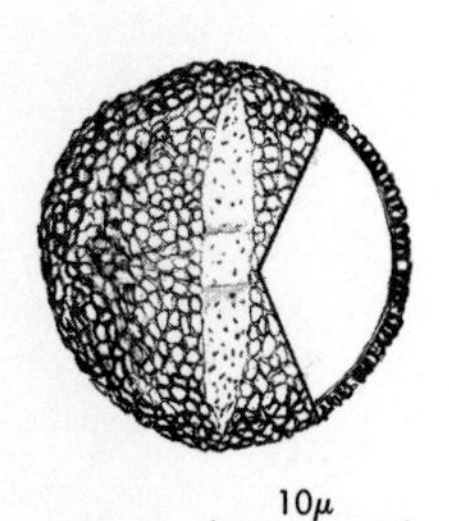

Fig. 284. *Sambucus pubens* Michx. Coll. Michigan; CAPRIFOLIACEAE.

Red-berried Elderberry

SIZE: Polar and equatorial axes about 19μ

RANGE: Newfoundland to Alaska, south to Oregon, Colorado, South Dakota, Iowa, Ohio and West Virginia

Shape spherical (P/E Index about 1.0). Furrows long, with persistent verrucate membranes. Pores elongated into obscure transverse furrows which completely encircle the equator. Exine reticulate, semitectate.

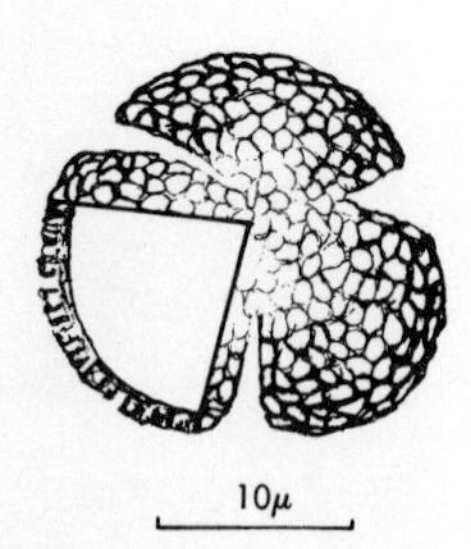

Fig. 285. *Viburnum acerifolium* L. Coll. Michigan; CAPRIFOLIACEAE.

Arrow-wood

SIZE: Polar axis about 19μ; equatorial diameter about 21μ

RANGE: Quebec to Minnesota south to Tennessee and Georgia

Shape spherical to oblate spheroidal (P/E Index about 0.9). Furrow long and polar area rather small (polar area index about 0.3 to 0.35). Pores elongated into transverse furrows which, though obscure, extend around the entire circumference at the equator. Exine perforate tectate forming a distinct reticulum. Reticulum finer adjacent to the furrows and at the poles than in the intercolpi.

36a Transverse furrows rounded at ends (Fig. 286)..*Celastrus pringlei* Rose.

Bittersweet

SIZE: Polar axis about 27μ; equatorial diameter about 24μ

RANGE: Mexico

Shape prolate spheroidal (P/E Index about 1.1); broadly rounded at the poles. Furrows narrow and relatively short, with distinct margos. Pore forms a short transverse furrow which is constricted at the furrow-crossing. Exine tectate with a distinct supratectate recticulum; columellae longer at the poles resulting in greater wall thickness in that region.

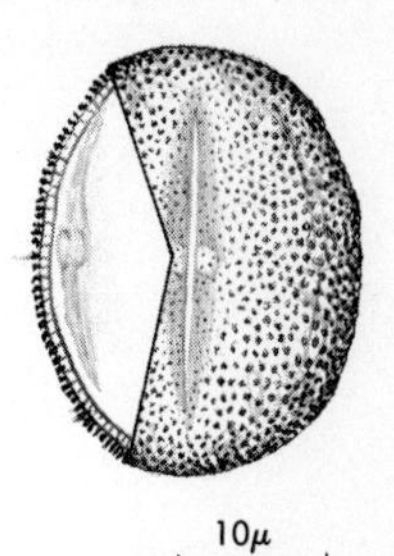

Fig. 286. *Celastrus pringlei* Rose. Coll. Mexico; CELASTRACEAE.

36b Transverse furrows tapered-pointed at ends..................37

37a Reticulum very coarse (large baculae and clavae), semitectate (Fig. 287)..............................*Viburnum lentago* L.

Nannyberry

SIZE: Polar axis about 32μ; equatorial diameter about 30μ

RANGE: Western Quebec to Manitoba south to Colorado, South Dakota, Missouri and New Jersey and upland south to Georgia

Shape prolate spheroidal (P/E Index about 1.1). Furrow long and slenderly tapered, transverse furrows abruptly narrow to attenuate extensions. Exine rather thick (about 4μ). Irregular and scattered baculae and clavae form an imperfect, coarse reticulum.

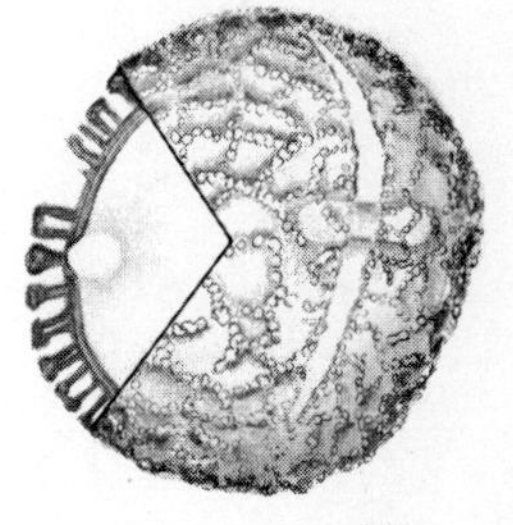

Fig. 287. *Viburnum lentago* L. Coll. Michigan; CAPRIFOLIACEAE.

37b Reticulum fine, apparently a structural feature; grain tectate (Fig. 288) . . . . ***Lysimachia terrestris*** **(L.) BSP.**

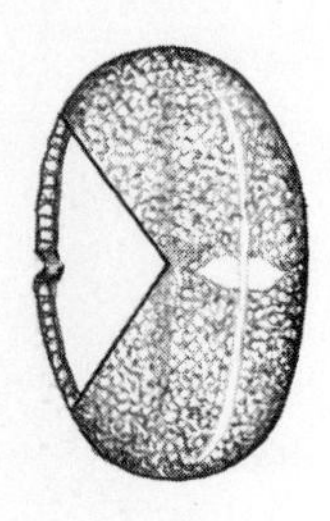

Fig. 288. *Lysimachia terrestris* (L.) BSP. Coll. Michigan; PRIMULACEAE.

Swamp Loosestrife

SIZE: Polar axis about 21μ; equatorial diameter about 17μ

RANGE: Newfoundland to James Bay and Minnesota south to Iowa, Kentucky and Georgia

Shape subprolate (P/E Index about 1.25), poles rounded. Furrows thin and long with small polar areas. Transverse furrows narrower at the furrow and apiculate or attenuate at ends. Exine tectate with distinct reticulum.

38a P/E Index 2.0 or greater, perprolate **45**

38b P/E Index 1.33 to 2.0, prolate **39**

39a Polar axis 20 to 30μ long **42**

39b Polar axis 35μ or greater **40**

40a Equatorial outline ovate **41**

40b Equatorial outline shows equatorial constriction, especially along the furrows . . . .
Heracleum maximum **Bartr.** **(see description, page 162, Figure 344)**

41a Polar axis approximately 60μ; P/E Index about 1.55 (Fig. 289) . . . .
. . . . ***Panax trifolium*** **L.**

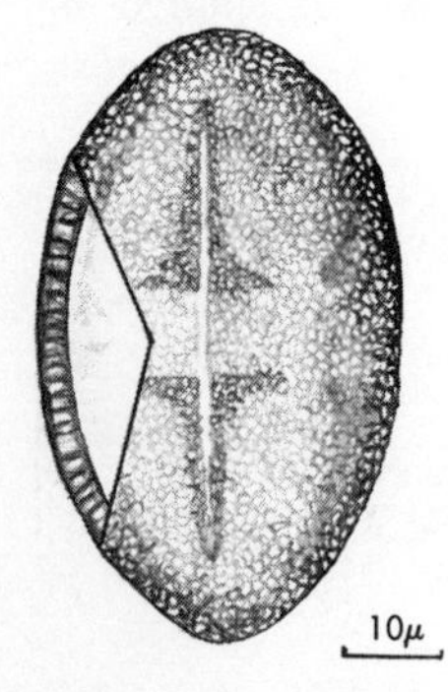

Fig. 289. *Panax trifolium* L. Coll. Michigan; ARALIACEAE.

Dwarf Ginseng

SIZE: Polar axis about 62μ; equatorial diameter about 40μ

RANGE: Prince Edward Island to Minnesota, south Nebraska and Iowa, Ohio, and New England

Shape prolate (P/E Index about 1.55). Furrows slender, with distinct margos which also border the transverse furrow. Exine tectate, reticulate due to the arrangement of the columellae.

41b Polar axis 40 to 55μ; P/E Index 1.3 to 1.45 (Fig. 290, 291) . . *Rhus* spp.

Dwarf Sumac

SIZE: Polar axis 56μ; equatorial diameter 39μ

RANGE: Florida to eastern Texas north to northern Illinois and southern Maine

Shape prolate (P/E Index about 1.4); diamond-shaped in equatorial view, triangular in polar view. Furrows long, somewhat contracted at the equator; transverse furrows long, rectangular or auriculately expanded at ends. Exine tectate, columellae distinct producing a subreticulate or vaguely striate pattern.

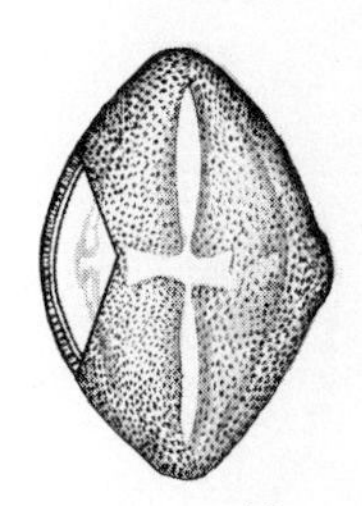

Fig. 290. *Rhus copallina* L. Coll. Oklahoma; ANACARDIACEAE.

Smooth Sumac

SIZE: Polar axis about 43μ; equatorial diameter about 32μ

RANGE: Maine and southwest Quebec and southern British Columbia south to northern Florida and Louisiana

Shape subprolate (P/E Index about 1.35). Furrows long, polar area small (polar area index 0.3); transverse furrows rectangular. Exine tectate, columellae form a subreticulate pattern.

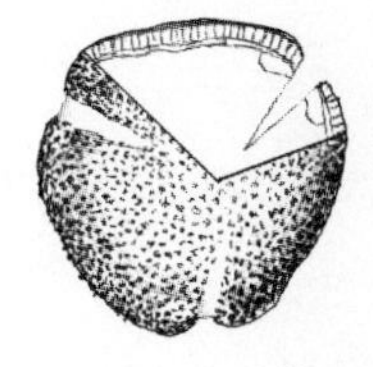

Fig. 291. *Rhus glabra* L. Coll. Michigan; ANACARDIACEAE.

42a Transverse furrow narrow and of uniform width, wall >2μ thick (Fig. 292) . . . . *Euphorbia strictospora* (Eng.) Small

Spurge

SIZE: Polar axis about 23μ; equatorial diameter about 16μ

RANGE: Kansas to Colorado south to Arizona, Mexico and Texas

Shape prolate (P/E Index about 1.4 to 1.45); poles somewhat pointed. Furrow somewhat constricted at the equator; transverse furrow very slender. Wall thick (>2μ), exine tectate with tall, distinct columellae which form a fine reticulum, surface verrucate roughened.

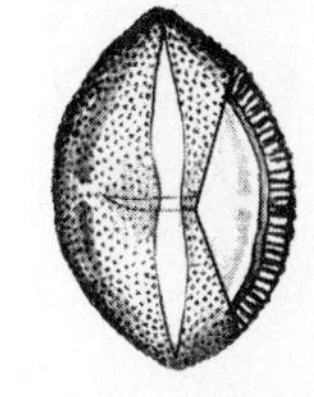

Fig. 292. *Euphorbia strictospora* (Eng.) Small Coll. Colorado; EUPHORBIACEAE.

42b Transverse furrow wider, usually of uneven breadth 43

43a Transverse furrow square or bifurcated at ends (Fig. 293)........ ....*Ptelea trifoliata* L.

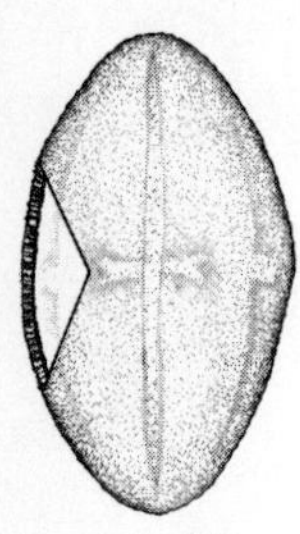

Fig. 293. *Ptelea trifoliata* L. Coll. Kansas; RUTACEAE.

Wafer Ash

SIZE: Polar axis 32μ; equatorial diameter 19μ

RANGE: Florida to Arizona and Mexico north to Nebraska, southern Michigan, southern Ontario and New England

Shape prolate (P/E Index 1.65 to 1.70). Furrows long, borders slightly irregular, membrane persistent and fine granular. Transverse furrows long and bifurcated at ends. Exine tectate, surface verrucate.

43b Transverse furrow pointed at ends..........................44

44a Furrow slightly to strongly constricted at equator (Fig. 294)........ ....*Xanthoxylum americanum* Mill.

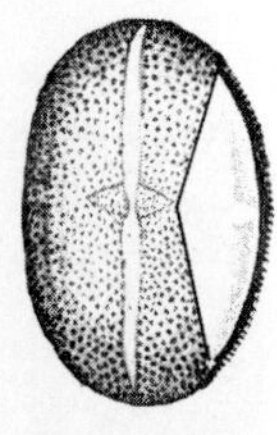

Fig. 294. *Xanthoxylum americanum* Mill. Coll. Kansas; RUTACEAE.

Northern Prickly Ash

SIZE: Polar axis about 24μ; equatorial diameter about 17μ

RANGE: Western Quebec to North Dakota south to Missouri, Oklahoma and Georgia

Shape prolate (P/E Index about 1.4), poles broadly rounded. Furrows slender and very long, constricted at the equator; polar area small. Transverse furrow elliptical, ends pointed. Exine fine reticulate, the diameter of the reticular openings is about 0.8μ.

44b Furrow with essentially parallel sides (Fig. 295)................. ....*Hypericum perforatum* L.

Common St. John's Wort

SIZE: Polar axis about 25μ; equatorial diameter about 17μ

RANGE: New Foundland to British Columbia and southward throughout the United States (naturalized from Europe)

Shape prolate (P/E Index 1.45 to 1.5), poles broadly rounded. Furrows long and gradually tapered to poles, polar area small; transverse furrows wide, somewhat indistinct. Exine tectate, the columellae forming a faint reticulate pattern; surface psilate.

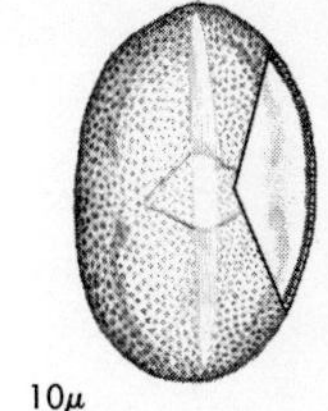

Fig. 295. *Hypericum perforatum* L. Coll. Michigan; GUTTIFERAE.

45a Transverse furrow with tapered-pointed ends (Fig. 296)..*Hydrophyllum appendiculatum* Michx.

Waterleaf

SIZE: Polar axis about 30μ; equatorial diameter about 15μ

RANGE: Ontario to Minnesota south to eastern Kansas, Tennessee and southwestern Pennsylvania

Shape prolate or perprolate (P/E Index about 2.0). Furrow slender and relatively short; transverse furrow elliptical. Exine tectate, thickest at the equator; columellae form a fine reticulate pattern; surface psilate.

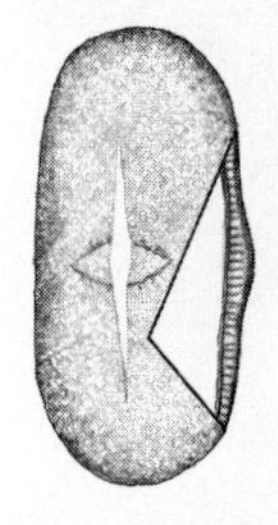

Fig. 296. *Hydrophyllum appendiculatum* Michx. Coll. Indiana; HYDROPHYLACEAE.

45b Transverse furrow with rounded ends (most of these grains actually have a psilate surface, but reveal a reticulate structural pattern in L/O analysis) (Fig. 297, 298, 299)................Umbelliferae

Spotted Cowbane

SIZE: Polar axis about 40μ; equatorial diameter about 19μ

RANGE: Gaspe Peninsula to Manitoba south to Missouri, Texas, Tennessee and Maryland

Shape perprolate (P/E Index about 2.1), somewhat contracted at the equator. Furrows long, slender, attenuate, but slightly expanded at the equator. Transverse furrows with rounded ends. Exine tectate, the columellae form an indistinct reticulum.

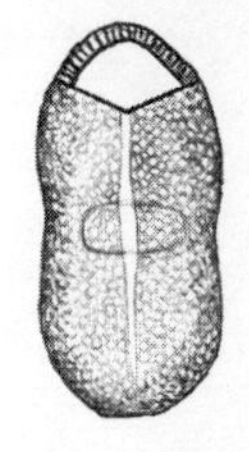

Fig. 297. *Cicuta maculata* L. Coll. Michigan; UMBELLIFERAE.

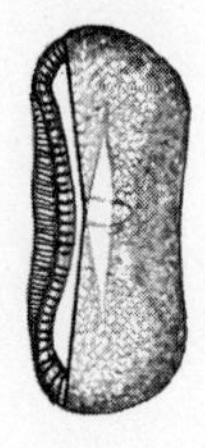
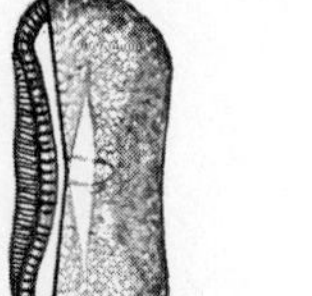

Fig. 298. *Conium maculatum* L. Coll. Wyoming; UMBELLIFERAE.

Poison Hemlock

SIZE: Polar axis about 37μ; equatorial diameter about 16μ

RANGE: Alabama to Texas north to Michigan, Ontario and New England. Also on Pacific Coast (naturalized from Europe)

Shape perprolate (P/E Index 2.25 to 2.3); grains contracted at the equator, especially at the apertures. Furrows relatively short and attenuate; transverse furrow elliptical. Exine tectate, columellae form an indistinct reticulum.

Fig. 299. *Osmorhiza longistylis* (Torr.) DC. Coll. Michigan; UMBELLIFERAE.

Anise Root

SIZE: Polar axis about 25μ; equatorial diameter about 12μ

RANGE: Gaspe Peninsula to southern Alberta, south to eastern New Mexico, Oklahoma, Missouri, and Virginia

Shape perprolate (P/E Index about 2.1), grains broadest near the poles, more or less contracted at the equator. Furrow rather short, and very slender; transverse furrow broad, round at ends and somewhat narrowed where it is crossed by the furrow. Exine tectate, with a distinct supratectate reticulum.

46a Shape class suboblate, spheroidal, or prolate spheroidal (P/E Index 0.8 to 1.14)..47

46b Shape class subprolate, prolate, or perprolate................54

47a P/E Index about 0.80 to 0.85..................................48

47b P/E Index about 1.0 to 1.14...................................49

48a Apertures three or four in number, pores about 3μ diameter (Fig. 300).................................*Rumex acetosella* L.

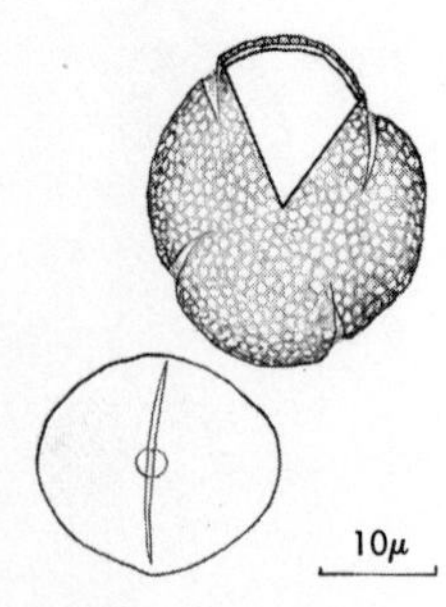

Fig. 300. *Rumex acetosella* L. Coll. Michigan; POLYGONACEAE.

Sheep Sorrel

SIZE: Polar axis about 19μ; equatorial diameter about 24μ

RANGE: Widespread, almost cosmopolitan in the United States and Canada. Naturalized from Europe

Shape suboblate (P/E Index about 0.8). Apertures three or four, furrows slender, pore round and much exceeding the furrow margins. Exine coarsely reticulate.

48b Apertures three, pores large, about 5μ diameter (Fig. 301)..*Cephalanthus occidentalis* L.

Buttonbush

SIZE: Polar axis about 18μ; equatorial diameter about 22.5μ

RANGE: Florida to Mexico to southern California, Oklahoma, southern Ontario and eastern Canada

Shape suboblate (P/E Index about 0.85). Furrows slender, pores large and round greatly exceeding the furrow margins. Exine tectate, reticulate; reticulum finer at poles.

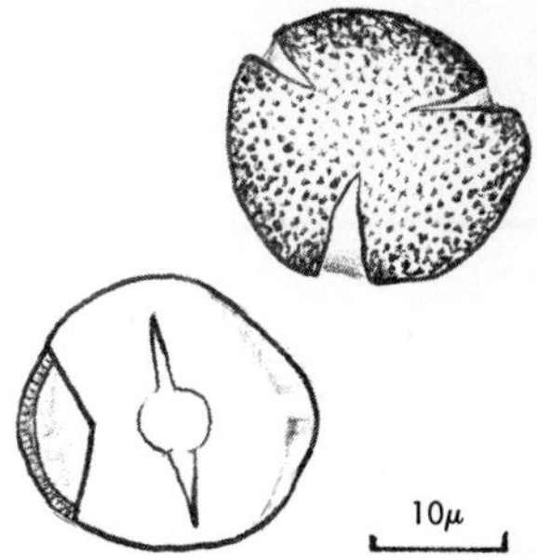

10μ

Fig. 301. *Cephalanthus occidentalis* L. Coll. Michigan; RUBIACEAE.

49a Exine subreticulate in appearance, spines short or obsolete..*Artemisia* spp.
(See description, page 153, Figure 321)

49b Exine with reticulate appearance..............................50

50a Size approximately 15μ (Fig. 302)........*Fendlera rupicola* Gray

Fendlera

SIZE: Polar and equatorial axes 14 to 15μ

RANGE: Southern Colorado to western Texas and Arizona

Shape spherical (P/E Index about 1.0). Furrows long, broadest at equator gradually tapering toward poles. Pore round, included within furrow margins. Exine tectate, columellae forming a reticulate pattern.

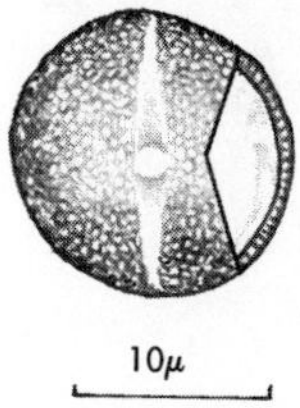

10μ

Fig. 302. *Fendlera rupicola* Gray Coll. New Mexico; SAXIFRAGACEAE.

50b Size greater than 20μ.......................................51

51a Size approximately 25μ......................................52

51b Size approximately 35μ......................................53

52a Furrow with distinct margo (Fig. 303)..... *Canotia holocantha* Torr.

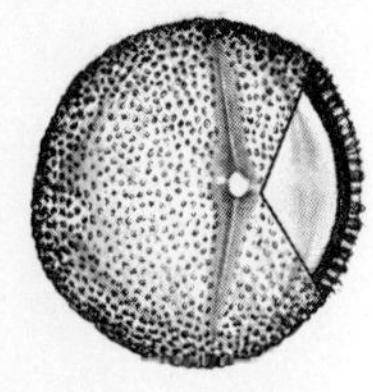

Fig. 303. *Canotia holocantha* Torr. Coll. Arizona; CELASTRACEAE.

Canotia

SIZE: Polar axis about 24μ; equatorial diameter about 22μ

RANGE: Southern Utah to Arizona, west to southern California and northwestern Mexico

Shape prolate spheroidal (P/E Index 1.1 to 1.15). Furrows with distinct margos; pore round, broader than furrow. Exine thick, perforate tectate, having a fine reticulum.

52b. Furrow lacking a distinct margo (Fig. 304)... *Verbascum virgatum* Stokes

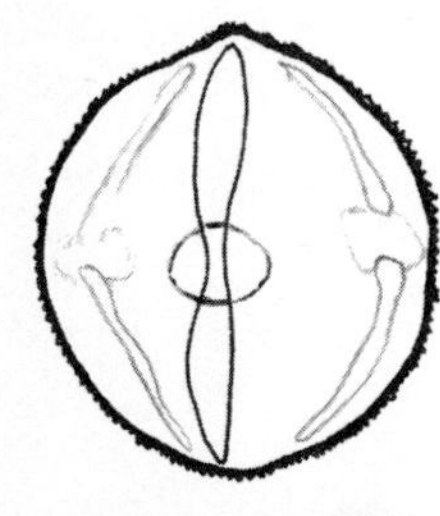

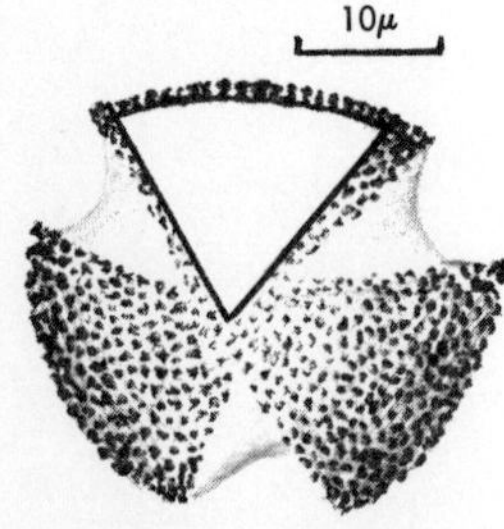

Fig. 304. *Verbascum virgatum* Stokes. Coll. Florida; SCROPHULARIACEAE.

Mullein

SIZE: Polar and equatorial axes about 28μ

RANGE: Eastern Canada to Ontario and Ohio south to South Carolina, coastal plain, Texas and California. Native of Europe

Shape approximately spherical (P/E Index about 1.0). Furrows long, contracted at the equator; polar area small (polar area index about 0.3). Pore round, exceeding the furrow margins. Exine perforate tectate, reticulate.

53a Pore large, broader than furrow (Fig. 305)... *Fouquieria splendens* Engelm.

Ocotillo

SIZE: Polar axis about 38μ; equatorial diameter about 35μ

RANGE: Western Texas to southeastern California and northern Mexico

Shape prolate spheroidal (P/E Index about 1.1). Furrows long with approximately parallel borders along most of extent, contracted at the equator. Pore round, exceeds the furrow margins. Exine tectate, reticulate; reticulum somewhat finer near the furrows.

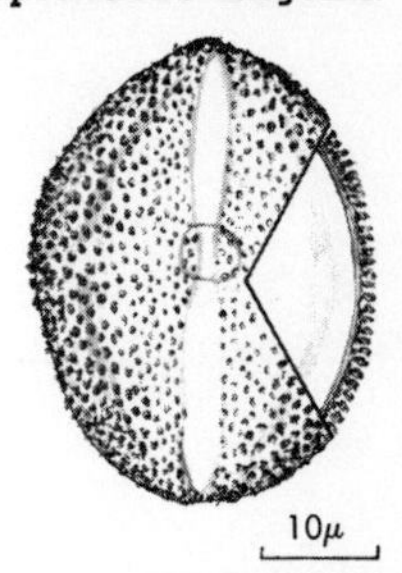

Fig. 305. *Fouquieria splendens* Engelm. Coll. New Mexico; FOUQUIERIACEAE.

53b Pore smaller, entirely within furrow borders (Fig. 306)... *Vitis linsecomii* Buckl.

Grape

SIZE: Polar axis about 36μ; equatorial diameter about 35μ

RANGE: Tennessee to Louisiana, Texas and Missouri

Shape spheroidal to prolate spheroidal (P/E Index about 1.05), subtriangular in polar view. Furrows long, gradually tapered from equator to poles, pore small, included within the furrow borders. Exine tectate (or perforate tectate), fine reticulate.

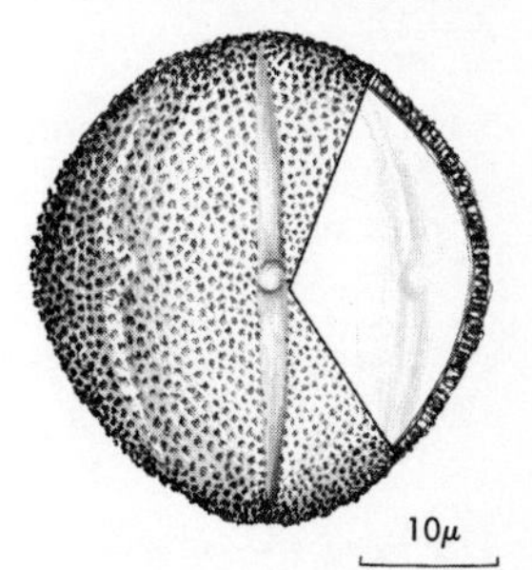

Fig. 306. *Vitis linsecomii* Buckl. Coll. Texas; VITACEAE.

54a Reticulum restricted to intercolpi; furrow border verrucate-scabrate (Fig. 307)................. *Parthenocissus inserta* (Kerner) Fritsch.

Woodbine

SIZE: Polar axis about 40μ; equatorial diameter about 32μ

RANGE: Quebec to Manitoba and Montana south to Pennsylvania, Kansas, Missouri, New Mexico and California

Shape subprolate (P/E Index about 1.25). Furrows long, with round furrow included within furrow borders. Exine perforate tectate; reticulate in the intercolpal areas, scabrate or verrucate near the furrows. Reticular meshes elongated in the meridional direction.

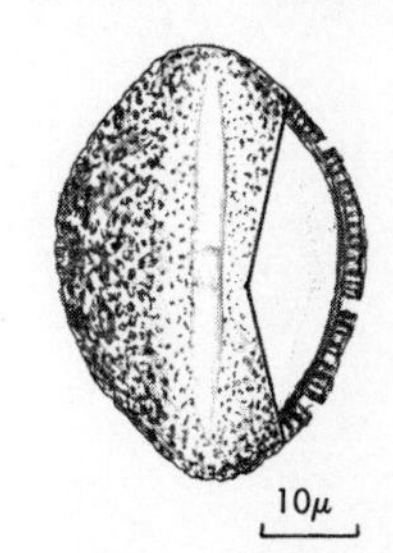

Fig. 307. *Parthenocissus inserta* (Kerner) Fritsch. Coll. Michigan; VITACEAE.

54b Reticulum throughout surface, but sometimes finer near the furrow 55

55a Shape subprolate (P/E Index 1.14 to 1.3). 56

55b Shape subprolate, prolate to perprolate. 59

56a Polar axis 45 to 50μ. 57

56b Polar axis about 25μ (Fig. 308). . . . . *Parnassia glauca* Raf.

Grass-of-Parnassus

SIZE: Polar axis about 25μ; equatorial diameter about 20μ

RANGE: Newfoundland to Manitoba, south to South Dakota, Iowa, Ohio and New England

Shape subprolate (P/E Index about 1.25). Furrows slender-attenuate, with distinct margos. Pore round considerably exceeding the furrow. Exine intectate, reticulate. Meshes of reticulum very coarse in the intercolpar area, becoming finer near the furrows.

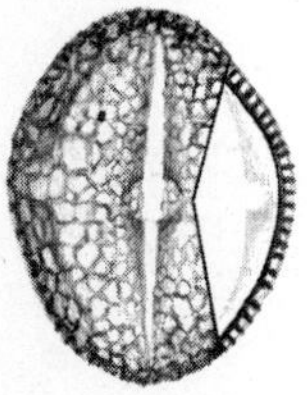

10μ

Fig. 308. *Parnassia glauca* Raf. Coll. Michigan; SAXIFRAGACEAE.

57a Furrows short, polar area large (Fig. 309). . . . . *Gentiana andrewsii* Griseb.

Closed Gentian

SIZE: Polar axis about 48μ; equatorial diameter 39μ

RANGE: Arkansas to Georgia north to Massachusetts, Quebec, Manitoba and Saskatchewan

Shape subprolate (P/E Index about 1.2). Furrow rather short, pore slightly elongate (meridionally), barely exceeding the furrow margins. Exine intectate to perforate tectate. Reticulum very coarse in the intercolpal areas, very fine near the furrows.

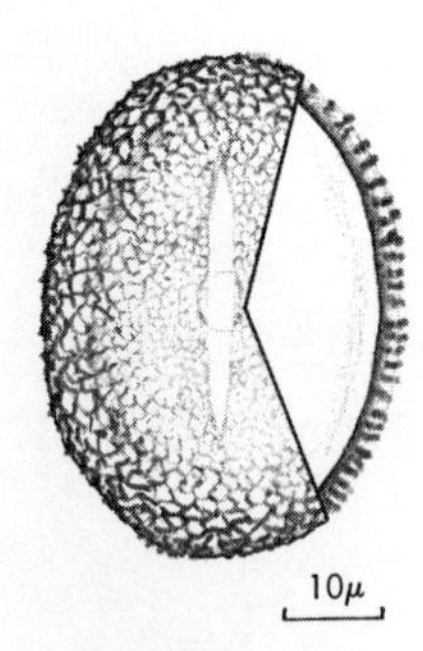

Fig. 309. *Gentiana andrewsii* Griseb. Coll. Michigan; GENTIANACEAE.

57b Furrow long, polar area small. 58

58a Exine perforate tectate; pore evident (Fig. 310)..*Staphylea trifolia* L.

Bladdernut

SIZE: Polar axis about 51μ; equatorial diameter about 44μ

RANGE: Southwestern Quebec to Minnesota south to Oklahoma and Georgia

Shape subprolate (P/E Index about 1.15). Furrows long, polar area small, pore round and included within furrow borders. Furrow membrane persistent, with a fine verrucate surface. Exine perforate-tectate, reticulate.

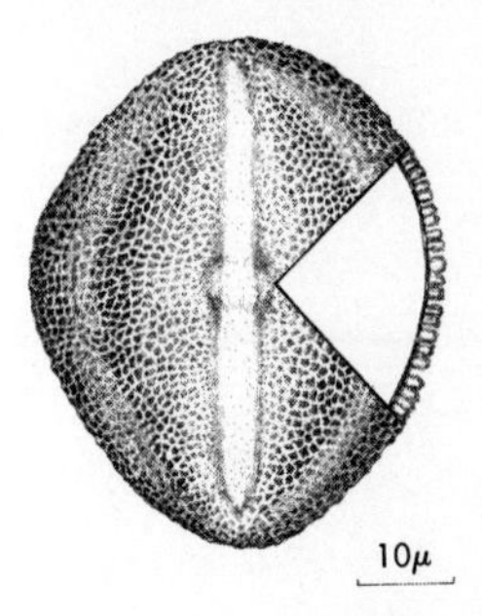

Fig. 310. *Staphylea trifolia* L. Coll. Michigan; STAPHYLACEAE.

58b Exine tectate; pore often obscure (Fig. 311)..*Gymnocladus dioica* (L.) K. Koch

Kentucky Coffee-tree

SIZE: Polar axis about 46μ; equatorial diameter about 40μ

RANGE: Central New York to South Dakota, south to Tennessee and Oklahoma; naturalized beyond

Furrow contracted at the equator, pore rather indistinct. Exine tectate with surface very finely reticulate (or scabrate ?).

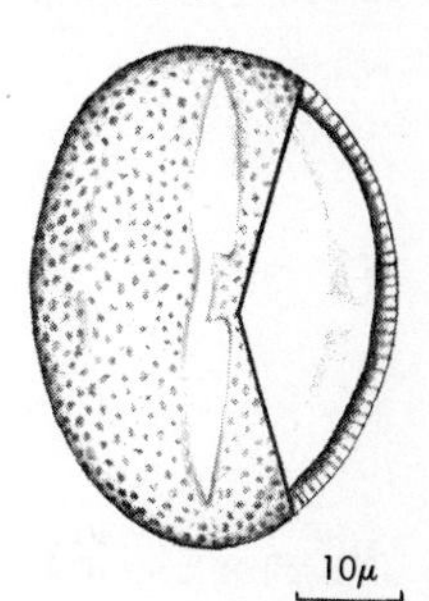

Fig. 311. *Gymnocladus dioica* (L.) K. Koch. Coll. Arn. Arb.; LEGUMINOSAE.

59a Reticulum finer adjacent to furrows and at poles, very coarse in the intercolpi (Fig. 312)........................*Salix nigra* Marsh.

Black Willow

SIZE: Polar axis about 21μ; equatorial diameter about 15μ

RANGE: New Brunswick to North Dakota, south to Arkansas, Alabama, Tennessee and North Carolina

Shape subprolate or prolate (P/E Index about 1.4). Pore elliptical (meridionally elongate), slightly exceeding the furrow borders. Exine intectate, reticulate; reticulum meshes finest near the furrows and at the poles.

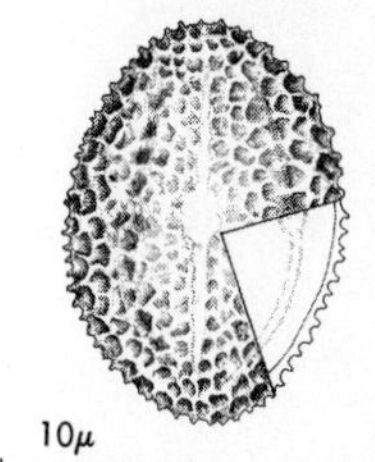

Fig. 312. *Salix nigra* Marsh. Coll. Michigan; SALICACEAE.

59b Reticulum uniform or finer only adjacent to the furrows, not at poles 60

60a Polar axis greater than 50μ, exine 4μ or thicker (Fig. 313) . . . . *Eriogonum chrysocephalum* A. Gray.

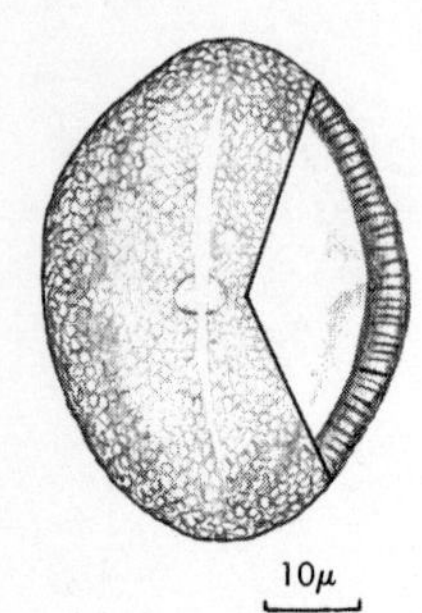

Fig. 313. *Eriogonum chrysocephalum* A. Gray. Coll. Utah; POLYGONACEAE.

Umbrella Plant

SIZE: Polar axis about 53μ; equatorial diameter about 40μ

RANGE: Nebraska and Colorado to Utah and Idaho

Shape prolate (P/E Index about 1.33). Exine thick (> 4μ), tectate, with stout, tall columellae which form a fine reticulate pattern. Furrow slender, gradually tapering from the equator; pore somewhat elliptical, equatorially elongate, exceeding the furrow margins.

60b Polar axis less than 50μ, exine thinner (Leguminosae, in part) . . . 61

61a Size 20 to 25μ (polar axis) 62

61b Size about 45μ (polar axis) 63

62a Pore indistinct, reticulum finer near furrow (Fig. 314) . . . . *Gleditsia tricanthos* L.

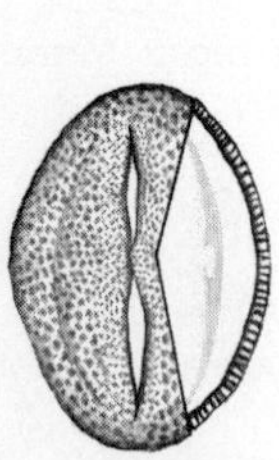

Fig. 314. *Gleditsia tricanthos* L. Coll. Ohio; LEGUMINOSAE, Caesalpinoideae.

Honey Locust

SIZE: Polar axis about 24μ; equatorial diameter about 18μ

RANGE: Florida to Texas north to Kansas and Ontario

Shape prolate (P/E Index about 1.33). Furrows with margos contracted at the equator, pore rather indistinct. Exine perforate tectate, reticulate. Reticular meshes widest in the intercolpar areas, finer near the furrows.

62b Pore distinct and large, reticulum uniform (Fig. 315)..*Melilotus officinalis* (L.) Lam.

Yellow Sweet Clover

SIZE: Polar axis about 23μ; equatorial diameter about 17μ

RANGE: Nearly throughout the United States and southern Canada; naturalized from Europe

Shape prolate (P/E Index 1.35), poles broadly rounded. Exine tectate, reticulate. Furrow slender, with a large circular pore. Similar in general characteristics to pollen of many related members of the Papilionoideae (Fabaceae).

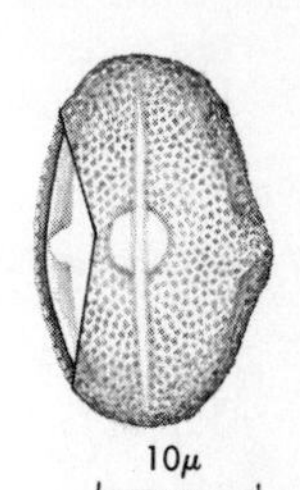

Fig. 315. *Melilotus officinalis* (L.) Lam. Coll. Michigan; LEGUMINOSAE.

63a P/E Index about 1.4; pore indistinct (Fig. 316)...*Trifolium pratense* L.

Red Clover

SIZE: Polar axis about 45μ; equatorial diameter about 32μ

RANGE: Introduced from Europe, widespread throughout the United States and southern Canada

Shape prolate (P/E Index about 1.4); pore indistinct, furrow often with merely roughened constriction at the equatorial limb. Surface coarsely reticulate. Probably indistinguishable from pollen of other species of clover or other legumes in the Papilionoideae.

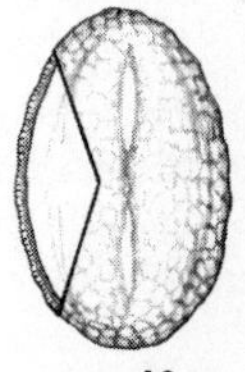

Fig. 316. *Trifolium pratense* L. Coll. Michigan; LEGUMINOSAE.

63b P/E Index about 2.2; pore distinct (Fig. 317).....*Vicia villosa* Roth.

Hairy Vetch

SIZE: Polar axis about 45μ; equatorial diameter about 20μ

RANGE: Widespread in eastern North America, naturalized from Europe

Shape perprolate (P/E Index about 2.25), polar ends broadly rounded. Furrows narrow, with distinct marginal thickenings. Pore round, greatly exceeding the furrow. Exine apparently intectate, with coarse reticulum.

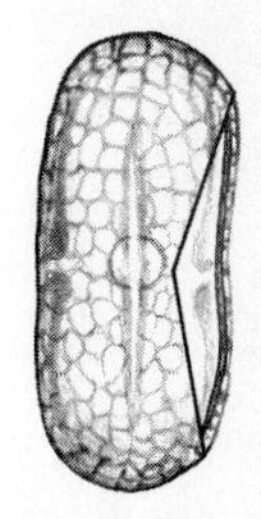

Fig. 317. *Vicia villosa* Roth. Coll. Michigan; LEGUMINOSAE.

64a Exine surface echinate, spines reduced or nearly obsolete in some genera; exine tectate with supratectate pila; Compositae........65

64b Exine surface striate-rugulate, verrucate-scabrate, or apparently so; exine intectate or tectate; if supratectate pila are present the grains are either prolate or lack transverse furrows............73

65a Furrows short, sometimes appearing as three elongate equatorial pores; pore more or less indistinct (Fig. 318, 319)................ ....*Ambrosia spp.* (and other Ambrosineae)

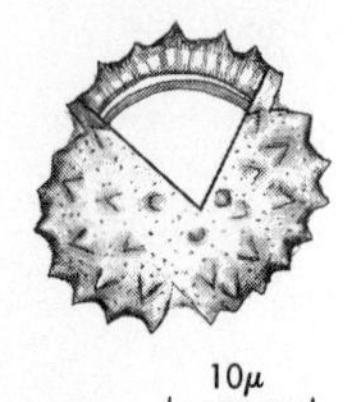

Fig. 318. *Ambrosia trifida* L. Coll. Indiana; COMPOSITAE, Tubuliflorae.

Great Ragweed

SIZE: Polar axis about 18.5μ; equatorial diameter about 20μ

RANGE: Southwestern Quebec to British Columbia south to Arizona and Florida

Shape spheroidal to oblate spheroidal (P/E Index about 0.92). Spines broad-based and short. Furrows short (about 2μ), sometimes almost poroid; pores obscure, apparently elongated as transverse furrows. Very similar to other species of *Ambrosia;* other genera in the Ambrosineae may be rather similar or quite different. *Xanthium pensylvanicum* is very similar to *Ambrosia,* but larger (to 25μ). *Iva xanthifolia* has exine sculpture similar to *Ambrosia,* but has very long furrows with distinct pores; *Iva ciliata* has short "ambrosioid" furrows, but the spines are longer, bulbous-based, and apiculate.

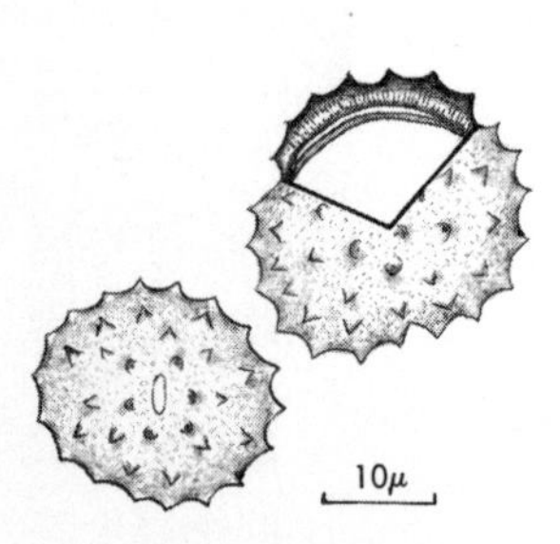

Fig. 319. *Ambrosia artemisiifolia* L. Coll. Michigan; COMPOSITAE, Tubuliflorae.

Common Ragweed

SIZE: Polar axis about 19μ; equatorial diameter about 22μ

RANGE: Widespread weed in the United States east of Rocky Mts. to southern Canada

Spines short and broad-based, supratectate. Endexine often shrunken or pulled away from ektexine. Shape suboblate, P/E Index about 0.86. Apertures actually tricolporate, but furrows very short (poroid).

65b Furrows long, polar area smaller than in Ambrosineae; pores usually distinct. Other morphological forms will be encountered in many Composites; these may not key to one of the types described; Wodehouse (1935) and Stix (1960) should be consulted for details. The key separations are based partly on the exine analysis of Stix 66

66a Exine surface undulate, apparently due to remnants of vestigial spine bases (Fig. 320)................................*Centaurea*

Star Thistle

SIZE: Equatorial diameter about 27μ

RANGE: Michigan to Washington south to Missouri, Texas and southern California

The *Centaurea*-type of Compositae pollen is subprolate, furrow membrane smooth to slightly verrucate. Exine about 3 to 7.5μ thick, thinner at the poles. Coarse subtectate baculae about 1.5 to 2.5μ long; fine, supratectate radial columellae fused distally forming a surface membrane. Surface undulate, apparently vestigial spines (Stix, 1960).

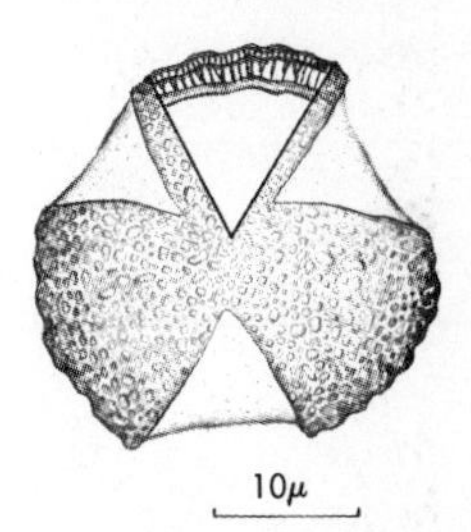

Fig. 320. *Centaurea picris* Pall. Coll. New Mexico; COMPOSITAE.

66b Exine surface echinate, spines short in *Artemisia*.............67

67a Spines 1μ or less in length, or essentially obsolete; pore sometimes indistinct (Fig. 321)..................................*Artemisia*

Wormwood

SIZE: Polar axis about 27μ, equatorial diameter about 25μ

RANGE: Florida to Texas and Arizona, north to Washington, Manitoba and southern Maine

Shape prolate spheroidal (P/E about 1.1); furrow broad and expanded, pore distinct or obscure. Exine distinctly stratified, the radial baculae outside the tectum are more or less fused at the distal ends, especially at the bases of spines. Surface subechinate or spines obscure; surface sometimes appears subreticulate.

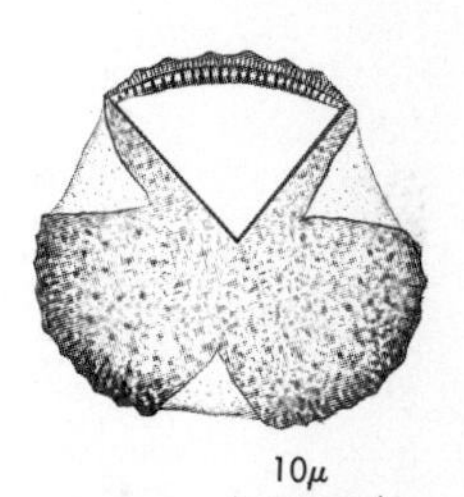

Fig. 321. *Artemisia caudata* Michx. Coll. Indiana; COMPOSITAE, Tubuliflorae.

67b Spines generally 2μ or longer; pore distinct, often a transverse furrow 68

68a Columellae present (Figure 323)............................71

68b Columellae absent (Figure 322).............................69

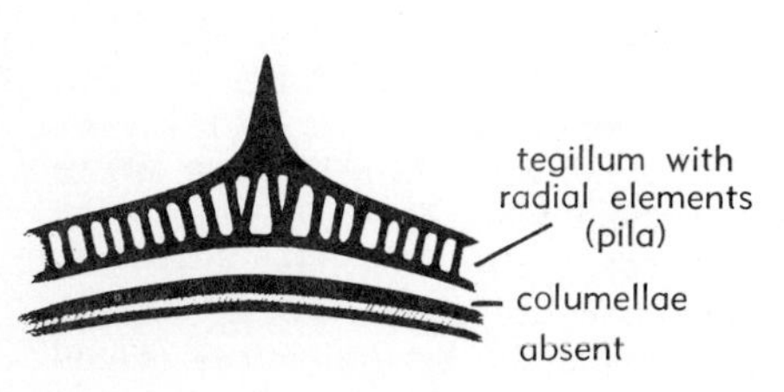

Fig. 322.

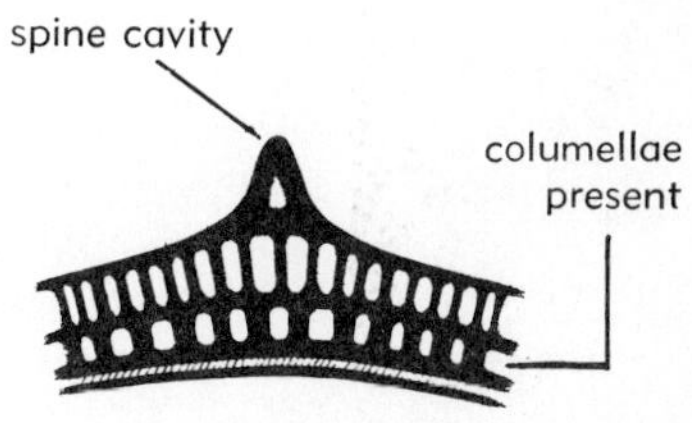

Fig. 323.

69a Transverse furrow sharply constricted at center of furrow, spine cavity often present; spines broad (Fig. 324)........*Carduus*-type

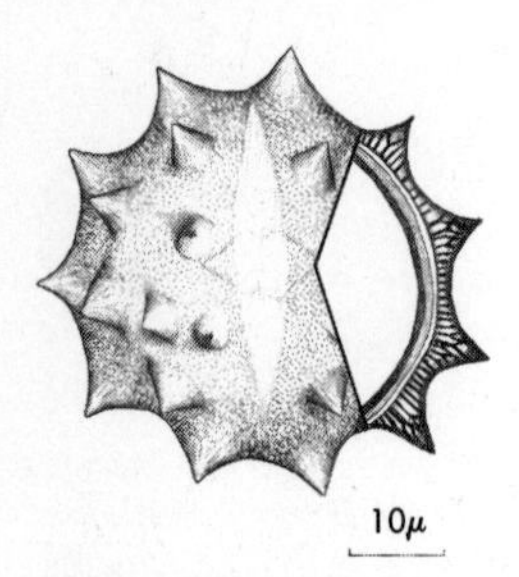

Fig. 324. *Carduus nutans* L. Coll. Michigan; COMPOSITAE, Tubuliflorae.

Musk Thistle

SIZE: Polar axis 42 to 43μ; equatorial diameter 45 to 47μ

RANGE: Southeastern Canada to Iowa south to Missouri and Maryland

Shape oblate spheroidal (P/E Index about 0.93). Furrow broad, with persistent smooth membrane. Transverse furrow abruptly contracted at center of furrow forming "notches." Exine about 8μ thick, including spines (about 3 to 4μ); spines rounded to pointed, usually with an internal cavity. Exine tectate; numerous, distally thickened, supratectate pila are more or less fused into a surface membrane.

69b Transverse furrow, if present, not constricted; spine cavity absent; spines narrow attenuate....................................70

70a Exine about 10μ thick, including spines (Fig. 325)...*Helianthus*-type

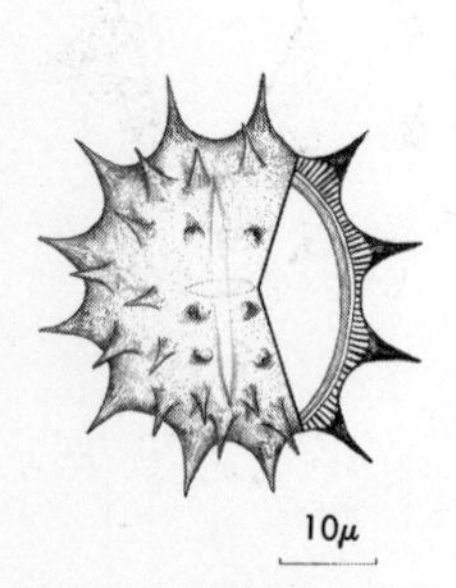

Fig. 325. *Helianthus Dalyi* Britt. Coll. Michigan; COMPOSITAE, Tubuliflorae.

Daly's Sunflower

SIZE: Polar axis about 40μ; equatorial diameter about 38μ

RANGE: Massachusetts to Long Island west to Michigan and Minnesota

Shape spherical to prolate spheroidal (P/E Index about 1.05). Furrows narrow and of intermediate length; transverse furrows narrowly elliptical with pointed ends. Exine tectate, about 10μ thick, including attenuate spines (5μ high, 3μ broad at base). *Bidens, Helenium, Polymnia, Tagetes,* and *Zinnia* are among the genera included with *Helianthus* in the *Helianthus* pollen-type by Stix (1960).

70b Exine about 4 to 5μ thick, including spines (Fig. 326)............
..*Baccharis*-type

Showy Goldenrod

SIZE: Polar axis about 34μ; equatorial diameter about 30μ

RANGE: North Carolina to Louisiana and Oklahoma north to southern Minnesota, Michigan and Massachusetts

Shape spherical to prolate spheroidal (P/E Index about 1.1). Furrows of intermediate length, ends rounded, transverse furrows pointed. Exine about 4.5μ thick, including 2.5 to 3.5μ long spines. Stix (1960) investigated pollen of one species of *Solidago* (*S. altissima*) and included it in the Baccharis-type.

10μ

Fig. 326. *Solidago speciosa* Nutt. Coll. Michigan; COMPOSITAE, Tubuliflorae.

71a Pore elongated transversely (transverse furrow)...............72

71b Pore elongated meridionally (Fig. 327)........................
............................*Cirsium altissimum* (L.) Spreng.

Tall Common Thistle

SIZE: Polar axis about 57μ; equatorial diameter about 60μ

RANGE: Northern Florida to Texas north to North Dakota, Michigan and southern New York

Shape oblate spheroidal (P/E Index about 0.95). Furrow narrow with attenuate ends; pore broader than furrow borders, meridionally elongate. Exine 8 to 9μ thick, including 4 to 5μ high spines. Columellae radially oriented between spine bases, oblique beneath spines. Supratectate pila bear a median thickening and are distally fused into a continuous membrane. Spines usually have an internal cavity. Stix (1960) describes the *Cirsium* pollen type (based on *C. palustre* (L.) Scop.) as 48 x 50μ, with transverse furrow.

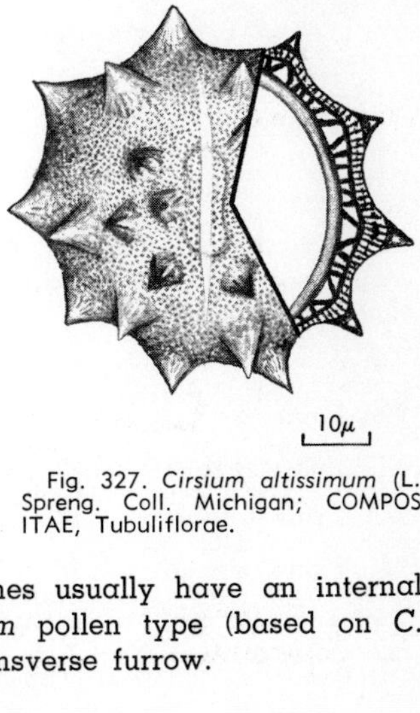

10μ

Fig. 327. *Cirsium altissimum* (L.) Spreng. Coll. Michigan; COMPOSITAE, Tubuliflorae.

72a Spines sharp-pointed or attenuate, usually with a spine cavity (Fig. 328) .... ***Eupatorium*-type**

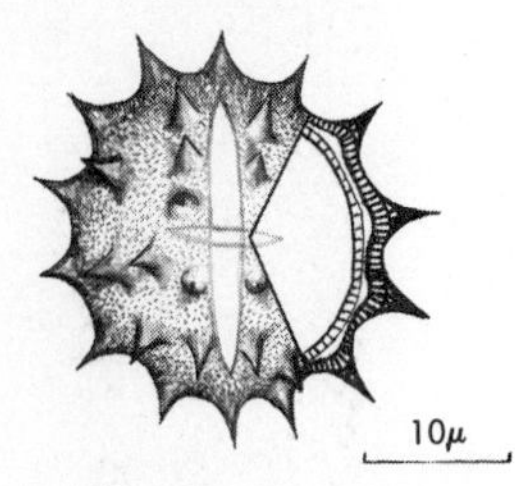

Fig. 328. *Eupatorium perfoliatum* L. Coll. Iowa; COMPOSITAE, Tubuliflorae.

Thoroughwort

SIZE: Polar and equatorial axes about 28μ

RANGE: Quebec to southeast Manitoba south to Texas, Louisiana, and Florida

Shape approximately spherical (Stix, 1960, describes the *Eupatorium* pollen-type as oblate spheroidal about 22 x 24μ). Furrow broad and long, with persistent, smooth membrane; transverse furrows have a very thin elliptical shape, with attenuate ends. Exine about 4 to 4.5μ thick, including 2.5μ long spines. A small cavity occurs between the exine layers beneath each spine. Stix (1960) also includes species of *Brickellia, Liatrus, Kuhnia, Ageratum* and *Mikania* within the *Eupatorium* pollen-type.

72b Spines blunt or at least not attenuate, spine cavity absent or extremely small (Fig. 329) .... ***Senecio*-type**

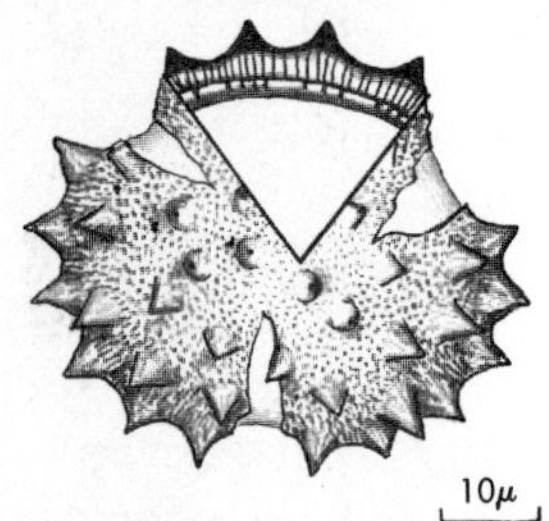

Fig. 329. *Senecio purshianus* Nutt. Coll. Wyoming; COMPOSITAE, Tubuliflorae.

Groundsel

SIZE: Polar axis about 48μ; equatorial diameter about 44μ

RANGE: Saskatchewan to British Columbia south to Utah and Texas

Shape prolate spheroidal (P/E Index about 1.1). Furrows rather long, polar area small (polar area index about 0.3); transverse furrows with pointed ends. Exine about 6μ thick, including the 2 to 3μ long spines. Exinous columellae stout, but scattered, the supratectate pila thin and elongate, fused distally to form a surface membrane.

73a Exine with striate or rugulate appearance **74**

73b Exine with a clavate or verrucate appearance **79**

74a Exine rugulate **75**

74b Exine striate **76**

75a Rugulate pattern restricted to intercolpi; furrow margins verrucate; pore distinct (Fig. 330) .... ***Shepherdia canadensis* (L.) Nutt.**

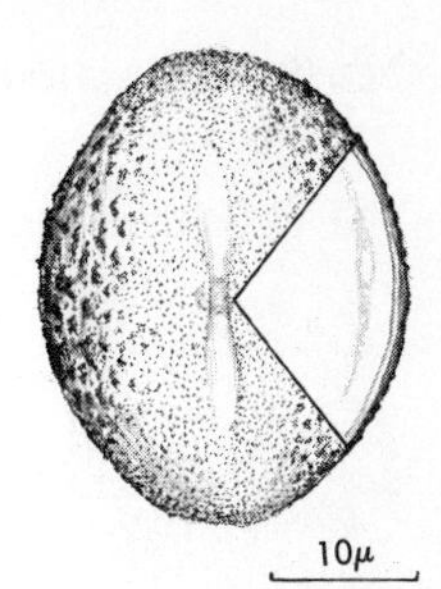

Fig. 330. *Shepherdia canadensis* (L.) Nutt. Coll. Michigan; ELAEAGNACEAE.

Soapberry

SIZE: Polar axis about 32μ; equatorial diameter about 25μ

RANGE: Newfoundland to Alaska south to South Dakota (New Mexico), northern Illinois, Ohio and New England

Shape subprolate (P/E Index 1.25 to 1.3). Furrows relatively short, contracted at the equator, pore round somewhat exceeding the furrow borders. Exine surface verrucate adjacent to the furrows and very coarsely rugulate in the intercolpate areas.

75b Rugulate pattern throughout; pore indistinct (Fig. 331)............ ....*Helianthemum canadense* (L.) Michx.

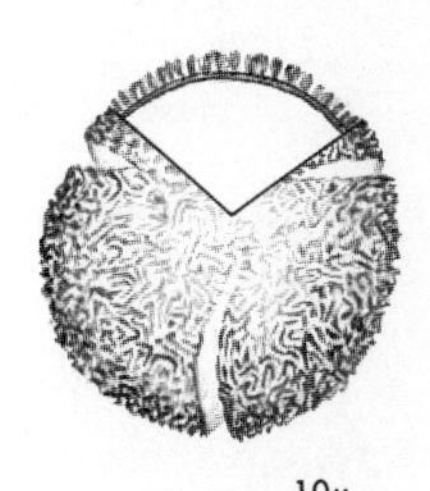

Fig. 331. Helianthemum canadense (L.) Michx. Coll. Michigan; CISTACEAE.

Frostweed

SIZE: Polar axis about 41μ; equatorial diameter about 45μ

RANGE: Nova Scotia to Wisconsin, south to Missouri and North Carolina

Shape oblate spheroidal (P/E Index about 0.9), pollen of some species of *Helianthemum* is spherical to prolate spheroidal. Furrows with an indistinct pore ("roughening") at the equator. Exine baculate and either intectate or perforate tectate. Surface pattern faintly reticulate to substriate. Some members of the Cistaceae are reticulate, others, including some species of *Helianthemum* are striate.

76a Striae very coarse (low magnification), arranged in blocks of parallel ridges which intersect other groups of striae at angles (Fig. 332)..................................*Menyanthes trifoliata* L.

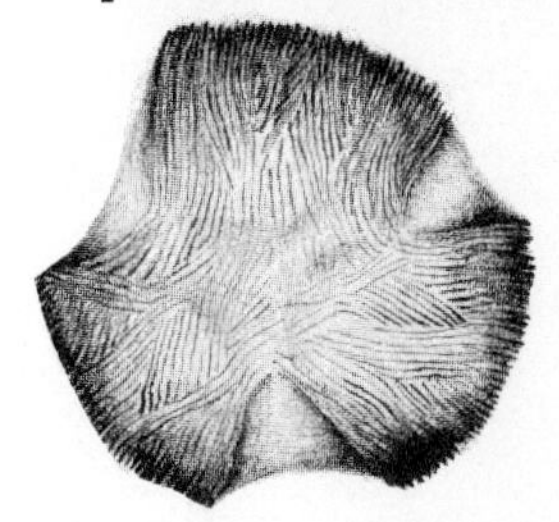

Fig. 332. *Menyanthes trifoliata* L. Coll. Michigan; GENTIANACEAE.

Buckbean

SIZE: Equatorial diameter about 30 to 32μ

RANGE: Labrador to Alaska south to Wyoming, Nebraska, Ohio and Maryland

Shape subspherical, furrows expanded often with pores indistinct. Wall surface distinctly striate, with blocks of striate intersecting each other in an angular pattern.

76b Striae more or less faint, arranged meridionally; Rosaceae, in part 77

77a Maximum dimension >30μ (Fig. 333)............*Pyrus malus* L.

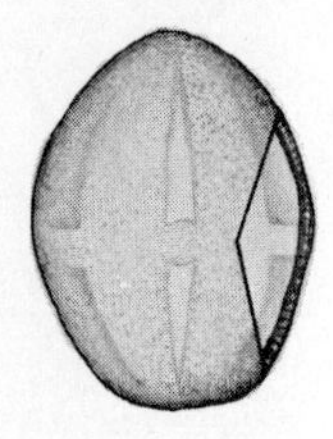

Fig. 333. *Pyrus malus* L. Coll. Michigan; ROSACEAE.

Apple

SIZE: Polar axis about 37μ; equatorial diameter about 29μ

RANGE: Cultivated, and spreading from cultivation throughout North America

Shape subprolate (P/E Index about 1.25). Furrows apparently interrupted at the equator by a weakly-developed pore. Exine thin, tectate; columellae forming a granular pattern which imparts a faint striate appearance to the wall surface.

77b Maximum dimension about 20 to 25μ.........................78

78a Shape subprolate, P/E Index about 1.25 (Fig. 334)............... ....*Geum canadense* Jacq.

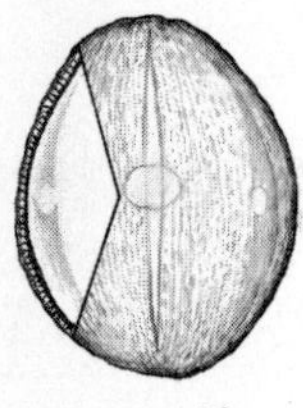

Fig. 334. *Geum canadense* Jacq. Coll. Michigan; ROSACEAE.

Canadian Avens

SIZE: Polar axis about 24μ; equatorial diameter about 19μ

RANGE: New Brunswick to Minnesota, its several varieties ranging south to Kansas, Texas and Georgia

Shape subprolate (P/E Index about 1.25). Furrow long and narrow, pore round exceeding the furrow borders. Exine tectate, with a distinctly striate pattern.

78b Shape approximately spheroidal (Fig. 335)....*Prunus virginiana* L.

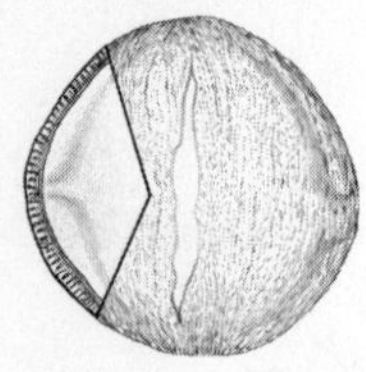

Fig. 335. *Prunus virginiana* L. Coll. Michigan; ROSACEAE.

Choke Cherry

SIZE: Polar and equatorial axes about 24μ

RANGE: Newfoundland to Saskatchewan, south to Kansas, Tennessee and North Carolina

Shape spheroidal (P/E Index about 1.0). Furrow margins irregular, pores indistinct or obscure, especially in face view. Exine tectate and under oil immersion is strongly striate.

79a Exine surface baculate or clavate (Fig. 336).. *Ilex* spp. (also *Nemopanthus*)

Cassena

SIZE: Polar axis about 43μ; equatorial diameter about 34μ (to 38μ)

RANGE: Florida to Texas northward to Arkansas and Virginia, primarily on coastal plain

Shape subprolate (P/E Index about 1.25). Furrow broad and long, polar area small (polar area index about 0.24). Furrow membrane persistent, fine verrucate; pore often indistinct or obscure. Exine intectate, with very coarse clavate elements.

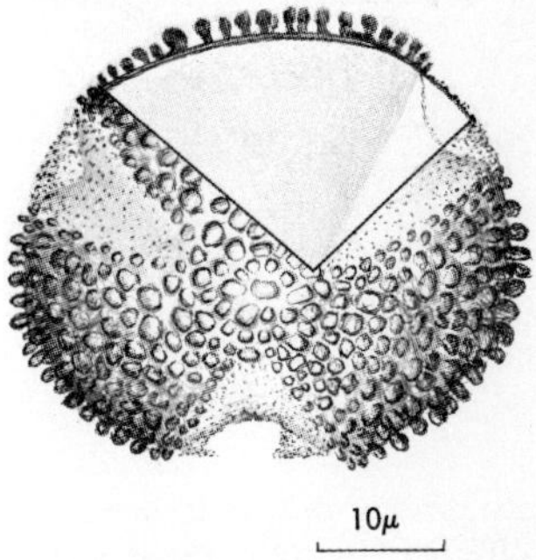

10μ

Fig. 336 *Ilex vomitaria* Ait. Coll. Texas; AQUIFOLIACEAE.

79b Exine surface scabrate-verrucate (in a few types keyed below the structural columellae give granular appearance which may be superficially like verrucae)..80

80a Shape approximately spheroidal..81

80b Shape subprolate, prolate, or perprolate..83

81a Transverse furrow present; exine probably psilate, granular appearance due to columellae (Fig. 337)........*Halesia carolina* L.

Silverbell Tree

SIZE: Polar axis about 37μ; equatorial diameter 34μ

RANGE: Florida to Texas north to Illinois and southwest Virginia

Shape prolate spheroidal (P/E Index about 1.05). Furrow of intermediate length (polar area index about 0.4), membrane fine-granular and persistent. Pore elliptical, elongated to form a short transverse furrow. Exine tectate, columellae very short, surface undulate to verrucate.

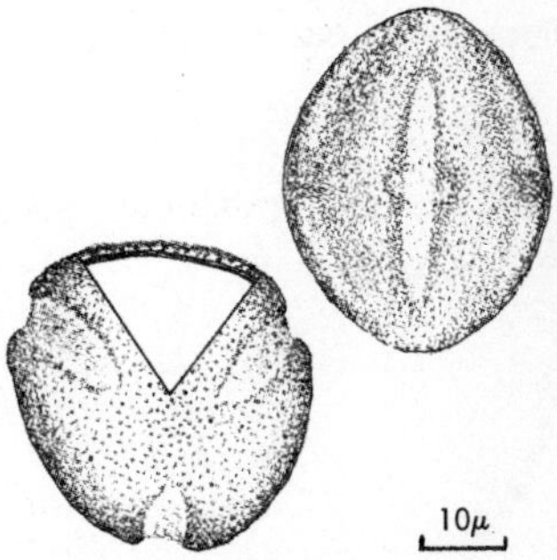

10μ

Fig. 337. *Halesia carolina* L. Coll. North Carolina; STYRACACEAE.

81b Transverse furrow absent..82

82a Furrows short, irregular; exine very coarsely verrucate (Fig. 338) ....*Leitneria floridana* Chapm.

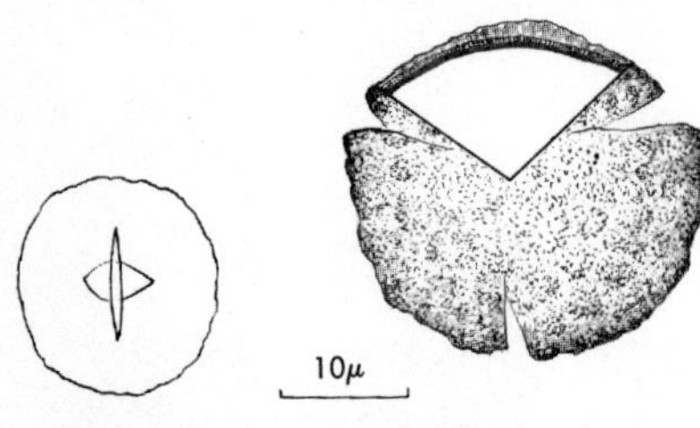

Fig. 338. *Leitneria floridana* Chapm. Coll. Florida; LEITNERIACEAE.

Corkwood

SIZE: Polar axis about 23μ; equatorial diameter about 24μ

RANGE: Coastal plain from Georgia and northern Florida to Texas

Shape oblate spheroidal (P/E Index about 0.96). Aperture number may occasionally vary from 4 to 6. Furrows, usually three, rather short (polar area index about 0.5). Transverse furrows broadly elliptical with pointed ends.

82b Furrows long, regular-margined and constricted at equator; verrucae fine (Fig. 339).......................*Lobelia siphilitica* L.

Blue Cardinal Flower

SIZE: Polar and equatorial axes about 25μ

RANGE: Maine to southern Manitoba, south to eastern Kansas, Texas, Louisiana, North Carolina and eastern Virginia

Shape approximately spherical. Furrow contracted at the equator, membrane fine granular (verrucate). Pore round, exceeding the furrow border. Exine with a fine-granular pattern. *L. cardinalis* is very similar but furrow membrane has a coarser verrucate surface.

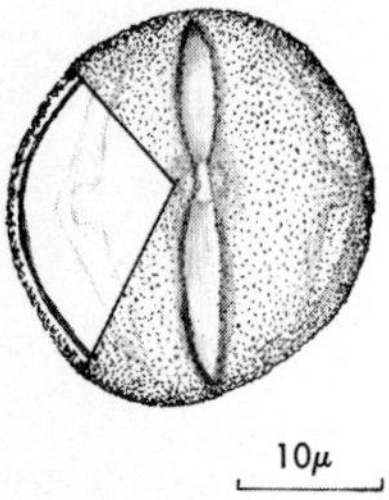

10μ

Fig. 339. *Lobelia siphilitica* L. Coll. Nebraska; CAMPANULACEAE.

83a Polar axis about 80μ long (Fig. 340).............*Stillingia texana*

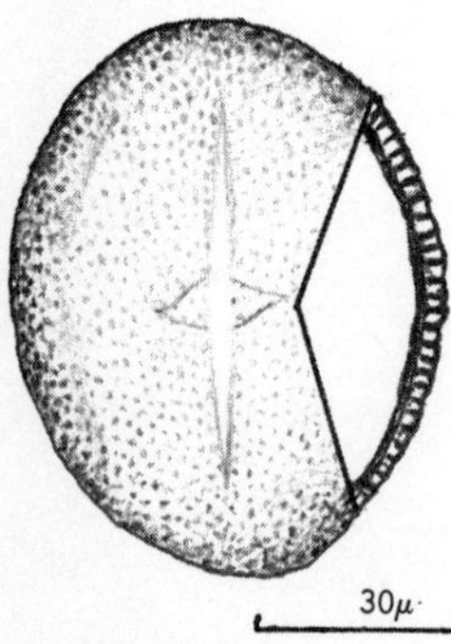

Fig. 340. *Stillingia texana*. Coll. Texas; EUPHORBIACEAE.

Queen's Delights

SIZE: Polar axis about 80μ; equatorial diameter about 66μ

Shape subprolate (P/E Index 1.15 to 1.2). Furrow thin and relatively short; transverse furrow elliptical with attenuate ends. Exine tectate with surface verrucae which impart a coarse granular appearance.

83b Polar axis 20 to 40μ **84**

84a P/E Index about 2.0 (Fig. 341) .... ***Sanicula gregaria* Bickn.**

Black Snakeroot

SIZE: Polar axis about 40μ; equatorial diameter about 20μ

RANGE: New Brunswick to Minnesota, south to eastern Kansas, Missouri, Alabama and northern Florida

Shape prolate or perprolate (P/E Index about 2.0). Furrows very slender, with narrow margos. Transverse furrows prominent with rounded ends and contracted at the furrows. Exine tectate with distinct columellae; surface psilate or scabrate.

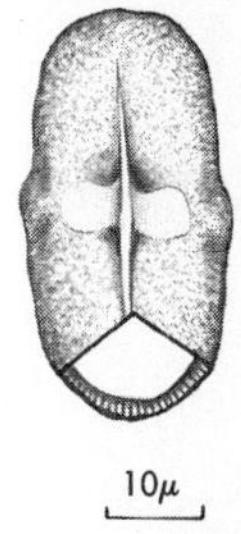

10μ

Fig. 341. *Sanicula gregaria* Bickn. Coll. Michigan; UMBELLIFERAE.

84b P/E Index 1.15 to 1.7 .... **85**

85a Transverse furrow present **86**

85b Transverse furrow lacking (Fig. 342) .... ***Cercis canadensis* L.**

Redbud

SIZE: Polar axis about 30μ; equatorial diameter about 21μ

RANGE: Florida to northeastern Mexico north to southern Wisconsin, southern Ontario, southeastern New York and Connecticut

Shape prolate (P/E Index about 1.4). Furrows contracted for about one-third their length at the equator; pores indistinct, especially in face view. Exine relatively thick (about 1.5μ), tectate, with a verrucate surface.

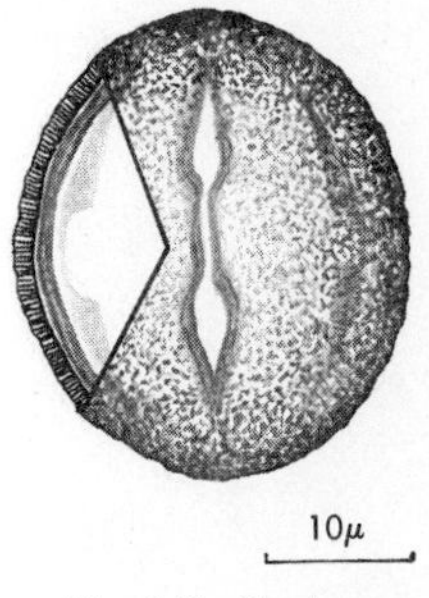

10μ

Fig. 342. *Cercis canadensis* L. Coll. Michigan; LEGUMINOSAE.

86a P/E Index 1.6 to 1.8 .... **87**

86b P/E Index less than 1.5; transverse furrows with rounded ends .. **88**

87a P/E Index about 1.65; transverse furrows bifurcated or square at ends (Fig. 343)................................*Ptelea trifoliata* L.

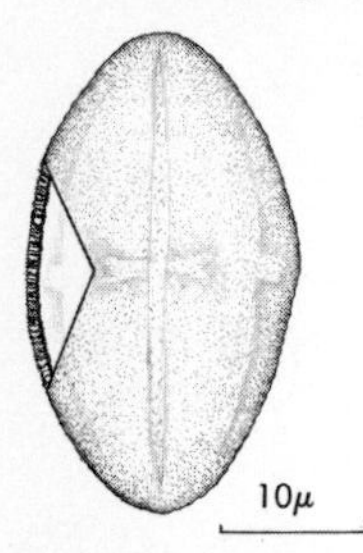

Fig. 343. *Ptelea trifoliata* L. Coll. Kansas; RUTACEAE.

Wafer Ash

SIZE: Polar axis 32μ; equatorial diameter 19μ

RANGE: Florida to Arizona and Mexico north to Nebraska, southern Michigan, southern Ontario and New England

Shape prolate (P/E Index 1.65 to 1.70). Furrows long, borders slightly irregular, membrane persistent and fine granular. Transverse furrows long and bifurcated at ends. Exine tectate, surface verrucate.

87b P/E Index about 1.8; transverse furrow with apiculate or rounded ends (Fig. 344)........................*Heracleum maximum* Bartr.

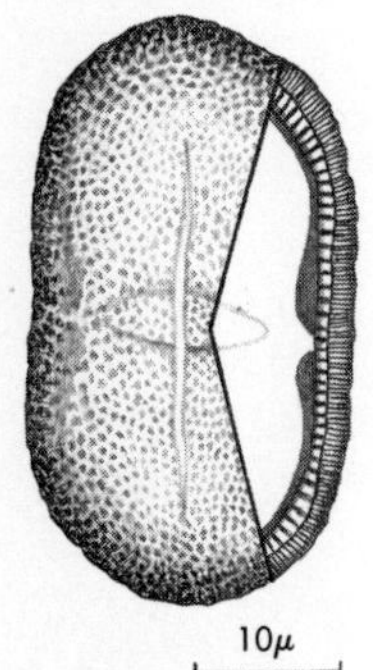

Fig. 344. *Heracleum maximum* Bartr. Coll. Michigan; UMBELLIFERAE.

Cow Parsnip

SIZE: Polar axis about 38μ, equatorial diameter about 21μ

RANGE: Labrador to Alaska south to California, New Mexico, Missouri, New England and Georgia

Shape prolate, P/E Index about 1.8, contracted at the equator; furrows slender, transverse furrows broad with tapered-rounded ends; exine thick, tectate, verrucate to faintly reticulate.

88a Shape class subprolate (Fig. 345)..........*Rosa palustris* Marsh

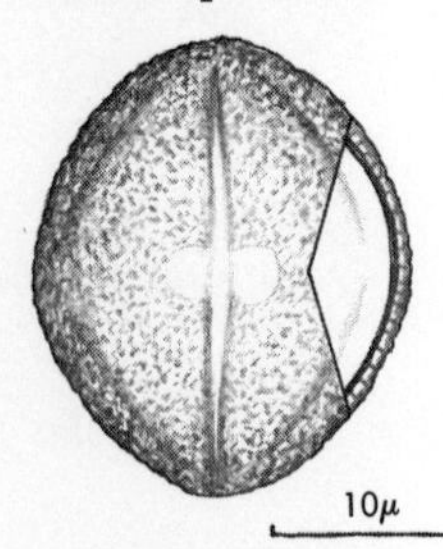

Fig. 345. *Rosa palustris* Marsh. Coll. Michigan; ROSACEAE.

Marsh Rose

SIZE: Polar axis 22μ; equatorial diameter about 19μ

RANGE: Florida to Arkansas, north to Minnesota and southeastern Canada

Shape subprolate (P/E Index 1.15 to 1.2). Furrows slender and narrowly attenuated. Transverse furrows with broadly rounded ends and contracted at the center. Exine tectate, with fine surface verrucae.

88b Shape class prolate (Fig. 346)........*Sedum stenopetalum* Pursh.

Stonecrop

SIZE: Polar axis about 22μ; equatorial diameter about 16μ

RANGE: Saskatchewan to Nebraska and New Mexico west to California and Alberta

Shape subprolate (P/E Index 1.35 to 1.4). Furrows very long, narrow attenuate near the poles and with margos near the equator. Furrow somewhat contracted at the equator; transverse furrows with rounded ends. Exine tectate; columellae form granular-striate pattern.

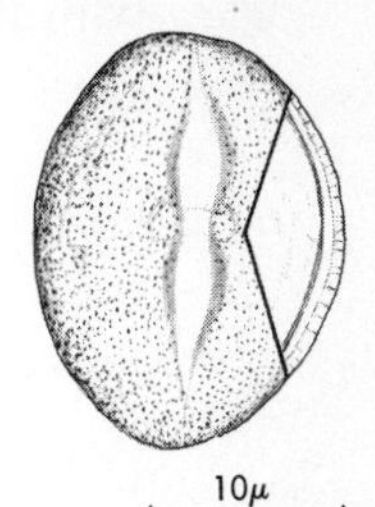

Fig. 346. *Sedum stenopetalum* Pursh. Coll. Montana; CRASSULACEAE.

STEPHANOCOLPORATE TYPES

Some tricolporate pollen types occasionally have four apertures; these exceptional types are not keyed in this section. *Pinguicula* and some members of the Boraginaceae also have pollen of this type, but are not illustrated.

1a Apertures 7, transverse furrows absent (Fig. 347)..*Sarracenia purpurea* L.

Pitcher Plant

SIZE: Polar axis 17.5μ, equatorial diameter 25μ

RANGE: Widespread in central and eastern Canada and the eastern United States

Stephanocolporate pollen with 7 to 8 apertures; pores circular and surpassing the furrow margins. Exine psilate, lacking discernible stratification. Shape oblate, P/E is 0.7.

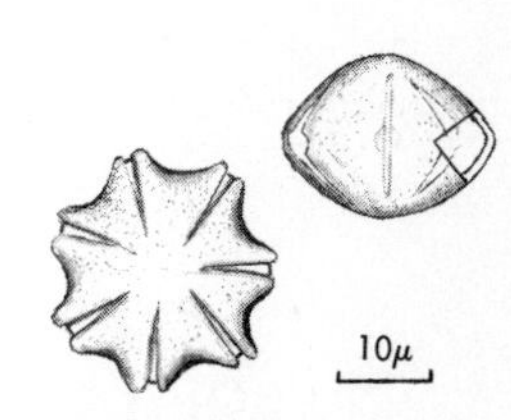

Fig. 347. *Sarracenia purpurea* L. Coll. Michigan; SARRACENIACEAE.

1b Apertures 9 or more, transverse furrows present...............2

2a Furrows 9 or more; grain with circular outline in equatorial view (Fig. 348)..*Polygala* spp.

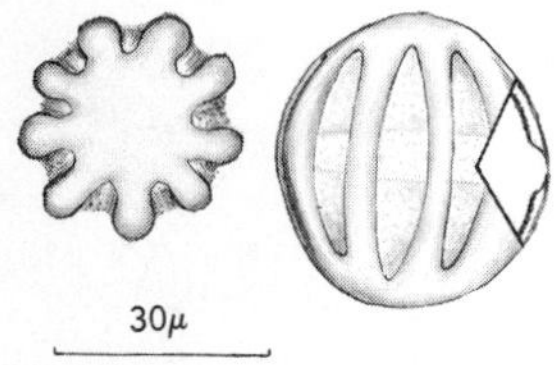

White Milkwort

SIZE: Polar axis 38μ, equatorial diameter 32μ

RANGE: Washington to Mexico, east to Minnesota, Nebraska, Kansas, Oklahoma and Texas

Nine furrows of moderate length, polar area large; some species have thin poroid areas in the polar cap. The transverse furrows associated with each pore are joined to form a girdle around the equator. The transverse furrow is not always obvious, especially in polar view; therefore, this type is keyed both in this category and with the stephanocolporate types, where it is properly placed. Some species of *Polygala* have 21-24-28 colpi.

Fig. 348. *Polygala alba* Nutt. Coll. Texas; POLYGALACEAE.

2b Furrows usually 14-15; grain with a distinctly swollen ridge at the equator (seen in equatorial view) (Fig. 349)........*Utricularia* spp.

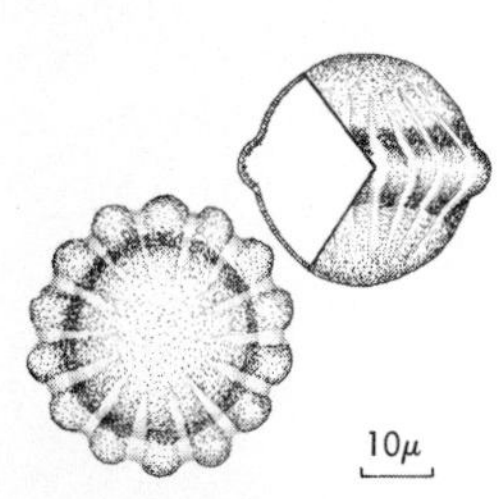

Fig. 349. *Utricularia minor* L. Redrawn from: Thanikaimoni, G. 1966. Pollen morphology of the genus *Utricularia*. *Pollen et Spores*, Vol. VIII(2):265-284 (Plate I, Figs. 3-4).

Bladderwort

SIZE: Longiaxe types (prolate), polar axis 33-36μ, equatorial diameter 28-30μ. Breviaxe types (oblate), polar axis 28-33μ, equatorial diameter 38-40μ

RANGE: Widespread in Northern United States, Canada and Eurasia

Pollen of aquatic species of *Utricularia* may have 11-28 colporate apertures. Since the equatorial girdle may not always be properly interpreted as fused transverse furrows (pores), this type is keyed with both stephanocolpate and stephanocolporate pollen types. This species may produce pollen of two shapes, "longiaxe" (prolate) and "breviaxe" types.

LOPHATE AND FENESTRATE TYPES

In fenestrate pollen (sensu stricto) the tectum is interrupted by a number of rather large openings (lacunae or windows). Thus, there are usually a number of wall-like or ridge areas (muri) in which the exine is very thick; the thinner lacunae may bear apertures (e.g. Veroniae). The Liguliflorae in the Compositae (Tribe Cichorieae) characteristically has lophate pollen. In this type the tegillum is not interrupted at the floor of the lacunae; instead, the tegillum extends over entire surface except apertures (Stix, 1960). Some species of cacti with pericolpate pollen may resemble typical fenestrate pollen although the exine of the muri may not be tectate.

1a Apertures pericolpate (18 to 20 short furrows) (Fig. 350)... *Opuntia* spp.

Prickly Pear Cactus

SIZE: 110 to 120μ diameter

RANGE: Widespread in plains and Rocky Mt. states

Exine of muri very thick (8μ), composed of clavae or baculae; exine of "windows" thin (1μ) reticulate. Shape of grain more or less octagonal. Pollen of O. *humifusa* Raf. is virtually identical with O. *polycantha*. Other genera of Cactaceae may be pericolpate (e.g. *Rhizophyllum, Neobesseya, Pereskia, Rhipsalis, Echinocactus*), tricolpate (e.g. *Epiphyllum, Echinocercus, Carnegia, Ferocactus*), or periporate (*Opuntia fulgida*, O. *pulchella*, O. *acanthocarpa*, O. *echinocarpa*, O. *exaltata*); all are moderately to very large (40 to 150μ).

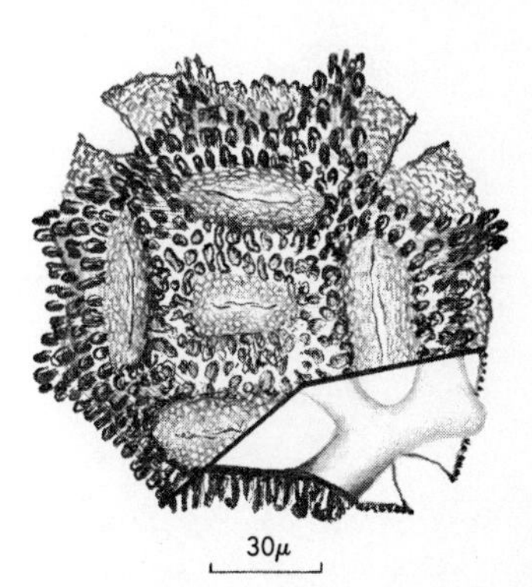

Fig. 350. *Opuntia polycantha* Haw. Coll. Montana; CACTACEAE.

1b Apertures tricolporate (Compositae)................................ 2

2a Tectum extends over lacunae (except apertures); lacunae number 9-11 (Fig. 351, 352).................................. Liguliflorae

Rattlesnake-Root

SIZE: About 60μ

RANGE: Quebec to Manitoba south to Georgia and Tennessee

Shape spheroidal to subtriangular. Number of lacunae appears to be nine, with two pole-fields absent (compare *Cichorium*).

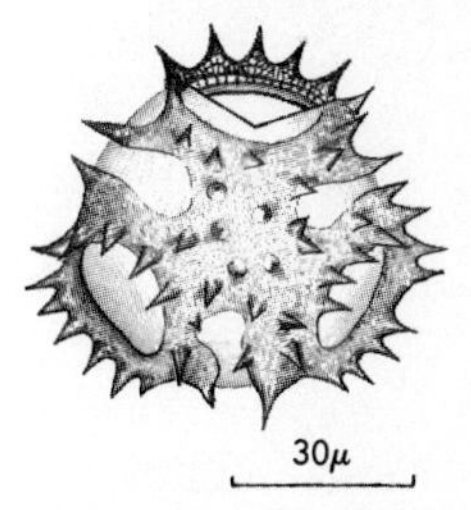

Fig. 351. *Prenanthes altissima* L. (polar view) Coll. New Hampshire; COMPOSITAE.

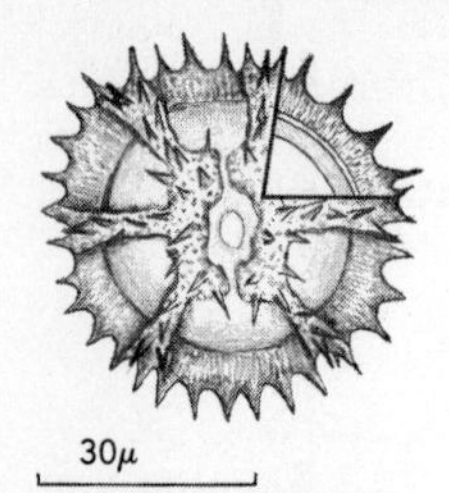

Fig. 352. *Cichorium intybus* L. (equatorial view) Coll. New Mexico; COMPOSITAE.

Chicory

SIZE: About 53μ

RANGE: Widespread in United States and southern Canada

Shape spheroidal. Number of lacunae eleven (*Taraxacum*-pattern, Stix, 1960). Aperturate lacunae 6-sided, the others 5-sided.

2b Tectum interrupted at lacunae; lacunae number 21 to 28; reticulate (Fig. 353, 354)..*Vernonia*

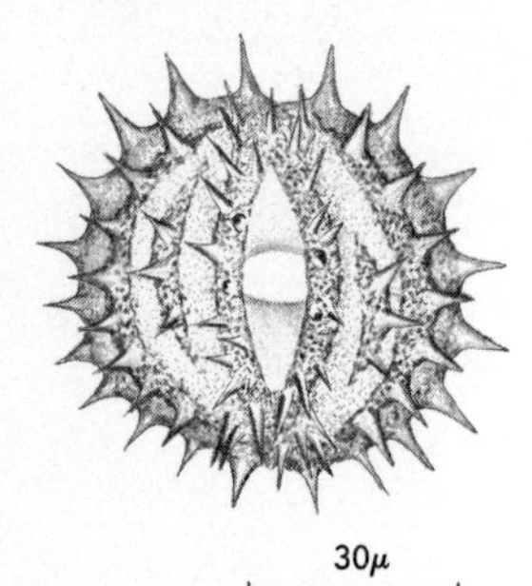

Fig. 353. *Vernonia altissima* Nutt. (equatorial view) Coll. Michigan; COMPOSITAE.

Ironweed

SIZE: 62μ x 69μ

RANGE: Georgia to Louisiana north to New York, Indiana and Missouri

Shape oblate spheroidal, P/E Index is 0.9. Spines slender. Exine obscurely reticulate due to pattern of columellae.

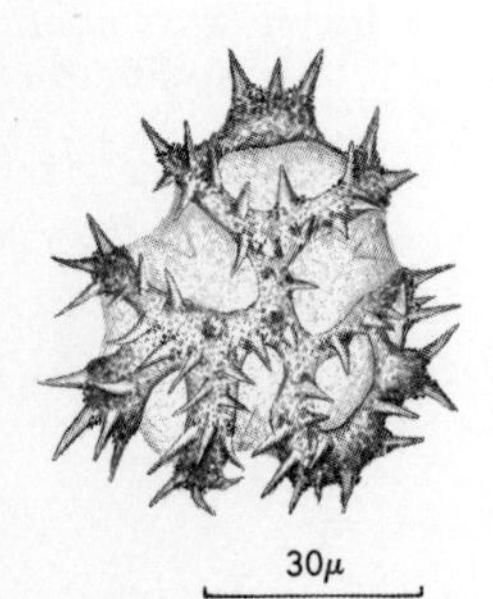

Fig. 354. *Vernonia lindheimeri* Engl. & Gray (polar view) Coll. Texas; COMPOSITAE.

Ironweed

SIZE: Diameter about 57μ

Spheroidal, P/E Index is 1.0. Spines broad-based attenuate. Exine faintly reticulate due to pattern of columellae.

DIPORATE TYPES

Besides the pollen of *Morus* and *Trema*, some triporate genera such as *Betula*, *Ostrya*, *Maclura*, *Myrica*, etc., may occasionally produce diporate grains. Two types of testaceous rhizopods (protozoa) are keyed with this group; they may be found in polleniferous sediments (especially in peat) and could be confused with pollen.

1a Wall composed of two or more layers........................2

1b Wall composed of a single layer..............................3

2a Pores protruding, endexine thickened at pore (Fig. 355)....*Trema* sp.

Trema

SIZE: 18-19.5μ (maximum diameter)

RANGE: Southern Florida and Keys

Essentially psilate exine with occasional evenly-spaced dots (verrucae?). Pores protruding due to ektexinous arc and endexinous thickening. Suboblate shape.

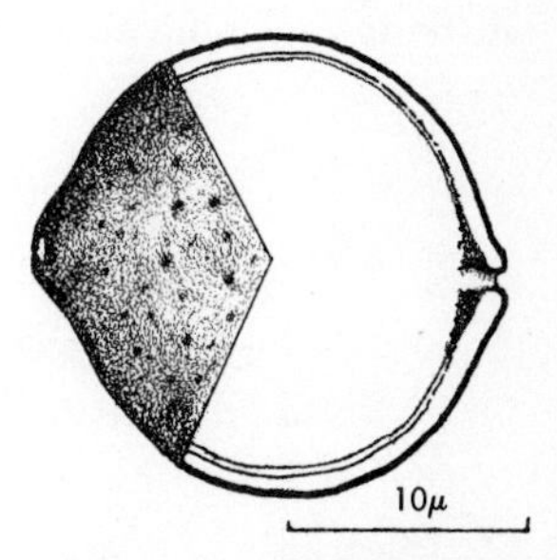

Fig. 355. *Trema floridana* Britt. Coll. Florida; ULMACEAE.

2b Pores non-aspidate, ektexine slightly thicker at the pore (Fig. 356).. ....*Morus alba* L.

Mulberry

SIZE: Polar axis 18-19.5μ, equatorial diameter 21.5-22.5μ

RANGE: Central and eastern U. S.; spread from cultivation

Exine psilate, pores non-aspidate, but vaguely annulate due to minor thickening of the endexine; pore margin sometimes irregular; shape suboblate to ellipsoidal.

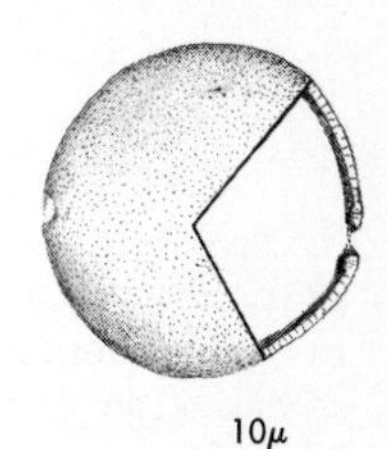

Fig. 356. *Morus alba* L. Coll. Michigan; MORACEAE.

3a Shape spherical, pores protruding and collared (Fig. 357).. *Amphitrema wrightianum*

Testaceous Rhizopod

SIZE: 90μ diameter

RANGE: Unknown. Habitat: moist rapidly growing peat moss (*Sphagnum*)

The tests of sphagnicolous rhizopods may be recovered with pollen from peat bog cores. They are generally well preserved and may in some cases be identified to species.

Fig. 357. *Amphitrema wrightianum* Coll. postglacial peat, Nova Scotia; PROTOZOA.

3b Shape cylindrical, with non-protruding terminal pores (Fig. 358).. *Amphitrema flavum*

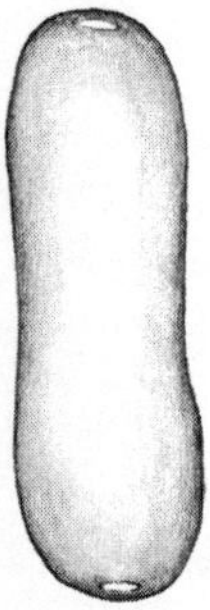

Testaceous Rhizopod

SIZE: 42μ x 160μ

RANGE: Unknown. Habitat: moist, rapidly growing *Sphagnum*

The cylindrical test is constricted near the middle. Unlike pollen grains, rhizopod tests lack wall stratification; the distinctive exine characteristics of pollen usually permits its separation from algal and protozoan remains.

Fig. 358. *Ditrema flavum* Coll. Nova Scotia, postglacial peat; PROTOZOA.

TRIPORATE TYPES

Triporate pollen is characteristic of the Betulaceae (Corylaceae), Myricaceae, Cannabinaceae, Urticaceae, and Onagraceae; it is found in some members of the Ulmaceae, Moraceae, Juglandaceae and is of scattered occurrence in a few other dicotyledonous families.

1a Exine echinate..2

1b Exine psilate or micro-reticulate....................................4

2a Pores round and annulate (Fig. 359)............................... ....*Sphaeralcea coccinea* (Nutt.) Rydb.

Red False Mallow

SIZE: Diameter about 60μ

RANGE: Manitoba to Texas and westward

Exine tectate, with broad-based spines about 5μ long. Pores annulate. Pollen of the Malvaceae is commonly polyporate, or in some genera 3-4 colporate.

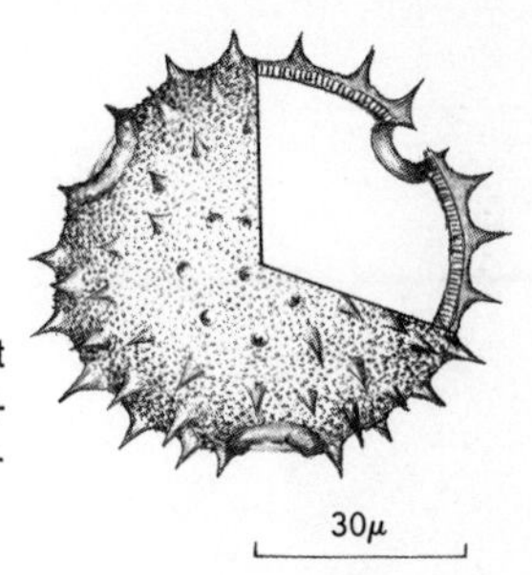

Fig. 359. *Sphaeralcea coccinea* (Nutt.) Rydb. Coll. Texas; MALVACEAE.

2b Pores elongate (possibly to be considered short furrows)........3

3a Shape of grain spherical, large, up to 100μ diameter (Fig. 360)..... ....*Dipsacus sylvestris* Huds.

Wild Teasel

SIZE: About 100μ diameter

RANGE: Southeastern Canada, northeastern United States south to North Carolina and Missouri

Apertures elongate, either short furrows or elongate pores; apertures with a strong margo (annulus). Exine very thick (endexine 2μ, ektexine 8.5μ, including 2μ echinae). Pore membranes bear large clavae and baculae.

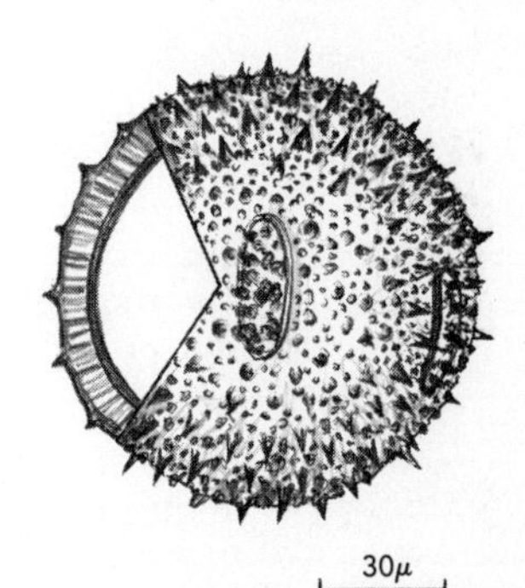

Fig. 360. *Dipsacus sylvestris* Huds. Coll. Michigan; DIPSACACEAE.

3b Shape of grain tri-lobed, small, less than 25μ (Fig. 361)............
...*Ambrosia* spp.

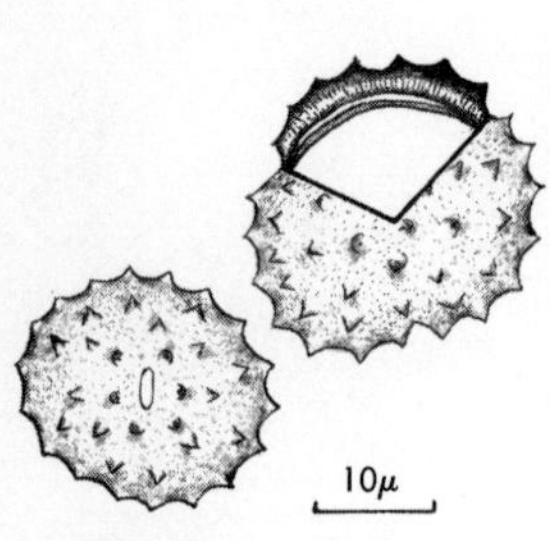

Fig. 361. *Ambrosia artemisiifolia* L. Coll. Michigan; COMPOSITAE.

Common Ragweed

SIZE: Polar axis about 19μ; equatorial diameter about 22μ

RANGE: Widespread weed in the United States east of Rocky Mts. to southern Canada

Spines short and broad-based, supratectate. Endexine often shrunken or pulled away from ektexine. Shape suboblate, P/E about 0.86. Apertures actually tricolporate, but furrows very short (poroid).

4a Apertures surrounded by thick dark, mesexinous thickenings......
(Fig. 362)...*Tilia*

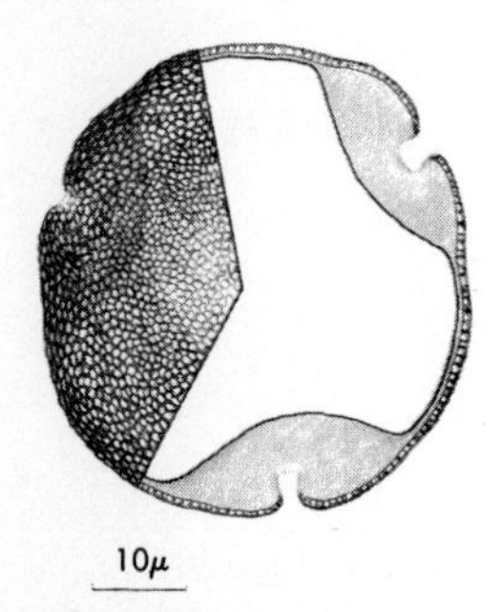

Fig. 362. *Tilia americana* L. Coll. New York; TILIACEAE.

Basswood

SIZE: About 45-47μ equatorial diameter

RANGE: Quebec to Manitoba south to Alabama and Texas

The three apertures (properly short furrows) look like gaping pores; they are surrounded by mesexinous thickenings which are dark after acetolysis. Exine surface is very finely reticulate.

4b Apertures otherwise 5

5a Shape of grain subtriangular; pores strongly protruding with large cylindrical or triangular pore cavities (vestibules) and usually internal collars or a series of ribs; viscid hairs usually present in fresh specimens..............................(Onagraceae) 6

5b Shape of grain subtriangular to round (polar view); pores aspidate, annulate, or simple, but lacking internal collars or ribs.........13

6a Vestibulum (pore cavity) with an internal endexinous collar......7

6b Vestibulum annulate or ribbed, but without a prominent endexinous collar 11

7a Size less than 65μ.. 8

7b Size greater than 100μ.. 9

8a Pore elongated and more or less slit-like (Fig. 363).. *Ludwigia alternifolia* L.

Seed Box

SIZE: Polar axis about 46μ, equatorial diameter about 60μ

RANGE: Florida to eastern Texas, north to Iowa and Massachusetts

Shape oblate, P/E about 0.75. Thin viscin threads present in fresh specimens.

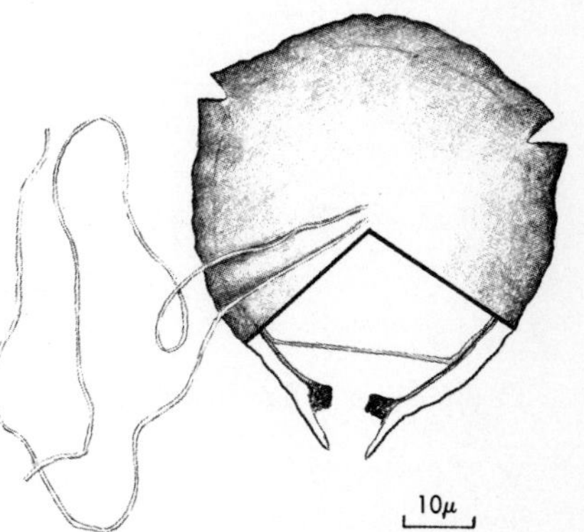

Fig. 363. *Ludwigia alternifolia* L. Coll. Virginia; ONAGRACEAE.

8b Pore round, often 4-pored (see also Fig. 396—Stephanoporate types) (Fig. 364).. *Circaea alpina* L.

Enchanter's Nightshade

SIZE: Polar axis 29μ, equatorial diameter 43μ

RANGE: Alaska and Canada, New England and northern United States, south in Appalachians to Georgia and in Rocky Mts. to Colorado and Utah

Both triporate and tetraporate grains common. Shape oblate, P/E Index about 0.66. Pores strongly aspidate with cylindrical or broadly triangular pore cavity. An endexinous collar with radial ribs lines the inner half of the pore cavity; circular ribs line the outer pore cavity.

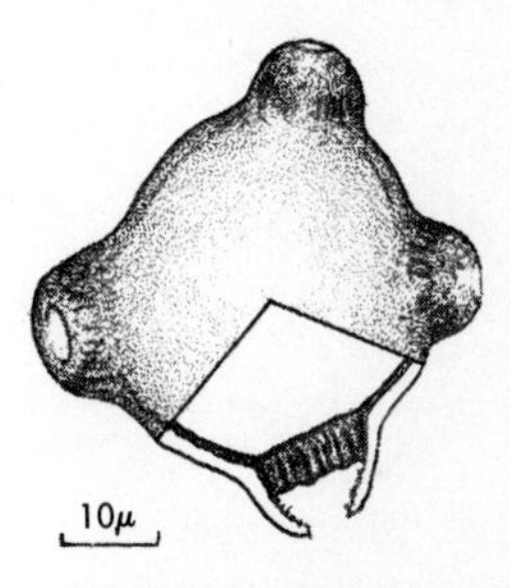

Fig. 364. *Circaea alpina* L. Coll. Michigan; ONAGRACEAE.

9a Endexine collar separate from the wall of the pore vestibulum (Fig. 365) .... ***Godetia bottae* Spach.**

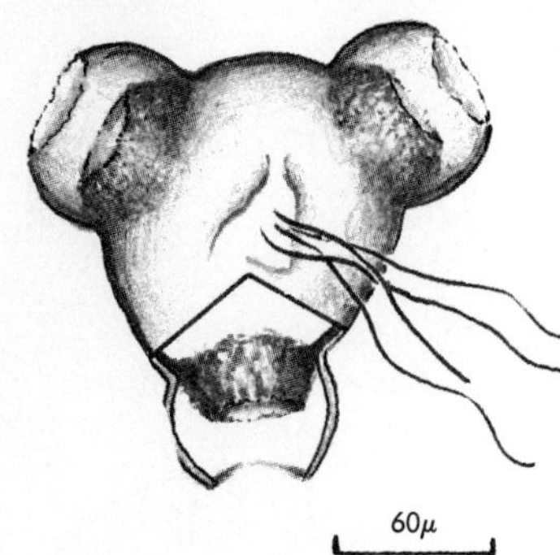

Fig. 365. *Godetia bottae* Spach. Coll. California; ONAGRACEAE.

SIZE: Equatorial diameter up to 160μ

Four or five viscin threads present in . fresh material. The complex pore structure of *Godetia* pollen seems to be very distinctive, permitting differentiation from other onagraceous pollen.

9b Endexine collar fused to the wall of the pore vestibulum **10**

10a Pore opening round (Fig. 366) .... ***Clarkia breweri***

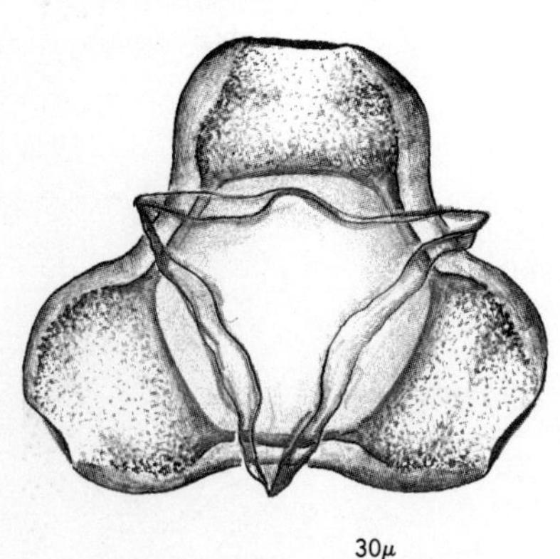

Fig. 366. *Clarkia breweri* Coll. California; ONOGRACEAE.

SIZE: Equatorial diameter approximately 135μ

Pores very broad and flattened at ends, ektexine of pores conspicuously thickened. A folded, triangular ektexinous wall layer seems to be consistently present around the center of the grain.

10b Pore opening frequently elongate and slit-like (Fig. 367)........... *Oenothera biennis* L.

Evening Primrose

SIZE: Equatorial diameter about 140μ

RANGE: Widespread east of Rocky Mts. in southern Canada and the United States

Shape (polar view) subtriangular with strongly aspidate squarish pores. Pore somewhat slit-like; viscin threads present in fresh specimens.

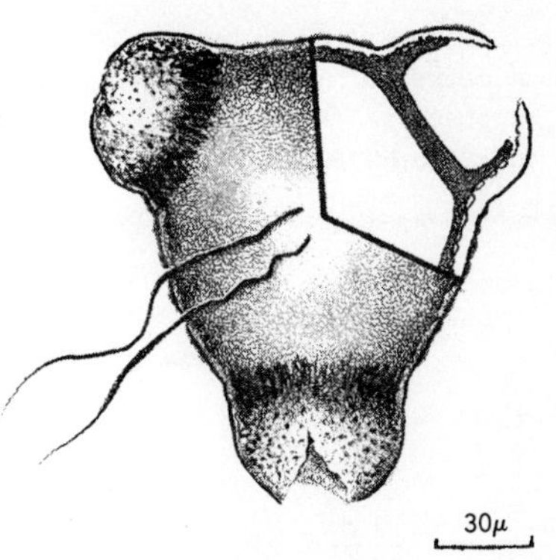

Fig. 367. *Oenothera biennis* L. Coll. Mich.; ONAGRACEAE.

11a Size less than 80μ (Fig. 368)....*Gayophytum racemosum* T. & G.

Baby's Breath

SIZE: Polar axis about 62μ; equatorial diameter about 75μ

RANGE: South Dakota to Colorado, west to California and Washington

Subtriangular shape (polar view) with rounded-aspidate pores, P/E Index about 0.8. Viscin threads not observed, but may be present.

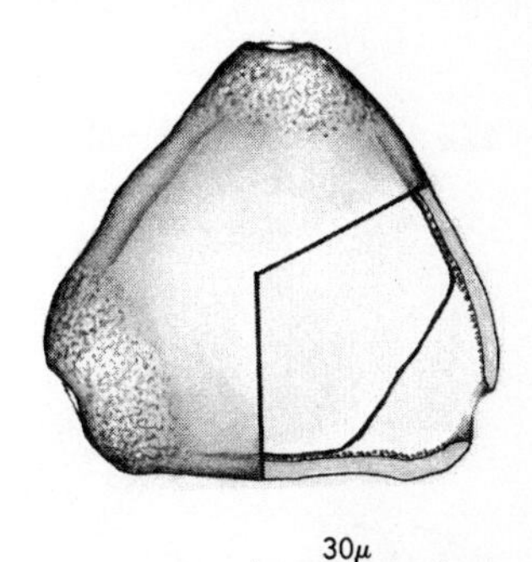

Fig. 368. *Gayophytum racemosum* T. & G. Coll. Nevada; ONAGRACEAE.

11b Size greater than 100μ......................................12

12a Shape subspherical (Fig. 369) ...... ***Zauschneria californica*** **Presl.**

SIZE: Equatorial diameter about 140μ

RANGE: West Coast of United States

Shape (polar view) rounded-triangular. Exine distinctly thickened at pores. Inner face of endexine very rough, surface of grain roughened (verrucate). The single thick viscin thread is composed of several intertwined strands.

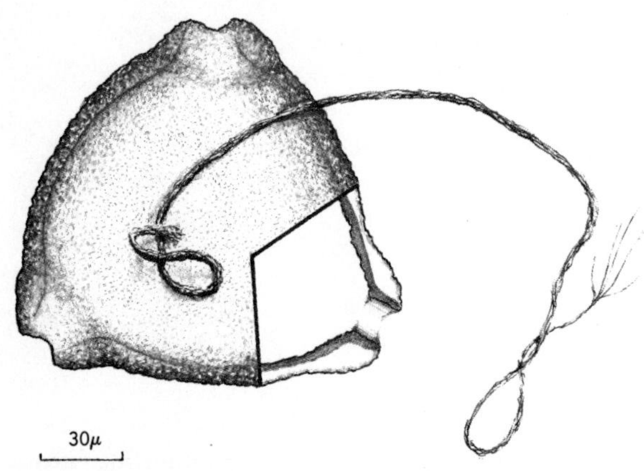

Fig. 369. *Zauschneria californica* Presl. Coll. California; ONAGRACEAE.

12b Shape subtriangular (Fig. 370) .... ***Gaura brachycarpa*** **Small**

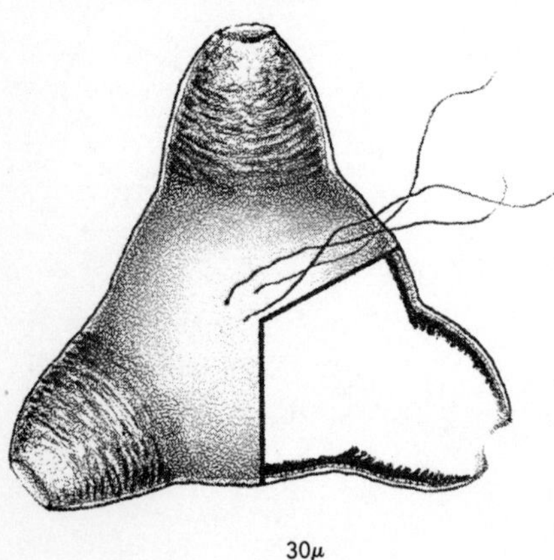

Fig. 370. *Gaura brachycarpa* Small Coll. Texas; ONAGRACEAE.

Gaura

SIZE: Equatorial diameter ranges from 115 to 160μ

Pores strongly aspidate; inner surface of pore cavity (vestibulum) with heavy ("muscular") ridges which appear as circular striae in surface view. Viscin threads thin. The sticky viscin threads hold masses of fresh pollen together, apparently making the insect pollen transfer more efficient.

13a Diameter about 14 to 17μ 14

13b Diameter greater than 20μ 15

14a Surface completely psilate; size 16 to 17μ (Fig. 371) .... *Acalypha hederacea* Torrey

Three-seeded Mercury

SIZE: Diameter 16 to 17μ

Shape spherical, with slightly aspidate pores. Exine at pore border separates to form a small pore cavity (*Betula*-like pore).

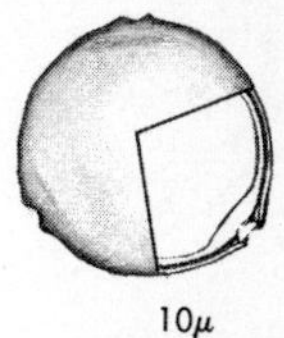

Fig. 371. *Acalypha hederacea* Torrey Coll. Texas; EUPHORBIACEAE.

14b Surface microverrucate to psilate; size 14 to 15μ (Fig. 372) .... *Urtica procera* Muhl.

Nettle

SIZE: 14 to 15μ diameter

RANGE: Eastern Canada to North Dakota, south to Louisiana and North Carolina

Shape spheroidal; pores three (occasionally 2). Exine tectate, with very fine granular or psilate surface.

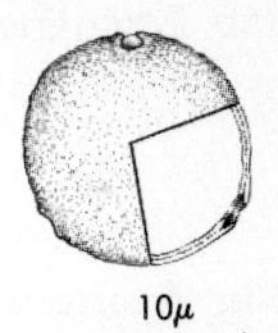

Fig. 372. *Urtica procera* Muhl. Coll. Minnesota; URTICACEAE.

15a Equatorial diameter 38 to 75μ, pores may be slightly off the equator in one hemisphere (heteropolar) (Fig. 373, 374) ...... *Carya* spp.

Pecan

SIZE: About 45μ equatorial diameter

RANGE: Indiana to Iowa, south to Alabama and eastern Texas

The pores are slightly aspidate in pecan pollen. The genera *Platycarya* and *Engelhardtia* also produce three-pored pollen (isopolar). *Pterocarya* (5-7 pored, 25-35μ diameter) is stephanoporate to periporate heteropolar. These predominantly Asian genera are represented in the North American Tertiary.

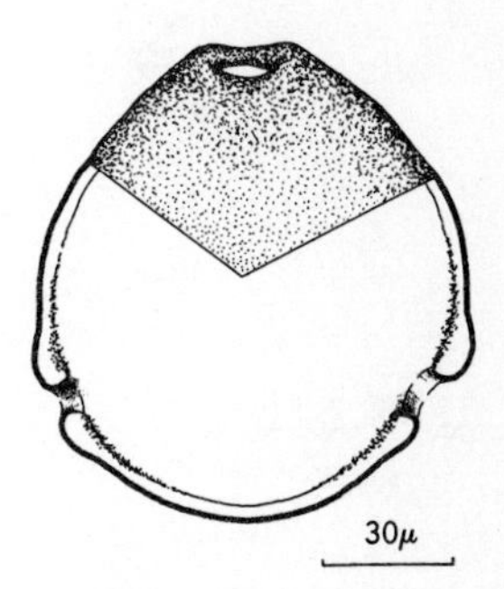

Fig. 373. *Carya illinoensis* (Wang.) K. Koch. Coll. Texas; JUGLANDACEAE.

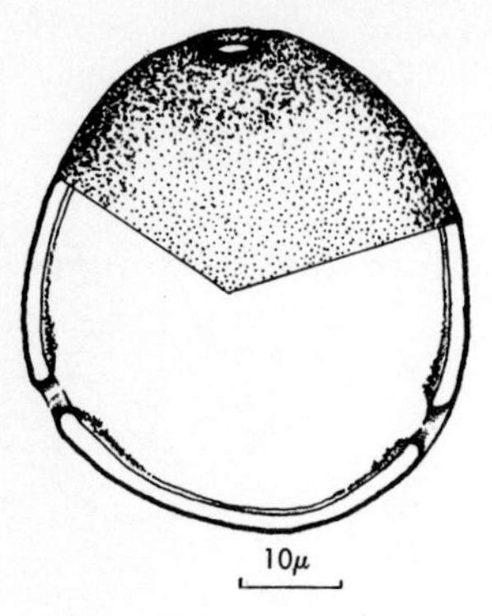

Fig. 374. *Carya aquatica* (Michx.f.) Nutt. Coll. Florida; JUGLANDACEAE.

Water Hickory

SIZE: 40-45μ, equatorial diameter

RANGE: Coastal Plains, Florida to Texas, north to Virginia and southeastern Missouri

The slightly heteropolar condition of hickory pollen is shown in this drawing of water hickory; one of the pores is displaced slightly from the equator into one hemisphere. The heteropolar nature of pollen of most Juglandaceae is more evident in walnut (*Juglans*), which is periporate.

15b Equatorial diameter less than 38μ, pores strictly equatorial....16

16a Apertures meridionally elongate (short furrows?) (Fig. 375)..*Monotropa*

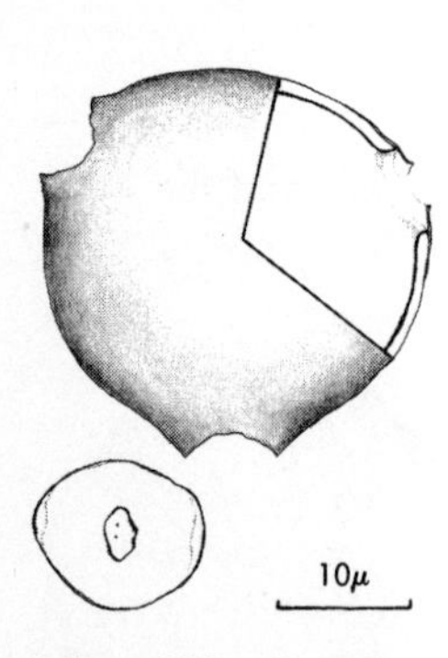

Fig. 375. *Monotropa uniflora* L. Coll. Michigan; PYROLACEAE.

Indian Pipe

SIZE: Polar axis about 22.5μ; equatorial diameter about 27μ

RANGE: Eastern United States and southern Canada to southeastern Alaska

Apertures are short furrows or pores about 5 x 7μ in size. Shape oblate, P/E Index about 0.8. *M. hypopitys* is similar in shape, size and apertures.

16b Apertures approximately circular................................17

17a Pores non-annulate and only slightly aspidate (Fig. 376) . . . . *Maclura pomifera* (Raf.) Schneid.

Osage Orange

SIZE: 22-25μ

RANGE: Southern New England to Iowa and southward

Pores usually 3, but frequently diporate in some collections, slightly aspidate. Pore slightly ellipsoidal, about 2.3μ in maximum diameter. Shape spherical to slightly oblate. Diporate *Maclura* may be very difficult to separate from *Morus;* triporate *Maclura* is very similar to *Celtis*. *Morus* pores are round with an irregular margin, shape of grains ellipsoidal; *Celtis* pores round and slightly annulate.

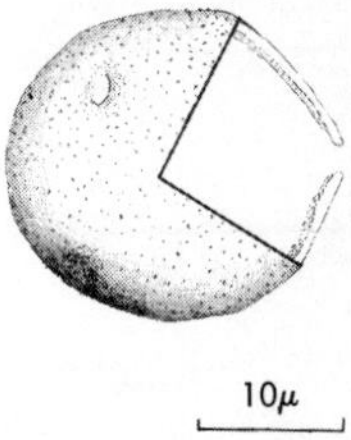

10μ

Fig. 376. *Maclura pomifera* (Raf.) Schneid. Coll. Texas; MORACEAE.

17b Pores annulate or aspidate 18

18a Grain spherical, pores annulate (Fig. 377) . . . . *Celtis* spp.

Hackberry

SIZE: 25-28μ diameter

RANGE: The several varieties range from southwestern Quebec to North Dakota and Utah southward to Florida, Louisiana and Oklahoma

Pores with a weak annulus, about 2.5μ diameter. Shape spherical.

10μ

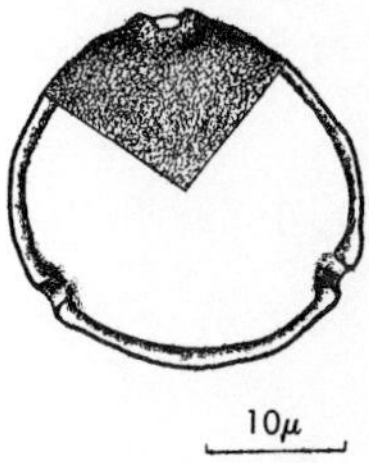

Fig. 377. *Celtis occidentalis* L. Coll. Kansas; ULMACEAE.

18b Grain subtriangular to spherical, pores aspidate (Betulaceae and Myricaceae) 19

19a Pore cavity present, formed by separation of the endexine and ektexine at the pore border. (Fig. 378, 379)............*Betula* spp.

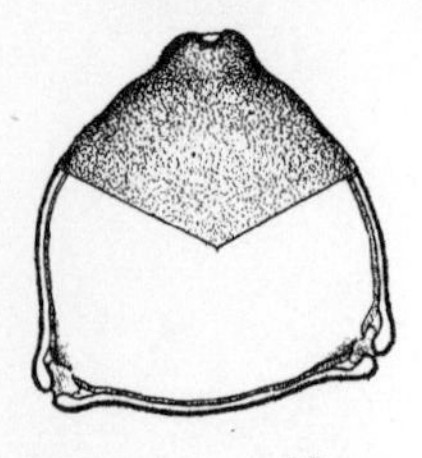

Fig. 378. *Betula papyrifera* Marsh. Coll. Michigan; CORYLACEAE.

Paper Birch

SIZE: About 27μ diameter

RANGE: The several varieties are widespread in Canada (to Alaska) and the northern United States

Shape spherical, P/E index about 1.0. Surface psilate, exine indistinctly tectate.

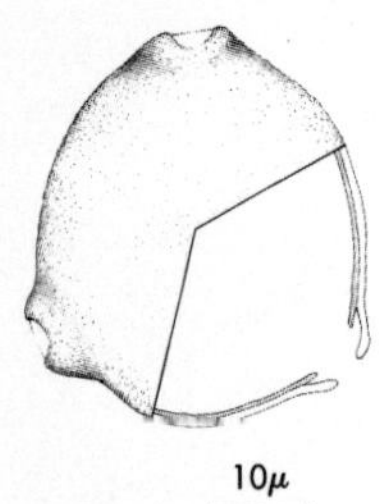

Fig. 379. *Betula occidentalis* Coll. California; CORYLACEAE.

Western Birch

SIZE: Equatorial diameter 32-38μ

RANGE: British Columbia to western Montana and Washington

Spherical with strongly aspidate pores. Surface psilate, exine indistinctly tectate.

19b Pore cavity absent.......................................20

20a Inner wall of pore roughened, ribbed, or jagged (Fig. 380, 381).... ....Myricaceae

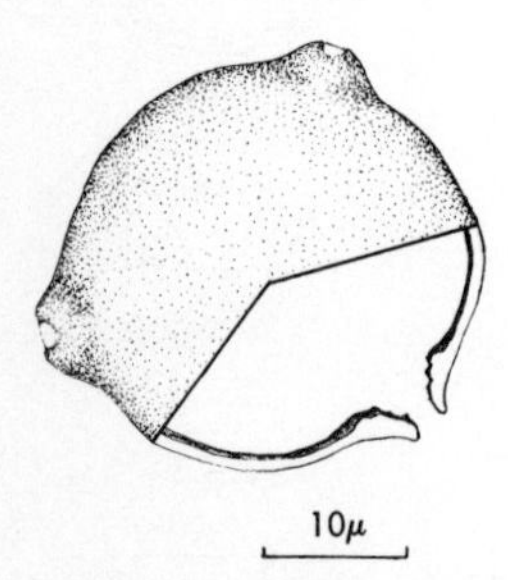

Fig. 380. *Myrica gale* L. Coll. Michigan; MYRICACEAE.

Sweet Gale

SIZE: 28-30μ diameter

RANGE: Widespread in Canada (to Alaska), the northern United States and Appalachian Mts. to North Carolina and Tennessee

Shape spherical with strongly aspidate pores. Inner wall of pore cavity (vestibulum) with "saw-toothed" or roughened surface.

Sweet Fern

SIZE: Equatorial diameter 27-29μ

RANGE: Southern Canada and the northeastern United States, westward to Manitoba and Minnesota and south to upland Georgia and Tennessee

Very similar and apparently indistinguishable from *Myrica*.

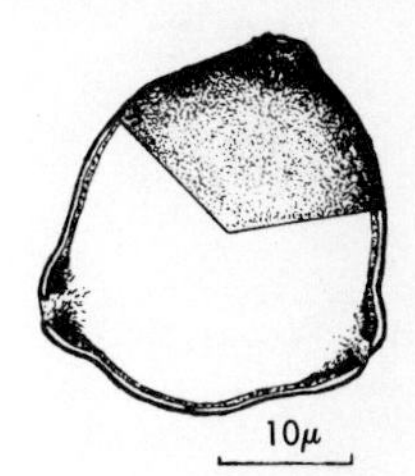

Fig. 381. *Comptonia perigrina* (L.) Coult. Coll. Michigan; MYRICACEAE.

20b Inner wall of pore smooth....................................21

21a Shape subtriangular, pore margin thickened and slightly aspidate (Fig. 382)..*Corylus*

Hazel

SIZE: 25-30μ diameter

RANGE: Maine to Saskatchewan south to Georgia and Oklahoma

Shape subtriangular in polar view. Surface psilate, exine indistinctly tectate.

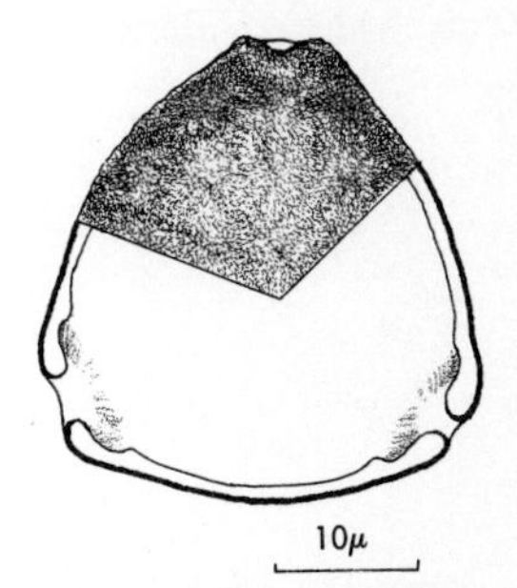

Fig. 382. *Corylus americana* Walt. Duke Univ. Herbarium; CORYLACEAE.

21b Shape spherical, pore margin of uniform thickness, distinctly aspidate (Fig. 383, 384)..........................*Ostrya* or *Carpinus*

Hop-Hornbeam

SIZE: About 30μ diameter

RANGE: Manitoba to Nova Scotia south to Georgia and Oklahoma

Equatorial limb circular with sharply protruding pores. Surface psilate, exine stratification obscure, indistinctly tectate.

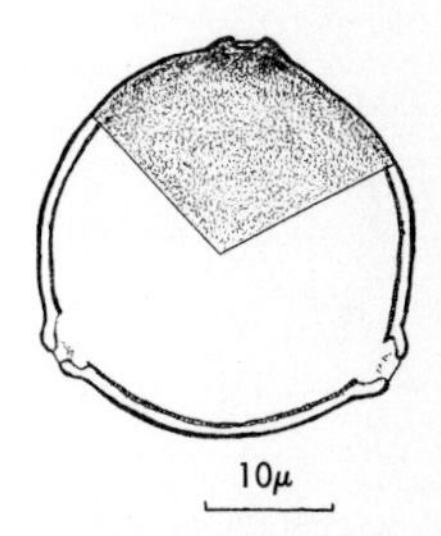

Fig. 383. *Ostrya virginiana* (Mill.) K. Koch Coll. Michigan; CORYLACEAE.

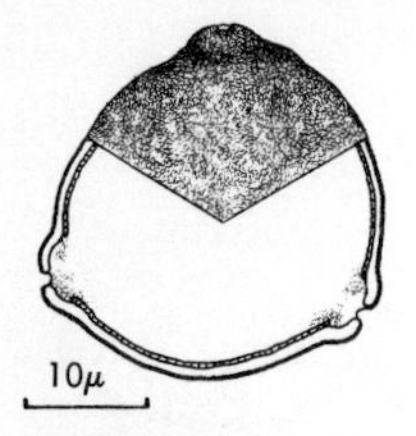

Fig. 384. *Carpinus caroliniana* Walt. Coll. Michigan; CORYLACEAE.

Blue Beech

SIZE: About 28μ diameter

RANGE: New England and Ontario to Minnesota south to North Carolina and Arkansas

Very similar to and apparently indistinguishable from *Ostrya* pollen.

STEPHANOPORATE TYPES

Occasionally pollen which is normally triporate will have 4 or 5 equatorial pores; if one of these extraordinary forms is encountered and cannot be satisfactorily identified in this section of the key, it should be sought in the triporate category.

1a Arcs extending between the pores (Fig. 385, 386)........*Alnus* spp.

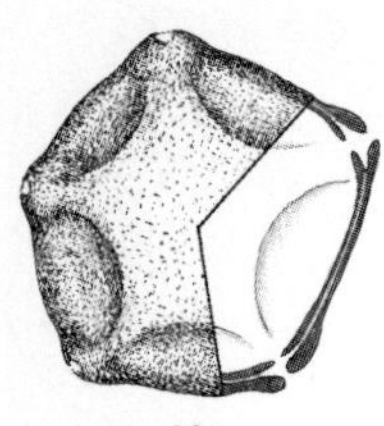

Fig. 385. *Alnus rugosa* (DuRoi) Spreng. Coll. Michigan; CORYLACEAE.

Speckled Alder

SIZE: Equatorial diameter about 30μ

RANGE: Northeastern United States and adjacent Canada

This alder has almost exclusively a 5-pored pollen (rarely 4 or 6 pored), shape (equatorial) angular, P/E Index about 0.7. Pores aspidate, ektexine separates from endexine at pore to form pore cavity.

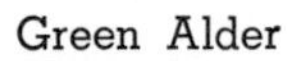

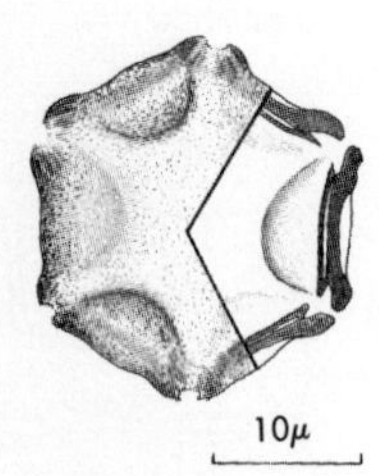

Fig. 386. *Alnus crispa* (Ait.) Pursh. Coll. Alaska; CORYLACEAE.

Green Alder

SIZE: Equatorial diameter about 22μ

RANGE: Widespread in Alaska and Canada, occasional in adjacent northeastern United States

Exine apparently intectate, ektexine thicker than endexine. In the samples examined, about 60% of the grains were 5-pored, the remainder 6-pored. It is probably impossible to separate species of alder pollen on basis of pore number and size; population statistics might permit such separation, however. For example, *A. alnobetula* (Ehrh.) K. Koch from Tennessee has large (28-40μ) almost exclusively 4-pored pollen, and that of *A. tennuifolia* Nutt from Idaho is smaller (22-27μ) and is 30% 4-pored, 70% 5-pored.

1b Arcs absent **2**

2a Surface convoluted, or rugulate (Fig. 387) .... ***Ulmus* spp.**

Slippery Elm

SIZE: Equatorial diameter about 37μ

RANGE: Western Florida to Texas, northward to N. Dakota and Quebec

Pores mostly 5 (occasionally 4 or 6), equatorial outline of grain polygonal. Shape oblate. The rugulate surface of elm pollen reminds some persons of the convoluted surface of the mammalian brain. Palynologists are not usually able to distinguish the pollen of the several species of elm.

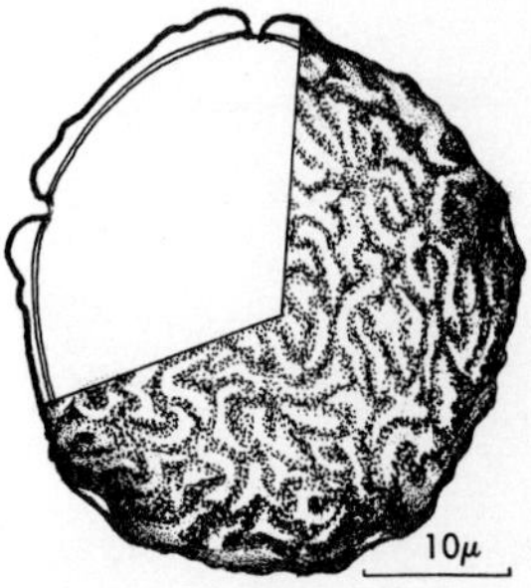

Fig. 387. *Ulmus rubra* Muhl. Coll. Michigan; ULMACEAE.

2b Surface psilate, echinate, or verrucate (not rugulate) **3**

3a Exine echinate **4**

3b Exine psilate or verrucate **5**

4a Pores large (about 5μ diameter); exine internally stratified, tectate (Fig. 388, 389) .... **Campanulaceae**

Venus' Looking Glass

SIZE: Diameter about 28μ

RANGE: United States and southern Canada east of Rocky Mts.

The four pores of pollen of Campanulaceae may not be perfectly equatorial; it is therefore keyed in both the stephanoporate or periporate pollen. Exine apparently tectate but columellae indistinct. Pore annulate due to endexine thickening. Spinule very small.

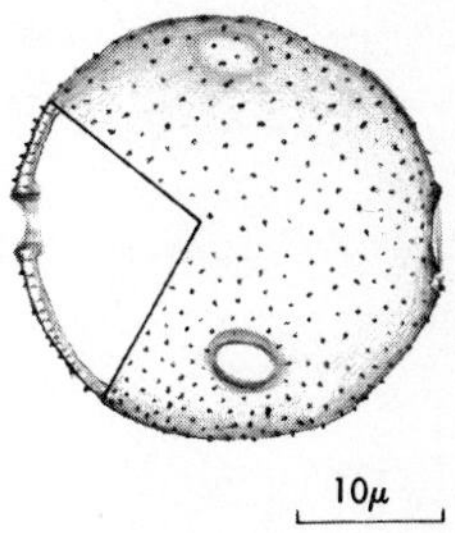

Fig. 388. *Specularia perfoliata* (L.) A. DC Coll. Texas; CAMPANULACEAE.

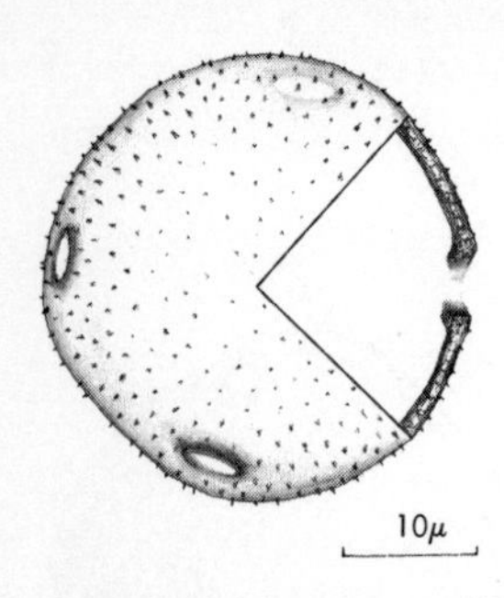

Fig. 389. *Triodanus coloradensis* (Buckl.) McVaugh Coll. Colorado; CAMPANULACEAE.

Colorado Bluebells

SIZE: Equatorial diameter 32μ

RANGE: Rocky Mountain area

Exine tectate with distinct columellae. Spinules small but distinct, and larger than in *Specularia perfoliata*. Pollen of this species is very similar to that of *Campanula* (cf. *C. petiolata*) except that the *Campanula* pollen is occasionally triporate, larger (33-46μ), and the spinules are larger, up to 0.5μ.

4b Pores small (2 to 3μ diameter), wall thick, with homogeneous structure (Fig. 390)........................*Uromyces fabae* (aeciospore)

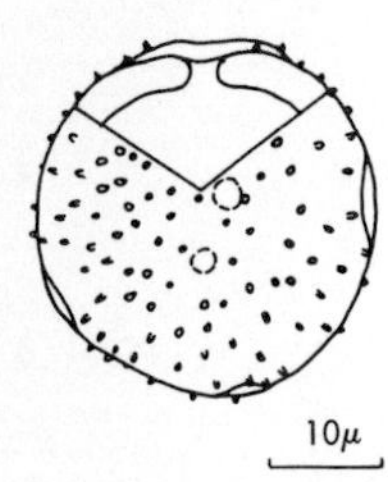

Fig. 390. *Uromyces fabae;* PUCCINIACEAE (aeciospore).

Broad Bean Rust

SIZE: About 27μ diameter

Wall very thick, not obviously stratified; pores generally in small concavities. Surface with short spines or baculae. Figure 133 illustrates the teliospore of this same species of rust.

5a Exine coarsely verrucate, with some gemmate and baculate elements (Fig. 391).......................*Proserpinaca palustris* L.

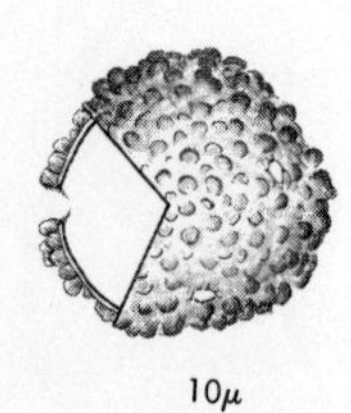

Fig. 391. *Proserpinaca palustris* L.; HALGORACEAE.

Mermaid's Weed

SIZE: Diameter about 21μ

RANGE: The several varieties extend from Nova Scotia to Minnesota south to Texas and Florida

Exine thick (about 1.5μ), densely covered with coarse verrucate, gemmate, baculate, and clavate elements. Mostly 5-pored (4-7 occasionally); pores either on equator or apparently periporate.

5b Exine psilate or faintly verrucate..............................6

6a Palynomorph elongate with a stalk scar at one end (Fig. 392)...... .... *Puccinia graminis* Pers.

Wheat Rust

SIZE: 14-17μ wide; 24-27μ long

RANGE: Nearly ubiquitous in North America

The spore is ellipsoidal with a flattened area on one end where the stalk was attached. Wall unevenly thickened, surface roughened. Germination pores 3-4.

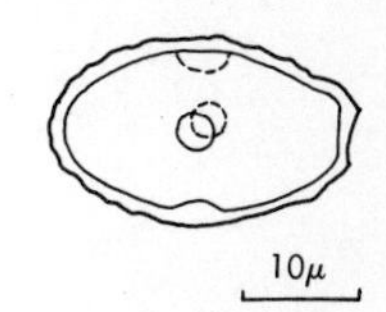

Fig. 392. *Puccinia graminis* Pers.; PUCCINIACEAE (uredospore).

6b Palynomorph more or less round in polar view.................7

7a Pores annulate..8

7b Pores aspidate, protruding, and with a distinct pore cavity........ .... Onagraceae (in part) 9

8a Diameter greater than 40μ (Fig. 393)........................... .... *Planera aquatica* (Walt.) Gmel.

Water Elm

SIZE: Equatorial diameter about 44μ

RANGE: Florida to Texas, north to North Carolina, Kentucky and southern Illinois

Grains 4 or 5-pored, in almost equal numbers in the samples examined. Pores slightly annulate. Surface psilate. Occasionally *Juglans cinerea* pollen will be stephanoporate with 4-5 pores; such pollen grains will key to *Planera*. The annulus in *Juglans* is different from *Planera*, however.

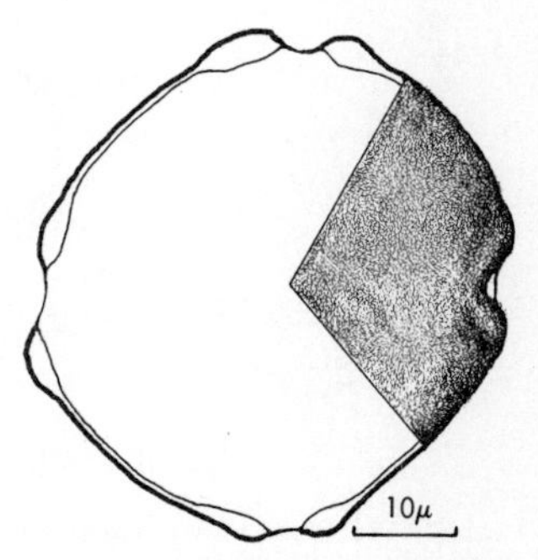

Fig. 393. *Planera aquatica* (Walt.) Gmel. Coll. Louisiana; ULMACEAE.

8b Diameter less than 30μ (Fig. 394) .... ***Myriophyllum***

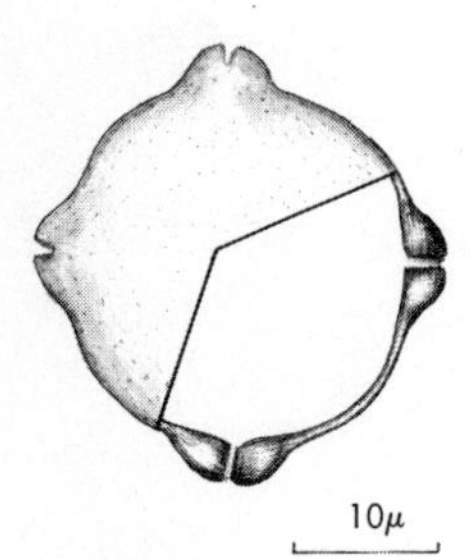

Fig. 394. *Myriophyllum spicatum* Michx. Coll. Michigan; HALGORACEAE.

Water Milfoil

SIZE: Polar axis 22μ, equatorial diameter 25μ

RANGE: Widespread in the eastern and central United States and Canada

Shape suboblate, P/E Index about 0.9. Apertures strongly annulate, either slightly elongate pores or short furrows. Number of apertures usually 4, occasionally 2-3 in the specimens examined. Some species of *Myriophyllum* are verrucate or slightly rugulate; *M. spicatum* is psilate. The genus *Pterocarya* (Juglandaceae) will key to this point also; its pollen is 3-7 pored, slightly aspidate, and often slightly heteropolar; this Asiatic genus may be encountered in Tertiary sediments of North America.

9a Diameter greater than 75μ (Fig. 395) .... ***Epilobium angustifolium* L.**

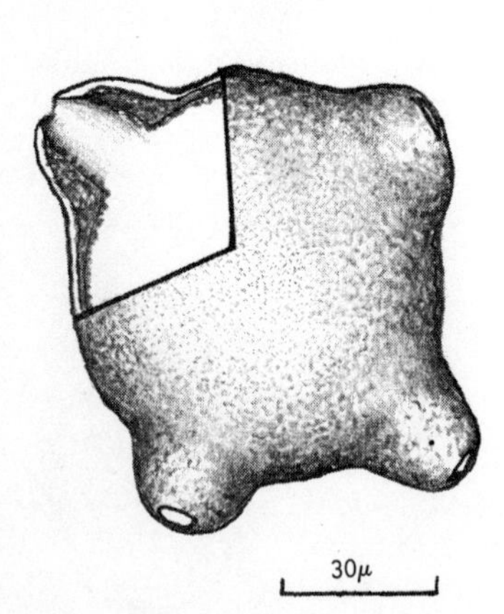

Fig. 395. *Epilobium angustifolium* L. Coll. Michigan; ONAGRACEAE.

Fire weed

SIZE: Equatorial diameter (including pores) about 80μ

RANGE: Widespread boreal species Alaska to Greenland, south to New England, northern plains states and Rocky Mt. region

This species may sometimes be triporate. Pores strongly aspidate, pore cavity large, cylindrical with internal ribs. Exine psilate. *E. glandulosum* Lehm. is similar to *E. angustifolium* (monads); several species (*E. strictum* Muhl., *E. leptophyllum* Raf., *E. adenocaulon* Haussk., *E. hirsutum* L., *E. coloratum* Biehler) frequently have tetrads, with occasional monads.

9b Diameter less than 50μ (Fig. 396)................*Circaea alpina* L.

Enchanter's Nightshade

SIZE: Polar axis 29μ, equatorial diameter 43μ

RANGE: Alaska and Canada, New England and northern United States, south in Appalachians to Georgia and in Rocky Mts. to Colorado and Utah

Both triporate and tetraporate grain common. Shape oblate, P/E Index about 0.66. Pores strongly aspidate with cylindrical or broadly triangular pore cavity. An endexinous collar with radial ribs lines the inner half of the pore cavity; circular ribs line the outer pore cavity.

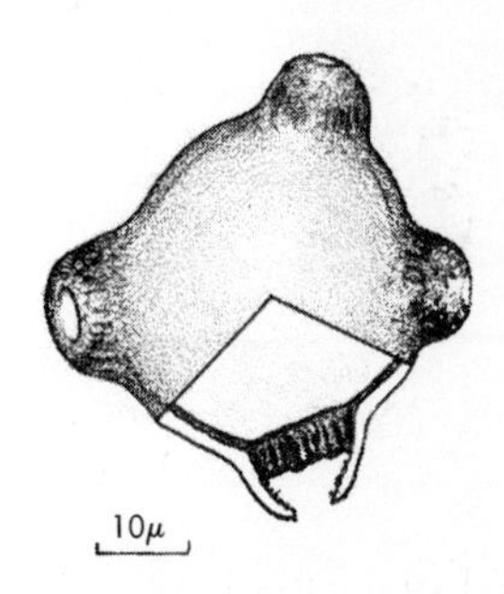

Fig. 396. *Circaea alpina* L. Coll. Michigan; ONAGRACEAE.

PERIPORATE TYPES

The Alismataceae, Amaranthaceae, Caryophyllaceae, Chenopodiaceae, Convolvulaceae, Cucurbitaceae, Plantaginaceae and Polemoniaceae produce periporate pollen. In addition, some genera or species in the Hamamelidaceae, Malvaceae, Polygonaceae, Ranunculaceae, Saxifragaceae and Zygophyllaceae have periporate pollen.

1a Surface echinate..2

1b Surface psilate, reticulate, baculate or verrucate................7

2a Spines short less than 1μ long...................................3

2b Spines longer, mostly 5μ or longer...........................4

3a Pores indistinct, 6-10 in number (Fig. 397)..........*Sagittaria* spp.

Arrowhead

SIZE: 24-25μ diameter

RANGE: Primarily in Great Lakes region

Pores indistinctly delimited and not always easily seen, 8-12 in number. Spines are short (about 1μ long) and evenly spaced. Pollen grains of *Alisma* are virtually the same, but may have more pores (up to 30). The pollen grains in *Alisma* (all species?) have 12-30 pores, may be reticulate, but lack spines.

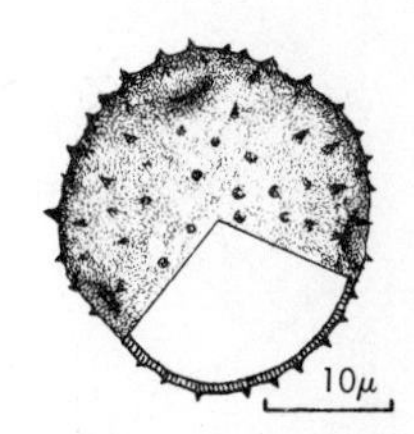

Fig. 397. *Sagittaria cristata* Engelm. Coll. Alaska; ALISMATACEAE.

3b Pores well defined and distinct, 15-16 in number, spines short and indistinct (Fig. 398) .... ***Sanguinaria canadensis*** **L.**

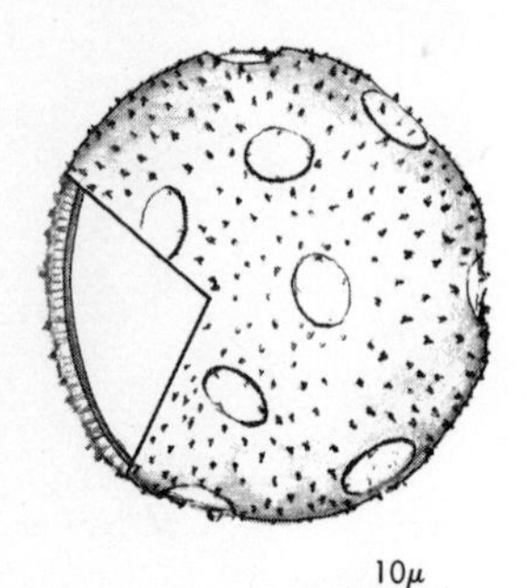

Fig. 398. *Sanguinaria canadensis* L. Coll. Michigan; PAPAVERACEAE.

Bloodroot

SIZE: (37-) 50-53μ diameter

RANGE: Quebec to Manitoba south to New England, Kentucky and eastern Kansas

Exine tectate with distinct columellae and a continuous tectum; spines short and uniformly spaced. Pores large (6-7μ diameter), approximately round, 15-16 in number. Surface may appear to be reticulate in some grains.

4a Pores operculate, spines on operculum **5**

4b Pores inoperculate, or if operculate the opercula without large spines **6**

5a Echinae long and blunt, apparently intectate (Fig. 399) .... ***Cucurbita*** **spp.**

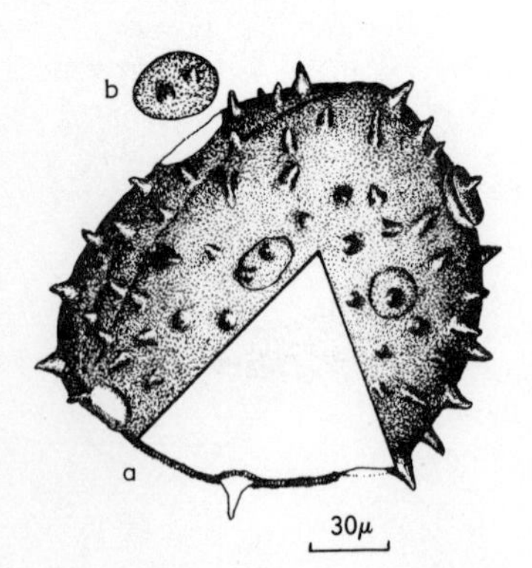

Fig. 399. *Cucurbita foetidissima* HBK. Coll. Kansas; CUCURBITACEAE.

Missouri Gourd

SIZE: About 160μ diameter

RANGE: Missouri and Nebraska to Texas, Mexico and California

Exine relatively thin, apparently intectate. Pores large, approximately 15-20 in number, operculate, with 1-2 spines on each operculum (b). Spines up to 12-15μ long. Similar in basic structure to other species of *Cucurbita*, *Cucumis* and *Cayaponia*. *Echinocystis* has stephanocolporate pollen; other cucurbits may be stephanocolpate.

5b Echinae broad-based and pointed (Fig. 400).. ***Malvastrum subtrifolium***

False Mallow

SIZE: Diameter 60-65μ

Four-five pores, perhaps stephanoporate in some specimens. Pores with thickened annulus at margin of operculum. Exine thick, with distinct columellae and complete tectum.

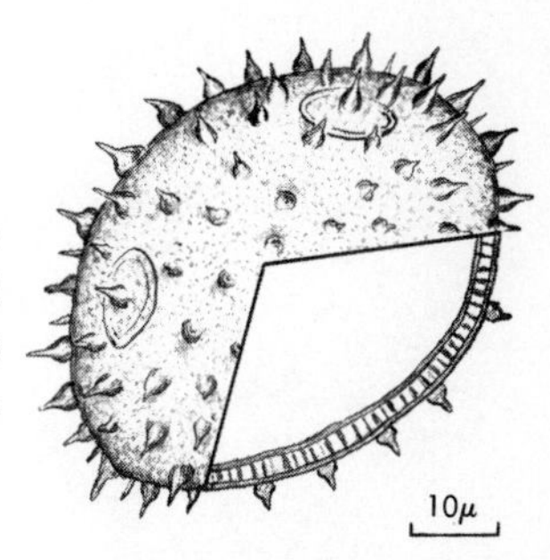

Fig. 400. *Malvastrum subtrifolium* Coll. Michigan; MALVACEAE.

6a Spines broad-based and blunt, exine very thick (Fig. 401).. ***Polygonum longistylum*** **Small.**

Long-styled Smartweed

SIZE: About 60μ diameter, spheroidal

RANGE: New Mexico to Louisiana north to southern Illinois and Kansas

Exine very thick (>7μ) with high ridges of exine (muri) surrounding numerous small lacunae. Muri appear to bear short broad-based spines. Pore number approximately 20, located in some of the lacunae. *Polygonum hydropiperoides* is also periporate but the exinous ridges lack coarse supratectate processes; 57μ in diameter. In contrast, *Polygonum amphilicum* is pericolpate (60μ, about 30 short furrows) and *P. californicum* is tricolporate with a transverse furrow, prolate (about 27 x 16μ). This genus and the family Polygonaceae is obviously eurypalynous.

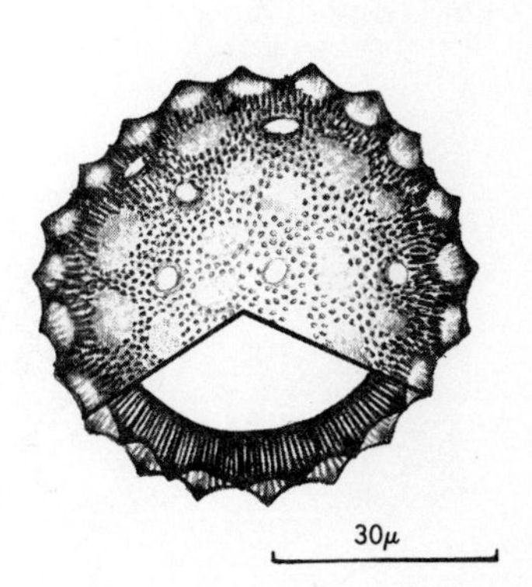

Fig. 401. *Polygonum longistylum* Small. Coll. Oklahoma; POLYGONACEAE.

6b Spines long, attenuate and bottle-shaped, exine thin (Fig. 402)..... ....*Hibiscus trionum* L.

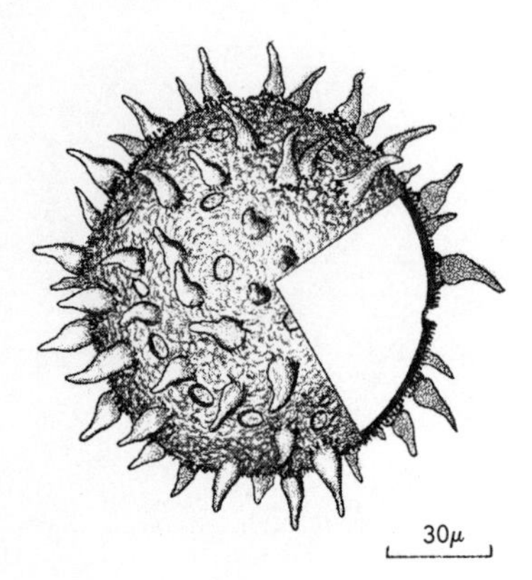

Fig. 402. *Hibiscus trionum* L. Coll. Michigan; MALVACEAE.

Flower-of-an-hour

SIZE: About 115μ diameter

RANGE: Widespread in eastern United States

Large, and spherical pollen with >30 pores. Exine surface closely beset with short baculae, and very large bottle-shaped spines 15-20μ long; each spine rests on a pedestal of tectum subtended by the baculate layer.

7a Shape subtriangular, pores indistinct (one terminal, 2-4 on "girdle") (Fig. 403)..Cyperaceae

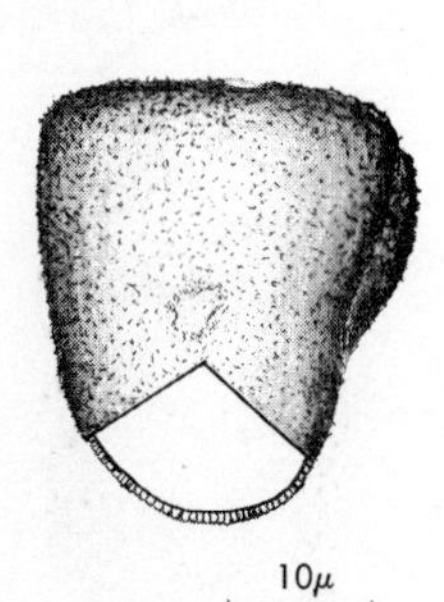

Fig. 403. *Scirpus cyperinus* (L.) Kunth. Coll. New York; CYPERACEAE.

Wool Grass or Bulrush

SIZE: Length 28-30μ, girth 24-26μ

RANGE: North Carolina to Oklahoma north to Iowa, Ohio and New England

Pollen of the Cyperaceae is very difficult (impossible?) to separate at the generic or specific level. The pollen of most North American genera is sub-triangular in shape with a pore at the broad end and 3 more-or-less obscure pores around the periphery, often about ⅓ of the distance from the broad end. The exine appears to be intectate verrucate. See also Fig. 119, *Eleocharis obtusa* which is keyed in the inaperturate section.

7b Shape spherical or oblate..8

8a Pollen grains heteropolar, pores on equator and on one hemisphere (Fig. 404, 405)..*Juglans* spp.

Butternut

SIZE: Equatorial diameter about 36μ (30-40μ)

RANGE: Eastern Canada to North Dakota, south to Arkansas and Georgia

Pollen grains heteropolar (pores on equator and one hemisphere); average pore number about 8, but varying from 2-15 (very exceptionally less than 5 or more than 10). Pores aspidate. According to Whitehead (1965), 10-35% of pollen of butternut has pores restricted to the equator (stephanoporate).

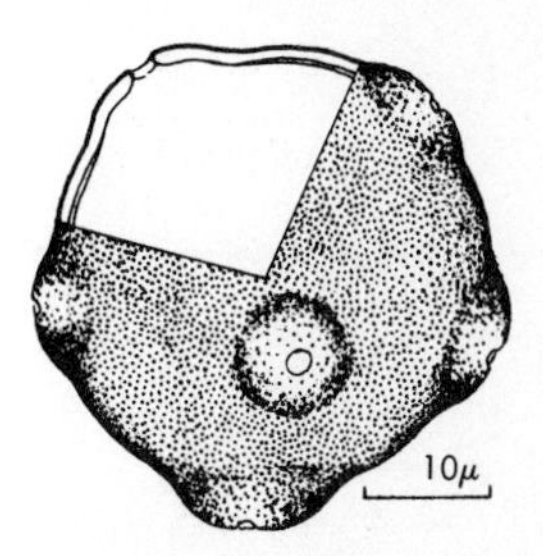

Fig. 404. *Juglans cinerea* L. Coll. Pennsylvania; JUGLANDACEAE.

Black Walnut

SIZE: Equatorial diameter about 35μ (30-40μ)

RANGE: Massachusetts to Minnesota, south to Florida and Texas.

Pollen grains usually periporate and heteropolar (pores on equator and one hemisphere); average pore number about 17, but varying from 9-37 (very exceptionally less than 13 or more than 22). Pores aspidate.

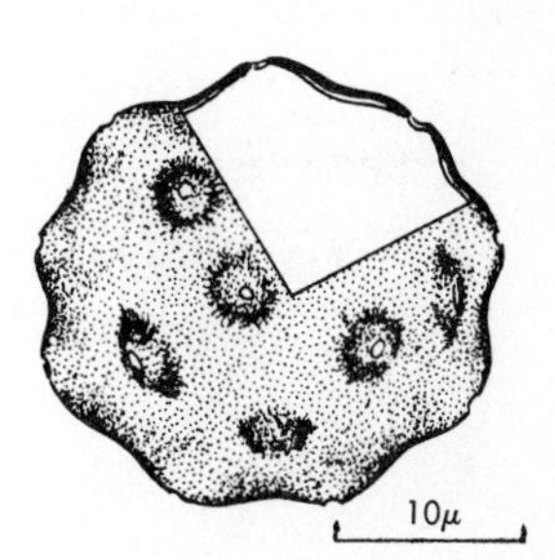

Fig. 405. *Juglans nigra* L. Coll. Michigan; JUGLANDACEAE.

8b Pollen grains isopolar..9

9a Size less than 20μ (Fig. 406) . . . . ***Thalictrum***

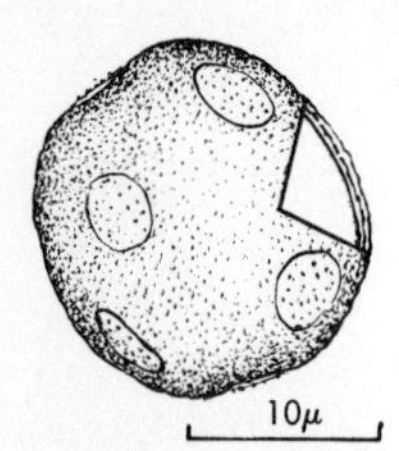

Fig. 406. *Thalictrum dioicum* L. Coll. Michigan; RANUNCULACEAE.

Early Meadow-Rue

SIZE: About 18μ diameter

RANGE: Georgia to Alabama north to North Dakota, southwestern Quebec and Maine

About 10 large pores, about 3.5-4μ diameter, pore membranes (opercula?) present, coarsely verrucate. In addition to *Thalictrum* several other genera in the Ranunculaceae have polyporate pollen (e.g. *Anemonella thalictroides*, *Coptis*).

9b Size greater than 20μ **10**

10a Diameter 70μ or larger (Fig. 407) . . . . ***Convolvulus sepium* L.**

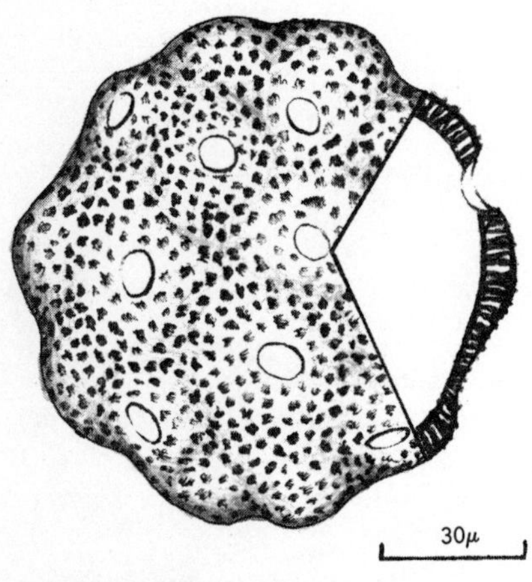

Fig. 407. *Convolvulus sepium* L. Coll. Michigan; CONVOLVULACEAE.

Hedge Bindweed

SIZE: 90-100μ diameter

RANGE: The several varieties range from Newfoundland to British Columbia south to Florida, New Mexico and Oregon

About 25* pores; exine tectate. Surface undulating with pores centered in the protruding areas. The grain illustrated is characteristic of the *Ipomoea*-type of Convolvulaceous pollen. Erdtman (1952) describes 10 species of *Convolvulus* as tricolpate! *Cuscuta* pollen is 3-4 colpate. Note: certain Cactaceae pollen (e.g. *Opuntia fulgida*, *O. pulchella*, *O. acanthocarpa*, *O. echinocarpa*, *O. exaltata*) may key here; for details see Tsukada, 1964.

---

*The microscopist attempting to count the number of pores by focussing on several optical sections through an isopolar, periporate pollen grain usually underestimates the actual pore number. The pore numbers given in the periporate section are determined in this way and are most likely too low. McAndrews and Swanson (1967) have proposed and tested a mathematical model for determining pore number by measuring the distance between adjacent pores (A-A′, the chord, C) and the diameter (D) of a pollen grain (Fig. 408). From the C/D ratio the pore number may be determined. In the key, the calculated pore number (N) is given in parentheses.

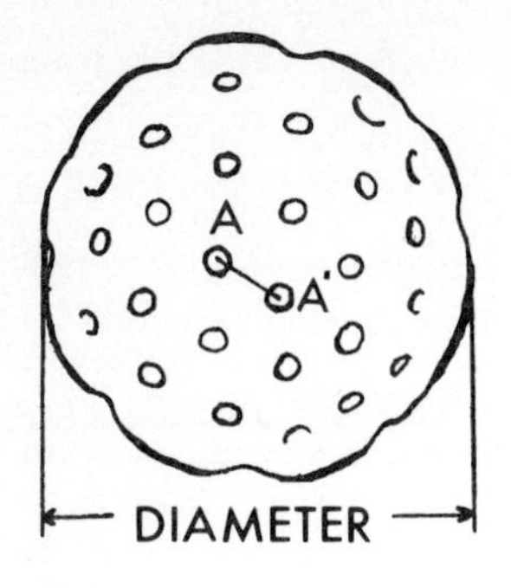

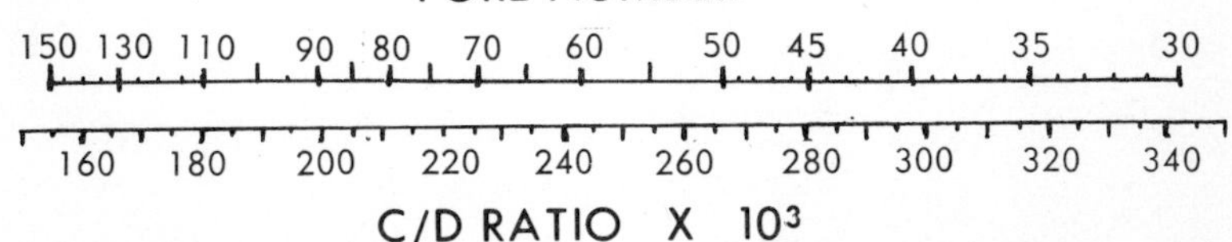

Fig. 408. Diagram of periporate (isopolar) pollen grain showing measurements for calculation of C/D ratio. A—A' is C, the chord, D is the diameter of the pollen grain. The scale permits determination of the pore number. Redrawn from: McAndrews, J. H. and A. R. Swanson. 1967. The pore number of periporate pollen with special references to *Chenopodium*. *Rev. Palaeobotany and Palynology*, Vol. 3:105-117 (Figs. 1 and 2).

10b Diameter mostly less than 50μ 11
11a Size 20-35μ 12
11b Size more than 35μ 16
12a Pores less than 20 in number 13
12b Pores usually more than 25 (Fig. 409, 410, 411) .... Amaranthaceae and Chenopodiaceae

Palynologists have been unable to separate most chenopod and amaranth pollen. Recently it has been shown by electron microscopy, using carbon surface replicas, that there is a different density of minexine spinules and a different ratio of minute holes to spinules per unit area (Fig. 473, 474). Pollen of the Amaranthaceae has more than 20 minute spinules per $4\mu^2$ and a ratio of holes to spinules of less than 1.5. Chenopod pollen has less than 20 spinules per $4\mu^2$ and a ratio (holes/spinules) greater than 1.6 (Tsukada, 1967).

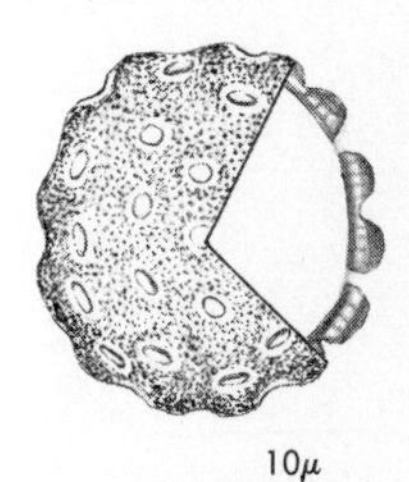

Fig. 409. *Atriplex powellii* Wats. Coll. Colorado; CHENOPODIACEAE.

Saltbush
SIZE: About 24-25μ diameter
RANGE: South Dakota and Montana south to New Mexico and Arizona

Periporate with about 40-45 pores. Exine thick, tectate; tectum thicker between pores.

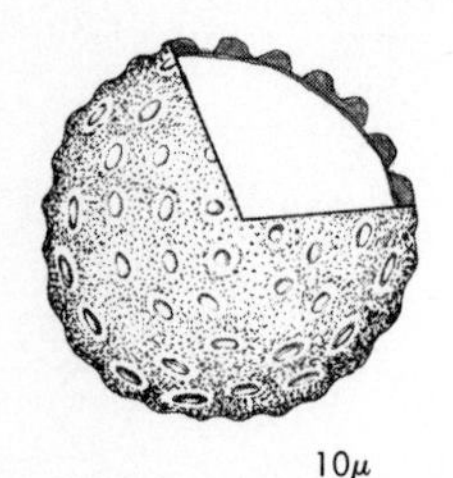

Fig. 410. *Chenopodium ambrosioides* L. Coll. Texas; CHENOPODIACEAE.

Goosefoot

SIZE: About 25μ diameter

RANGE: Widespread throughout most of United States; introduced from tropical America

Pores very numerous, approximately 70-75. Exine tectate with tectum thickened and bulging between the pores.

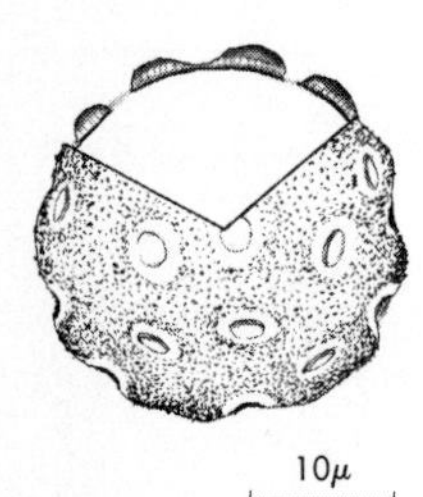

Fig. 411. *Salsola pestifer* A. Nels. Coll. Colorado; CHENOPODIACEAE.

Russian-thistle

SIZE: About 25μ diameter

RANGE: Extensively naturalized in the western United States, introduced from Eurasia

Pollen grains have about 39-40 pores.

13a Apertures annulate and aspidate, 14-16 in number (Fig. 412)..*Sarcobatus vermiculatus* (Hook) Torr.

Greasewood

SIZE: Diameter 23-26μ

RANGE: North Dakota to Alberta south to western Texas and California

Pores 14-16 in number, aspidate and annulate. Grain looks like *Juglans* on cursory examination but is isopolar, having pores evenly distributed over the spherical surface. Surface psilate; exine indistinctly tectate. *Sarcobatus* is one of the few genera in the Chenopodiaceae or Amaranthaceae which has distinctive pollen identifiable with the light microscope.

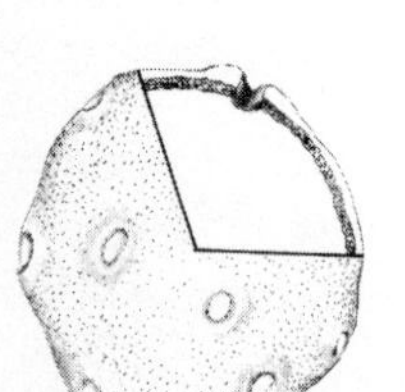

Fig. 412. *Sarcobatus vermiculatus* (Hook.) Torr. Coll. Idaho; CHENOPODIACEAE.

13b Apertures unthickened or annulate, but not aspidate...........14

14a Exine warty, very coarsely verrucate or gemmate (Fig. 413)...... ....*Proserpinaca palustris* L.

Mermaid Weed

SIZE: Diameter about 21μ

RANGE: The several varieties extend from Nova Scotia to Minnesota south to Texas and Florida

Exine thick (about 1.5μ) densely covered with coarse verrucate, gemmate, baculate and clavate elements. Mostly 5-pored (4-7 occasionally); pores either on equator or apparently periporate.

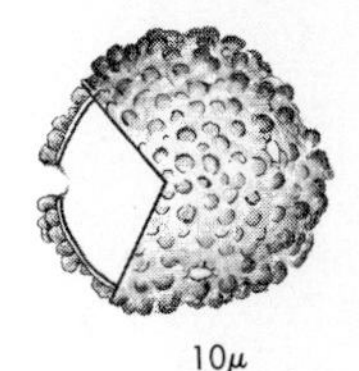

Fig. 413. *Proserpinaca palustris* L.; HALGORACEAE.

14b Exine psilate or verrucate..................................15

15a Pores 9-10, located on flattened or concave faces (Fig. 414)....*Ribes*

Wild Black Currant

SIZE: About 24-27μ diameter

RANGE: New Brunswick to Alberta south to Delaware, western Virginia, Missouri, Nebraska and New Mexico

Pores 9 or 10, located in flattened concave areas of the pollen grain. Outline of grain more or less polygonal. Exine thin, indistinctly tectate.

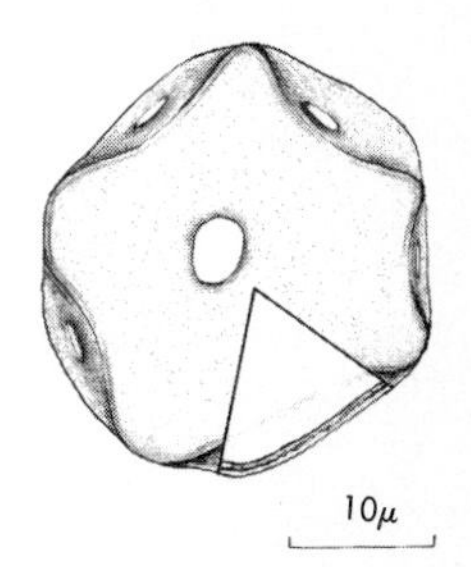

Fig. 414. *Ribes americanum* Mill. Coll. Michigan; SAXIFRAGACEAE.

15b Pores 4-11, located on surface of spheroidal grain (Fig. 415)...... ....*Plantago* spp.

Buckhorn

SIZE: Spheroidal about 27μ diameter

RANGE: Widespread in United States; naturalized from Europe

Pollen grains 6-10 (12) porate, with pores about 3μ diameter, annulate. Exine surface with scattered low verrucae or sometimes with a vaguely rugulate appearance. *Plantago coronopus* (4-7 porate) resembles *P. lanceolata* in having annulate pores. *P. major*, and *P. media* have poorly delimited pores with the margins unthickened.

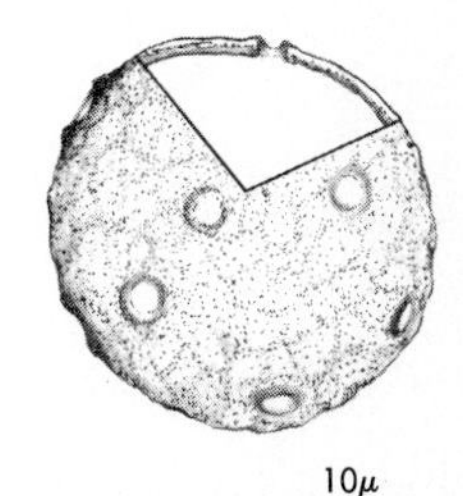

Fig. 415. *Plantago lanceolata* L. Coll. Michigan; PLANTAGINACEAE.

16a Pores very numerous (more than 50) and small (Fig. 416)..*Tribulus terrestris* L.

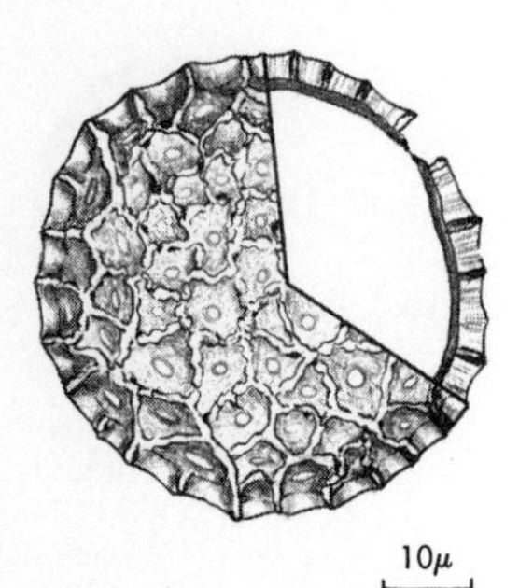

Fig. 416. *Tribulus terrestris* L. Coll. New Mexico; ZYGOPHYLLACEAE.

Caltrop

SIZE: Diameter about 45-48μ

RANGE: Extensively naturalized in the United States, especially on roadsides; native in southern Europe

Pollen grain with numerous (more than 50) small pores. Exine surface coarsely reticulate due to baculate and blunt-echinate elements which are more or less joined by a thin tectum to produce a per-reticulate to frustillate surface.

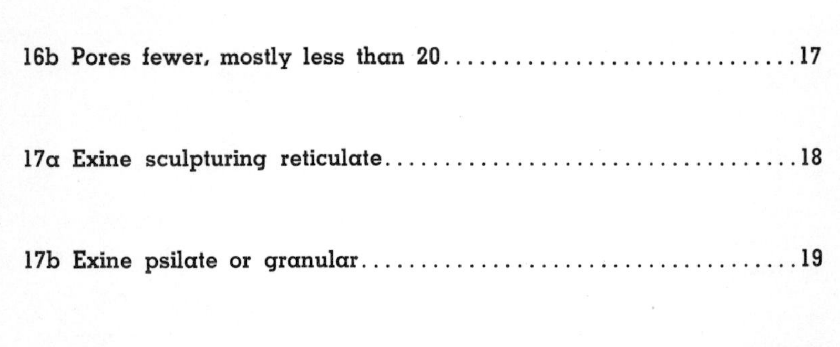

16b Pores fewer, mostly less than 20..............................17

17a Exine sculpturing reticulate..............................18

17b Exine psilate or granular..............................19

18a Reticulum very coarse, forming high muri (fenestrate?), with both large and small lacunae (Fig. 417)..............................*Phlox*

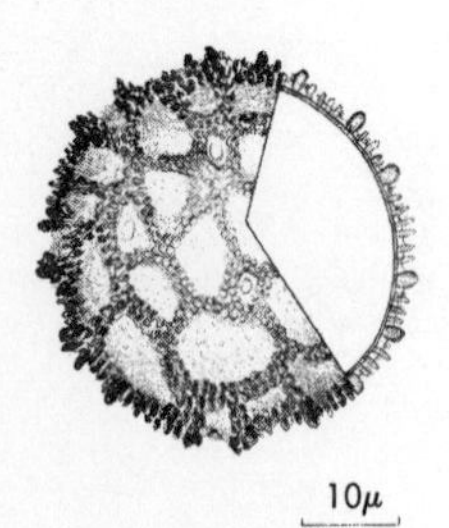

Fig. 417. *Phlox pilosa* L. Coll. Michigan; POLEMONIACEAE.

Hairy Phlox

SIZE: Diameter 38-42μ

RANGE: Florida to Texas north to eastern Kansas, southern Michigan and Connecticut

Pollen grain polyporate with (8-) 10-12 pores. Large baculae and clavae form muri (very coarse reticulate appearance). The "windows" (lacunae) are of two types: the larger are more or less hexagonal, the smaller are sub-pentagonal and each has a pore. *Phlox divaricata* pollen grains have virtually the same features, but are smaller (26-34μ diameter).

18b Reticulum fine, perforate tectate, lacking muri and lacunae (Fig. 418) .... ***Dianthus armeria* L.**

Deptford Pink

SIZE: Diameter about 38-41μ

RANGE: Southern Quebec and Ontario south to Georgia, Kentucky and Missouri, naturalized from Europe

Pollen grains with about 14 pores. Baculae and clavae form a coarse per-reticulate surface. Pore with a weak annulus. The pollen of the Caryophyllaceae may be distinguished from other polyporate types by a low number of pores in combination with rather thick tectate exine; pores often annulate or exine surface reticulate. The genus *Spergula* has colpate (tricolpate or pericolpate) pollen; *Corrigiola* pollen is triporate.

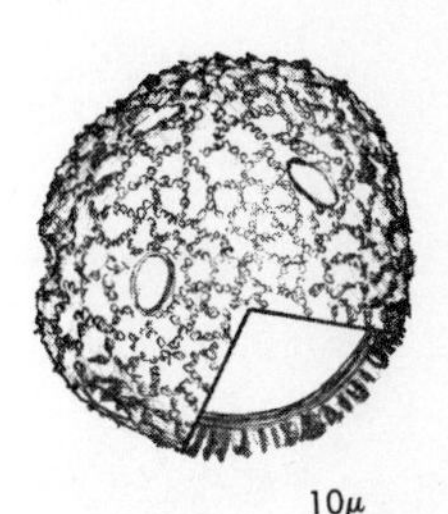

Fig. 418. *Dianthus armeria* L. Coll. Michigan; CARYOPHYLLACEAE.

19a Pores 7-8 in number (Fig. 419) .... ***Saponaria officinalis***

Bouncing-Bet

SIZE: Diameter 48-51μ

RANGE: Nova Scotia to Florida westward to New Mexico and Colorado

Grain with 7-8 very large (7μ diameter) pores. Exine clearly tectate with coarse, stout columellae. Pore membranes (opercula?) persistent, typically with a few large baculae or clavae projecting from them. In addition to *Dianthus* and *Saponaria*, polyporate genera of the Caryophyllaceae include: *Scleranthus, Sagina, Stellaria, Lychnis, Silene* and *Cypsophila*. Size of grain and number of pores have been used to separate pollen of the European genera; such study should be attempted among North American caryophylls.

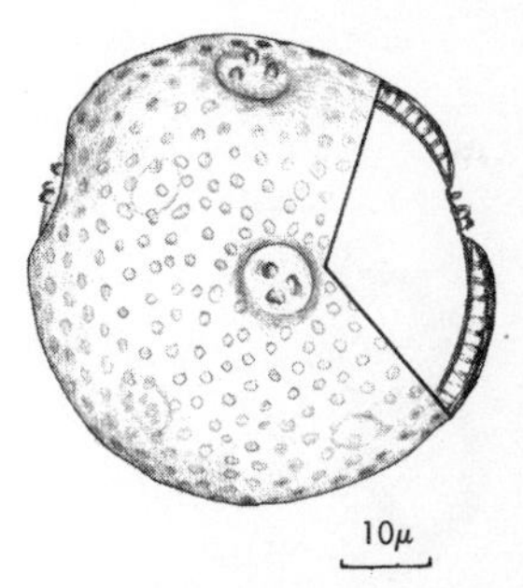

Fig. 419. *Saponaria officinalis* L. Coll. Michigan; CARYOPHYLLACEAE.

19b Pores 12-20 in number (Fig. 420) .... ***Liquidambar***

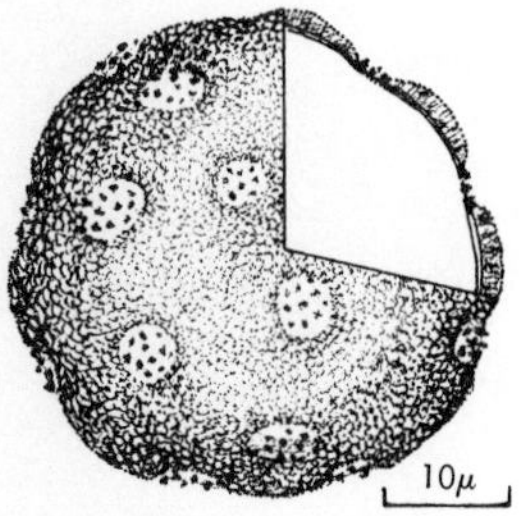

Sweet Gum

SIZE: Diameter 34-38μ

RANGE: Florida to Texas north to southeastern Missouri, southern Illinois, eastward to southern Connecticut

Grain with 12-20 circular pores. Pore membranes persistent with conspicuous verrucae. *Hamamelis* pollen is very different (tricolpate and about 19μ).

Fig. 420. *Liquidambar styraciflua* L.; HAMAMELIDACEAE.

DYAD TYPES

The monocot *Scheuchzeria palustris* var. *americana* is distinctive among the flowering plants in having inaperturate, finely reticulate pollen grains united in dyads. (It is not illustrated here.) Several types of fungal spores produce dyads; three of the commonest are keyed below.

1a Dyad straight **2**

1b Dyad curved and fusiform .... ***Sticta* sp.**

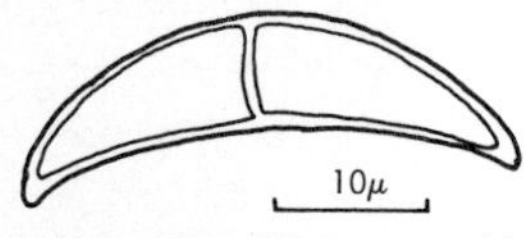

Fig. 421. *Sticta* sp.; LICHEN (ascospore).

SIZE: 7 x 30μ

Dyad crescent-shaped, each cell gradually tapered.

2a Wall uniformly thickened, not stalked .... ***Pullularia* sp.**

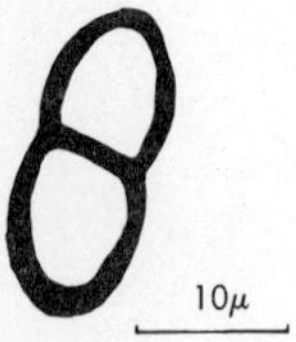

Black Yeast

SIZE: 8 x 19μ

Wall uniformly thickened; dark colored.

Fig. 422. *Pullularia* sp.; Moniliales (Deuteromycete fungus, Fungi Imperfecti), conidia.

2b Wall thickest at distal end, with a proximal stalk..................
.......................................*Puccinia graminis* Pers.

Wheat Rust

SIZE: 19 x 60μ

RANGE: Nearly ubiquitous in North America

A portion of the stalk of the teliospore is always in evidence. The terminal cell is usually broader than the basal cell.

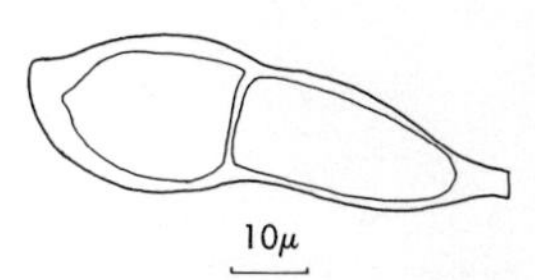

Fig. 423. *Puccinia graminis* Pers.; PUCCINIACEAE (teliospore).

TETRAD TYPES

In addition to the plants which regularly shed pollen as tetrads, one should expect to find occasional tetrads of most other pollen and spore types. Since both spores and pollen grains are produced in groups of four by meiosis, immature or aberrant tetrads, rather than the usual monads, may be encountered.

1a Tetrad with tetrahedral shape, one of the four grains often obscured by the other three in cursory examination.......................2

1b Tetrad with tetragonal or irregular shape, all four grains visible in same plane..11

2a Grains inaperturate, or with indistinct or obscure apertures, the numerous pores near the lines of attachment in the tetrad........3

2b Grains tricolporate, although the pore and aperture number may be difficult to discern (Ericaceae, Empetraceae, and Pyrolaceae)...6

3a Pollen grains with echinate and/or clavate ektexine sculpture....4

3b Pollen grains with rugulate or coarsely verrucate sculpture.......5

4a Exine with large blunt clavate warts and smaller spines (Fig. 424)..
.......................................*Dionaea muscipula* Ellis

Venus Flytrap

SIZE: Tetrad: >110μ diameter. Individual grains 70-80μ equatorial diameter

RANGE: North Carolina

The grains have 7-12 more or less elongate poroid, and somewhat obscured, apertures just below the equatorial rim. The apertures can be seen most readily in loose tetrads. The exine is ornamented with numerous dense baculae or spines and a few large, irregular, projecting baculae.

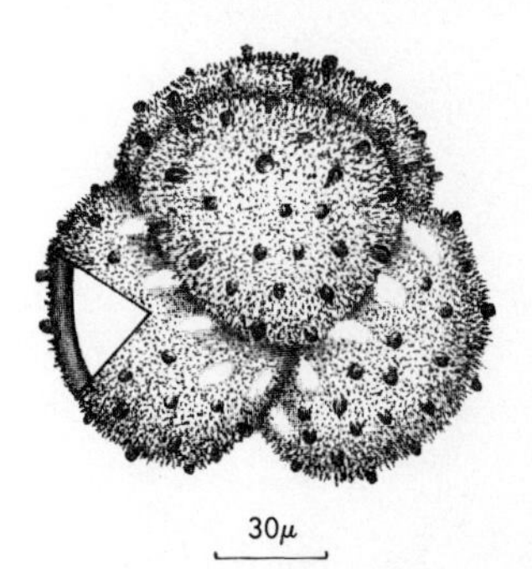

Fig. 424. *Dionaea muscipula* Ellis Coll. North Carolina; DROSERACEAE.

4b Exine with scattered large projecting spines and smaller spinules or clavae between them (Fig. 425b).......... *Drosera rotundifolia* L.

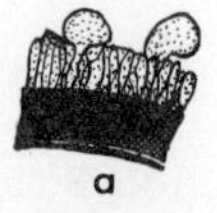

Fig. 425. a. Wall section of *Dionaea muscipula* Ellis; b. *Drosera rotundifolia* L. Redrawn from: Chanda, S. 1965. The pollen morphology of Droseraceae with special reference to taxonomy. *Pollen et Spores,* Vol. VII(3): 509-528 (Fig. 1).

Round-leaved Sundew

SIZE: Tetrad diameter about 60μ, individual grains about 38μ

RANGE: Widespread in United States and Canada on peaty or moist acid soils. Circumpolar

Tetrad similar to *Dionaea,* with about 10 apertures on proximal faces of each grain partially obscured by other members of the tetrad. Large pointed or blunt macro-spinules have small (about 0.5μ) microechinae and clavae between them (Fig. 425b wall section of *Drosera rotundifolia* L.).

5a Tetrad over 100μ in diameter (Fig. 426)... *Asimina triloba* (L.) Dunal.

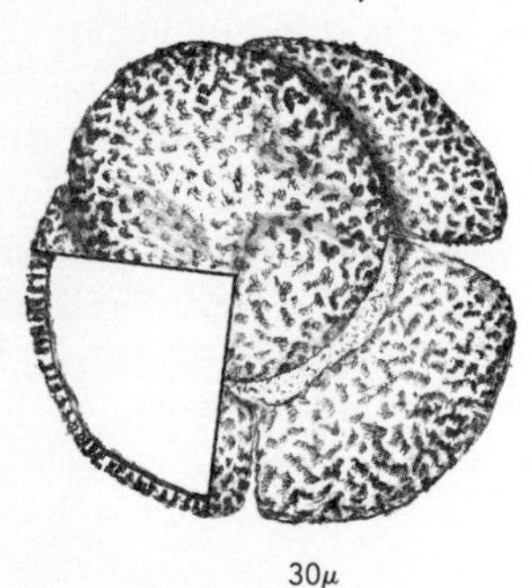

Fig. 426. *Asimina triloba* (L.) Dunal. Coll. Illinois; ANNONACEAE.

Common Paw-paw

SIZE: Tetrad diameter 133-155μ; individual grain 80 x 93μ

RANGE: Southern New York to Florida and west to Nebraska and Texas

Tetrads loose and monads sometimes encountered, indistinctly monocolpate. Exine coarsely rugulate on the distal hemisphere, with a distinct line of demarcation from the psilate or scabrate-granular proximal faces.

5b Tetrad smaller, less than 50μ (Fig. 427)... *Luzula campestris* (L.) DC.

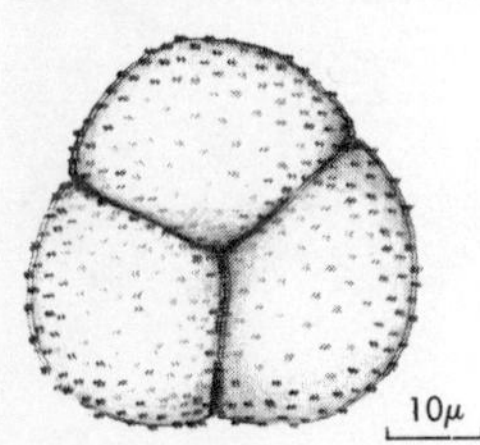

Fig. 427. *Luzula campestris* (L.) DC.; JUNCACEAE.

Woodrush

SIZE: Tetrad diameter 30-45μ

RANGE: European, locally established in eastern Canada and New England

Tetrads tetrahedral. Exine very thin, apparently possessing a thin "poroid" area at the distal pole (very obscure!).

6a Tetrads large, greater than 60μ diameter........................7

6b Tetrads smaller, generally less than 50μ diameter...............8

7a Furrows long and polar area small, polar area index less than 0.3 (Fig. 428)................................*Arbutus menziesii* Pursh.

Arbutus

SIZE: Tetrads 76μ diameter

RANGE: California

Tetrad sub-triangular. Individual grains tricolporate with distinct transverse furrows; exine psilate. The Ericaceae is a stenopalynous family; a complete analysis would reveal additional genera which would be practically indistinguishable from those enumerated in this key.

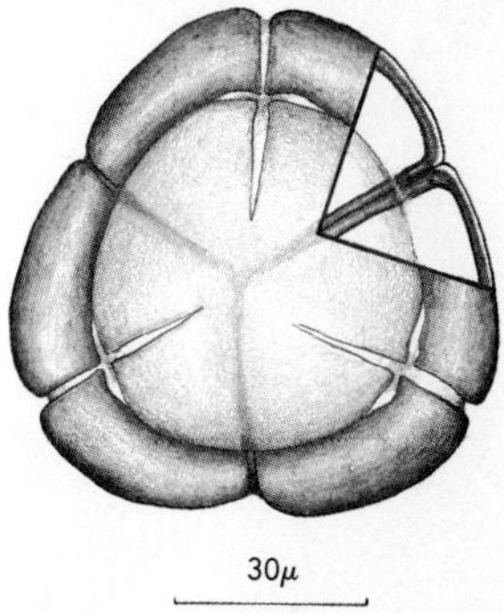

30μ

Fig. 428. *Arbutus menziesii* Pursh Coll. California; ERICACEAE.

7b Furrows shorter and polar area longer, polar area index greater than 0.5 (Fig. 429)....................*Rhododendron maximum* L.

Rosebay Rhododendron

SIZE: Tetrad 72μ diameter; individual grains 40-45μ diameter

RANGE: Central eastern United States

Tricolporate with distinct slender transverse furrows. Exine psilate.

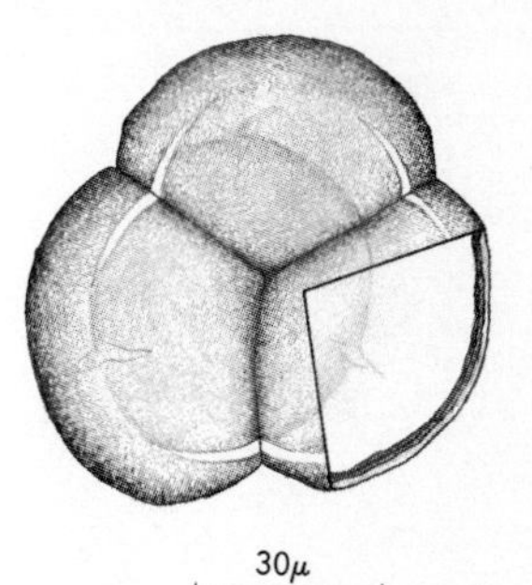

Fig. 429. *Rhododendron maximum* L. Coll. Duke Univ. Herb.; ERICACEAE.

8a Ektexine with distinct columellae.............................9

8b Ektexine lacking distinct columellae..........................10

9a Surface psilate, but coarsely "dotted" or granular due to thick columellae; transverse furrow present and distinct (Fig. 430).......... ....*Lyonia mariana* (L.) D. Don.

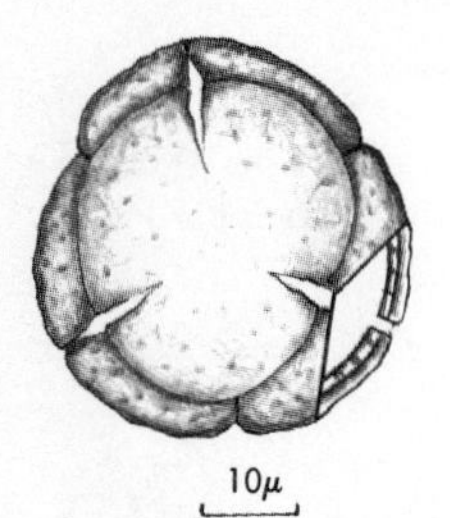

Fig. 430. *Lyonia mariana* (L.) D.Don. Coll. Univ. Mich. Herb.; ERICACEAE.

Stagger-bush

SIZE: Tetrad about 38μ diameter

RANGE: Southeastern United States north to New England

Tetrad rather spherical in outline, angles between the individual grains indistinct. Exine psilate and indistinctly tectate; the stout columellae give the wall a scattered dotted appearance.

9b Surface slightly verrucate; pores round or only slightly transversely elongate (Fig. 431)..........................*Pyrola rotundifolia* L.

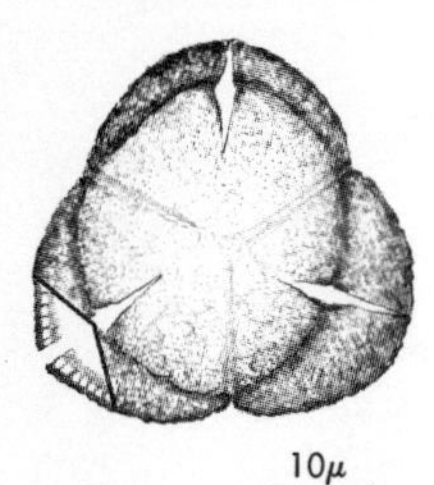

Fig. 431. *Pyrola rotundifolia* L. Coll. Michigan; PYROLACEAE.

Shinleaf

SIZE: Tetrad 33-38μ diameter

RANGE: Eastern Canada and the United States

Tricol(por)ate, pores indistinct and possibly slightly elongated into short transverse furrows. Exine tectate with distinct columellae; surface slightly verrucate. Other members of the Pyrolaceae have a tendency toward lose tetrads (*Chimaphila umbellata*) or free pollen grains (*Monotropa*).

10a Furrows short and polar area large, polar area index ≃ 0.55 (Fig. 432).................................*Epigaea repens* L.

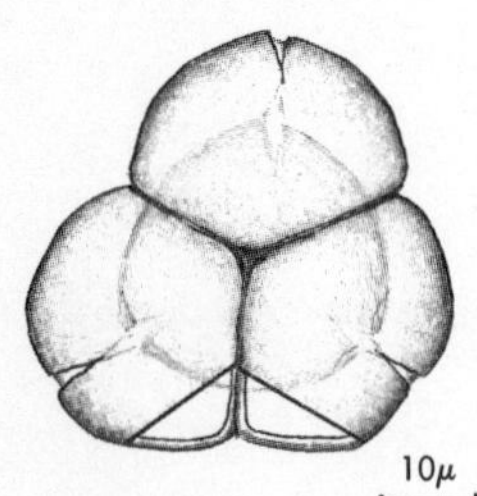

Fig. 432. *Epigaea repens* L. Coll. New Hampshire; ERICACEAE.

Trailing Arbutus

SIZE: Tetrad about 39μ diameter

RANGE: Eastern United States and Southeastern Canada

Tricolporate with short transverse furrows, exine psilate and apparently intectate. Tetrad subtriangular in outline.

10b Furrows longer and polar area reduced, polar area index $\simeq$ 0.3 (Fig. 433) . . . . ***Vaccinium corymbosum* L.**

Highbush Blueberry

SIZE: Tetrad 42μ diameter

RANGE: Southeastern Canada and eastern United States

Tricolporate pollen grains with somewhat obscure transverse furrows. Exine surface psilate, apparently intectate. Tetrads sometimes appear spherical, not as strongly lobed as some Ericaceae.

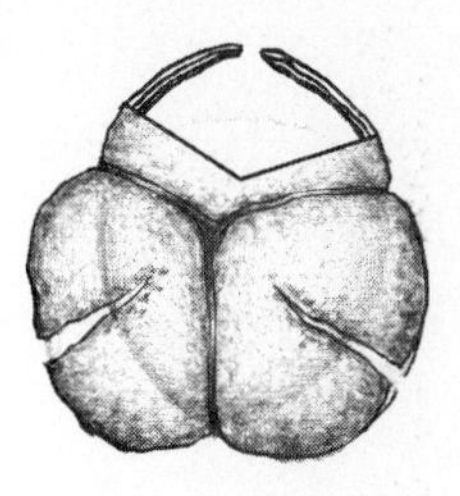

Fig. 433. *Vaccinium corymbosum* L. Coll. Michigan; ERICACEAE.

11a Pollen inaperturate **12**

11b Pollen grains with pores **14**

12a Tetrad, and its member grains, irregularly angular or trapezoidal in shape (Fig. 434) . . . . ***Calopogon pulchellus* (Salisb.) R. Br.**

Grass Pink

SIZE: Tetrads variable in size, commonly about 45μ x 60μ

RANGE: Eastern United States, Quebec and Newfoundland

The loose pollinia are composed of strongly coherent tetrads (occasionally triads). Tetrads and individual grains are irregularly polygonal or trapezoidal. Surface has coarse rugulate folds and verrucate patches.

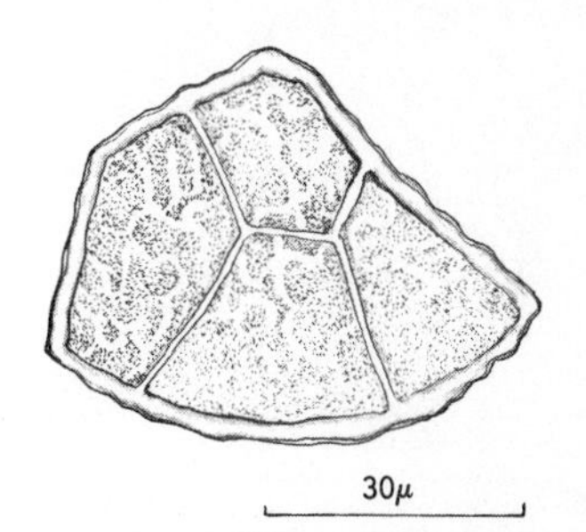

Fig. 434. *Calopogon pulchellus* (Salisb.) R. Br. Coll. Michigan; ORCHIDACEAE.

12b Tetrad, and its member grains, more or less square............13

13a Exine psilate (Fig. 435)............................*Mimosa* spp.

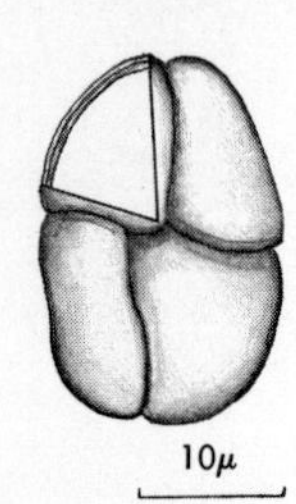

Fig. 435. *Mimosa biuncifera* Benth. Coll. Texas; LEGUMINOSAE.

Sensitive Plant

SIZE: Tetrad 17μ x 23.5μ

RANGE: Western Texas to Arizona

Exine is psilate and apparently intectate. Other North American species of *Mimosa* are apparently very similar; *M. lemmonii* Gray has been observed to be virtually indistinguishable.

13b Exine faintly reticulate (Fig. 436)................*Schrankia* spp.

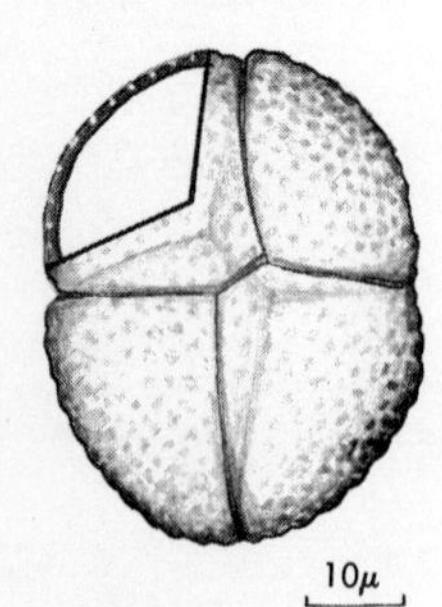

Fig. 436. *Schrankia occidentalis* (W & S.) Standl. Coll. Texas; LEGUMINOSAE.

Sensitive Brier

SIZE: Tetrad about 40 x 50μ

Tetrad quadrate to ovate. Exine intectate or perforate tectate; surface with abundant broad gemmae or coarse verrucae, surface appearance faintly reticulate.

14a Exine reticulate, a single, indistinct distal pore present (Fig. 437) .... *Typha latifolia* L.

Common Cat-tail

SIZE: Tetrad about 45μ broad, individual grains about 20 x 25μ

RANGE: Throughout the United States and southern Canada to Alaska

The tetrad is normally arranged as shown, but may occasionally be linear, rhomboidal, or of irregular shape. The exine is reticulate and bears a single pore which may be indistinct. The pore may be variously positioned on the grains, not necessarily at the distal pole. In many cases all four pores are on the same face of the tetrad. The tetrads sometimes separate into dyads or monads. When single, the pollen of *Typha latifolia* cannot be distinguished from *T. angustifolia* and species of *Sparganium*.

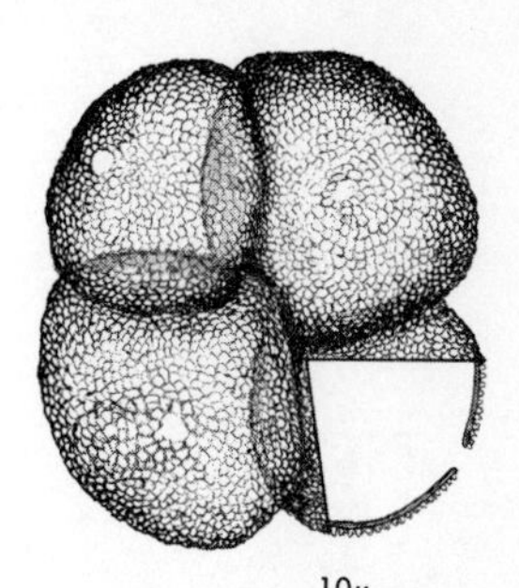

Fig. 437. *Typha latifolia* L. Coll. Michigan; TYPHACEAE.

14b Exine psilate; triporate (Fig. 438)....... *Apocynum cannabinum* L.

Indian Hemp

SIZE: Tetrad 24-26μ diameter, individual grains about 13μ

RANGE: Southern Canada and Washington, New England, eastern and gulf coastal states and southwest

Tetrad more or less spherical to quadrate. Exine intectate and psilate. Apertures are 3-4 irregular-sized pores, these slightly annulate.

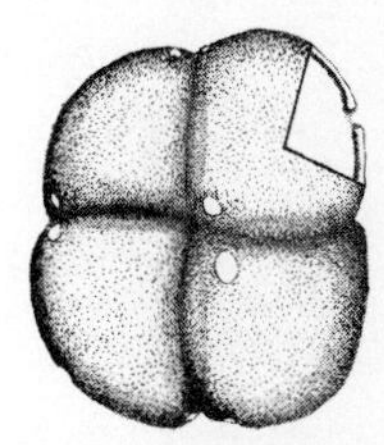

Fig. 438. *Apocynum cannabinum* L. Coll. Michigan; APOCYNACEAE.

POLYAD TYPES

The colonial alga *Pediastrum* is keyed with this pollen type because it is sometimes encountered in polleniferous sediments and could be confused with palynomorphs.

1a Cell mass circular or spherical 2

1b Cell mass elongate, sometimes pointed or twisted 5

2a Shape spherical or slightly discoidal 3

2b Shape a regular disc, one (or occasionally two) cells thick (Fig. 439, 440) . . . . ***Pediastrum* spp.**

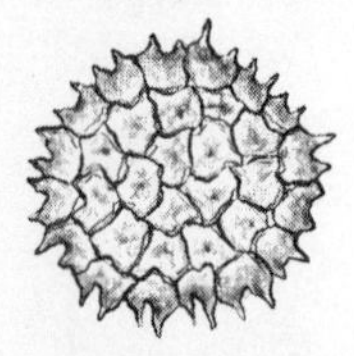

Fig. 439. *Pediastrum* sp. Coll: Tertiary of California; CHLOROPHYTA, HYDRODICTYACEAE. Redrawn from: Wilson, L. R. and W. S. Hoffmeister. 1953. Four new species of fossil *Pediastrum*. *Amer. Jour. Sci.*, Vol. 251:753-760 (Fig. 1).

Genus *Pediastrum*

SIZE: Colonies are usually 100-150μ in diameter

RANGE: Widespread in fresh-water lakes, pools and sluggish streams

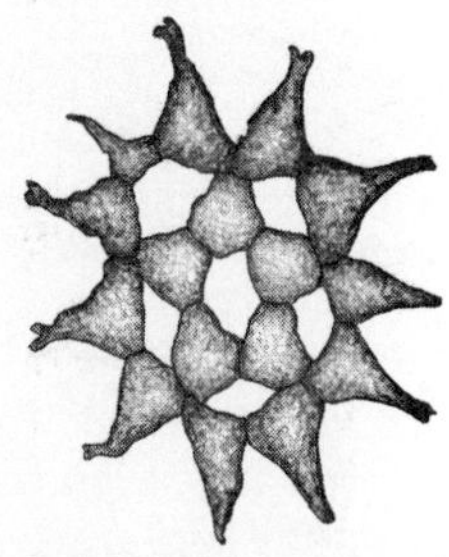

Fig. 440. *Pediastrum kajartes* Wilson and Hoffmeister. Coll: Cretaceous of Sumatra; HYDRODICTYACEAE, Redrawn from: Evitt, W. R. 1963. Occurrence of freshwater alga *Pediastrum* in Cretaceous marine sediments. *Amer. Jour. Sci.*, Vol. 261:890-893 (Fig. 4).

These colonial green algae produce multicellular colonies of 2, 4, 8, 16, 32, 64 or 128 polygonal cells. Most of the marginal cells have forked or double projections. Some fossil species (Fig. 440) have loosely arranged cells. Two fossil specimens are illustrated. For details see: How to Know Freshwater Algae, G. W. Prescott.

3a Clump of cells small, about 15μ (Fig. 441) . . . . ***Epicoccum* sp.**

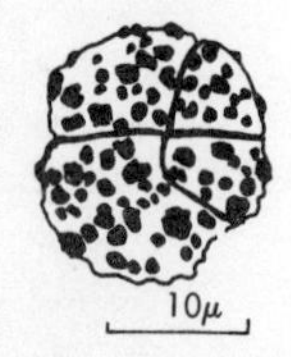

Fig. 441. *Epicoccum* sp.; Moniliales (Deuteromycete fungus, Fungi Imperfecti), conidia.

SIZE: Entire mass about 15μ

Multicellular conidium, usually 5 or more cells. Inaperturate, but may show attachment scar. Surface verrucate, the warts uneven in dimension. Mass more-or-less spheroidal; dark-colored when fresh.

3b Clump of cells large, usually greater than 50μ..................4

4a Component pollen grains quadrangular or pentangular, usually 16 in number; psilate (Fig. 442)........................*Acacia* spp.

Acacia

SIZE: Entire polyad 50μ diameter

RANGE: Texas to Nevada, southeastern California and northern Mexico

Polyad of 16 cells round and discoidal; two (or more) cells thick at middle. Most species of *Acacia* have 16 cells and are very similar to *A. greggii* (see list below); however, two species (*A. farinosiana* and *A. carenia*) with 48-celled polyads are in my collection. *Albizzia* appears to have the same polyad pattern as *Acacia*, but the component cells are more loosely attached. North American species of *Acacia* with 16 celled polyads include: *A. wrightii*, *A. vernicosa*, *A. roemeriana*, *A. farensiana*, *A. constricta*, *A. berlanderi*, *A. amentacea*, *A. pennatula*.

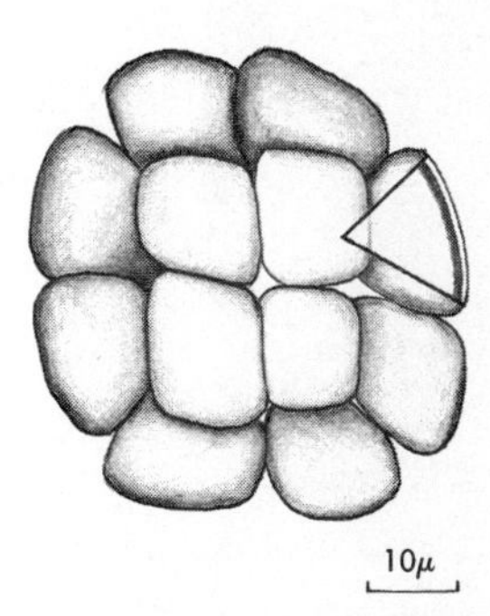

Fig. 442. *Acacia greggii* A. Gray Coll. Texas; LEGUMINOSAE.

4b Component pollen grains spherical; wall with coarse ridges forming a reticulum (Fig. 443).........*Proboscidea louisianica* (Mill.) Thell.

Proboscis Flower

SIZE: Separate pollen grains 20μ diameter

RANGE: Southwestern U. S. and southern plains

Pollen grains adhere in large spherical or circular masses approximately 100μ in diameter. Individual pollen grains very coarsely reticulate, the muri separated by areas of thin, psilate exine.

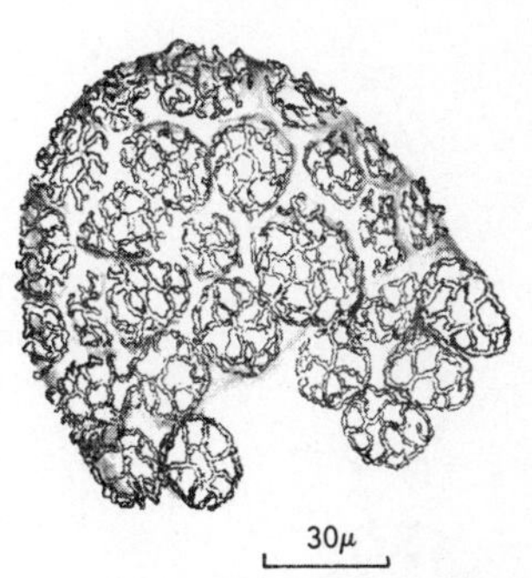

Fig. 443. *Proboscidea louisianica* (Mill.) Thell. Coll. Michigan; MARTYNIACEAE.

5a Size (maximum dimension) usually 500-1500μ (true pollinia, the adherent contents of an entire pollen sac may be present)........6

5b Size (maximum dimension) usually less than 100μ...............7

6a Individual pollen grains pentagonal, pollinium tear-drop shaped (Fig. 444)..*Asclepias* spp.

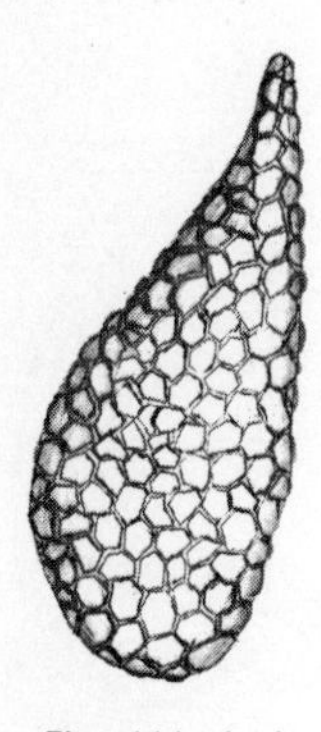

Fig. 444. *Asclepias syriaca* L. Coll. Michigan; ASCLEPIADACEAE.

Common Milkweed

SIZE: Pollinium 590 x 1350μ; individual pollen grains 55-65μ diameter

RANGE: Southern Canada and eastern United States

The pollinia are extracted from the pollen sac of the flower in pairs in most species of *Asclepias.* The individual pollinium of this species is estimated to contain in excess of 250 pollen grains. The individual grains are quadrate to hexagonal in shape. Apertures appear to be lacking in this species, although Erdtman illustrates *A. purpurascens* with grains united in linear tetrads, each bearing three pores.

6b Individual pollen grains quadrate or irregularly angular, sometimes clearly grouped in tetrads within the pollinium....................... (Fig. 445)..................*Habenaria blephariglottis* (Willd.) Hook.

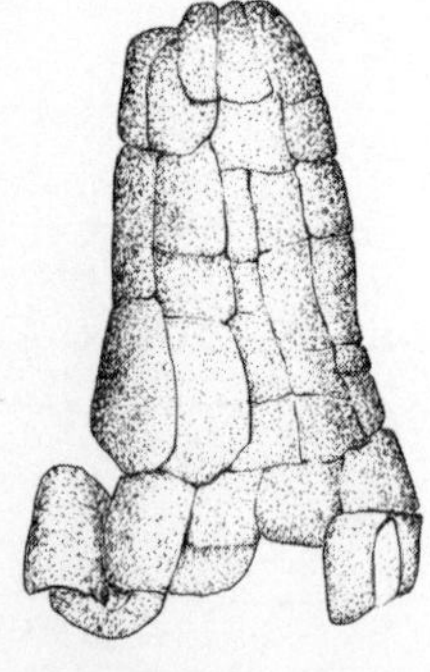

Fig. 445. *Habenaria blephariglottis* (Willd.) Hook. Coll. Michigan; ORCHIDACEAE.

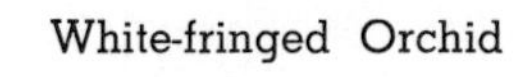

White-fringed Orchid

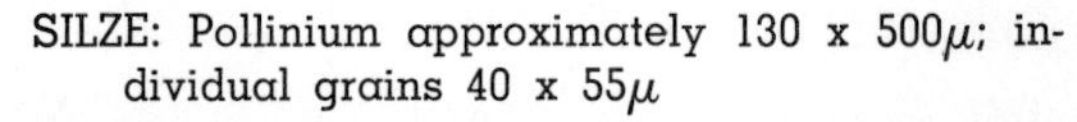

SILZE: Pollinium approximately 130 x 500μ; individual grains 40 x 55μ

RANGE: Northeastern U. S. and southeastern Canada

The individual pollen grains are more or less rectangular. The grains are obscurely monocolpate. The exine has a fine reticulum (oil!). Some species of orchids produce pollen in distinct tetrads within pollinia; some pollinia disintegrate when subjected to acetolysis (see Fig. 434 *Calopogon pulchellus*).

7a Multiseriate, more than 2-cells wide...............................8

7b Uniseriate or biseriate..9

8a Surface verrucate, 6 to 16 cells (Fig. 446)..........*Stemphylium* sp.

SIZE: 17 x 26μ

RANGE: Common parasite of vascular plants.

Multicellular (6-16 cells), inaperturate. Exine surface scabrate to verrucate; brown when fresh. Similar to *Alternaria* (Fig. 448), but never in chains and not "beaked."

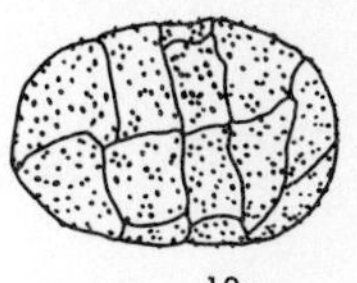

Fig. 446. *Stemphylium* sp.; Moniliales (Deuteromycete fungus, Fungi Imperfecti), conidia.

8b Surface psilate, about 32 cells (Fig. 447)............*Pleospora* sp.

SIZE: 13 x 36μ

RANGE: Worldwide

The individual ascospores are inaperturate, clumps of about 32 multicells. Surface psilate; brown in color when fresh.

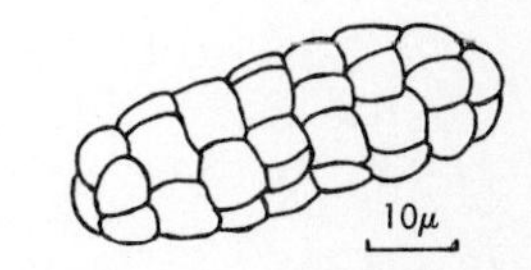

Fig. 447. *Pleospora* sp.; PLEOSPORACEAE (ascospore).

9a Biseriate or uniseriate, minutely verrucate (Fig. 448)..*Alternaria* sp.

SIZE: 20 to 40μ long

RANGE: Widespread plant saprophyte or weak parasite

Conidia multicellular (2 to 10 or more cells); pore present at point of attachment. Surface rough, minutely verrucate. Conidia stalked and the terminal cell usually "beaked." Fresh spores brown. An important fungal cause of hayfever.

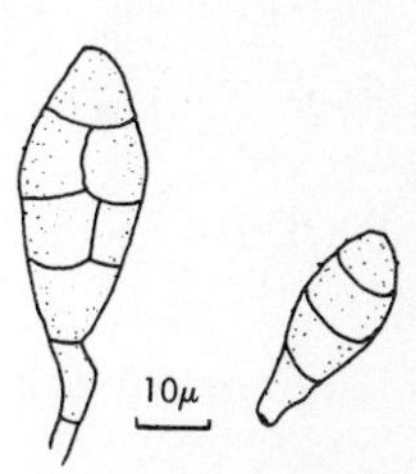

Fig. 448. *Alternaria* sp.; DEMATIACEAE (Deuteromycete fungus, Fungi Imperfecti), conidia.

9b Uniseriate, psilate..10

10a Chain of cells, usually 8 or 9 cells long, subterminal cells similar in diameter to others...11

10b Chain four to five cells, subterminal cell largest................. (Fig. 449).......................................*Curvularia* sp.

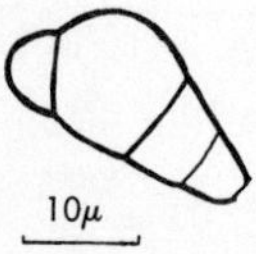

Curvularia leaf spot

SIZE: 7.5 x 22μ

RANGE: Common parasite of vascular plants

Inaperturate conidia composed of about four cells; the subterminal cell distinctly larger than the others. Surface psilate. Quite similar to *Brachysporium.*

Fig. 449. *Curvularia* sp.; Moniliales (Deuteromycete fungus, Fungi Imperfecti), conidia.

11a Spore chain coiled (Fig. 450).......................*Helicoma* sp.

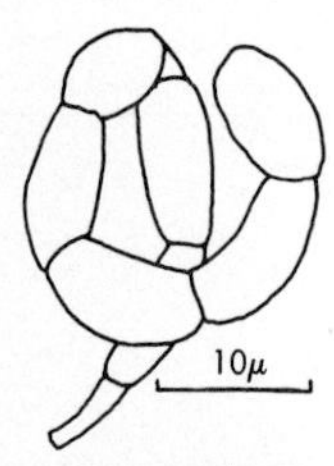

Fig. 450. *Helicoma* sp.; Moniliales (Deuteromycete fungus, Fungi Imperfecti), gi Imperfecti), conidia.

SIZE: 5μ x 80μ

Inaperturate, multicellular, coiled, with a basal stalk. Surface psilate; color hyaline or dark. These conidia are destroyed by acetolysis, but may be encountered in unacetolyzed water samples.

11b Spore chain straight.......................................12

12a Terminal cells tapered, but straight........*Helminthosporium* sp.

SIZE: 9 x 40μ

RANGE: Widespread plant parasite

Conidia composed of (3-) 5-9 cells. Shape cylindrical; wall very thick, with thin cross-walls. Surface smooth. Dark colored. Similar to *Heterosporium* which differs by being coarsely verrucate to minutely echinate.

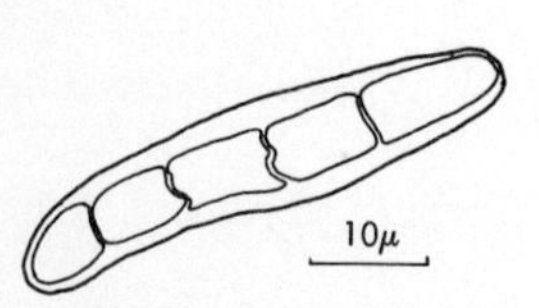

Fig. 451. *Helminthosporium* sp.; DEMATIACEAE (Deuteromycete fungus, Fungi Imperfecti), conidia.

12b Terminal cells curved (Fig. 452)....................*Fusarium* sp.

SIZE: 5μ x 70μ

RANGE: Widespread plant parasite

Conidia usually 9-celled inaperturate. Tips tapered (fusiform), one end has a notch-like scar (point of attachment) which is difficult to see. Psilate; hyaline when fresh.

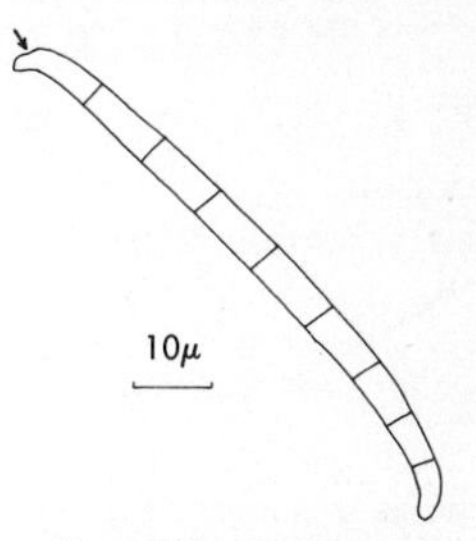

Fig. 452. *Fusarium* sp.; TUBERCULARIACEAE (Deuteromycete fungus, Fungi Imperfecti), conidia.

SYSTEMATIC LIST OF SPECIES OF WHICH SPORES AND POLLEN ARE ILLUSTRATED

Morphological types among pollen and spores do not follow taxonomic groups very consistently. There is, therefore, no convenient way to detect the palynological diversity or uniformity of a particular taxonomic group of plants in a key such as this. In a few stenopalynous families, all of the species produce similar pollen; these will occur in the same section of the key. A great many families are eurypalynous, often including genera or species whose pollen or spores occur in widely separated parts of the key. The reader who wishes to examine the array of palynomorphic types of a selected taxon may do so by means of the following list. There has been no attempt to be *inclusive* of all species (or in some cases even genera) in a plant family. This is especially true of taxa in which pollen grains of several genera or species are practically indistinguishable. I have attempted to include virtually all trees of the United States, as well as representatives of most genera of shrubs and anemophilous plants. In addition, selections have been made from remaining groups to provide as broad as possible a sampling of palynomorphic and taxonomic diversity. Coverage of non-vascular plants is especially incomplete, partly because of the paucity of reference specimens in most pollen-spore collections. In the following list, those types which are encountered frequently in air or sediment samples are marked with an asterisk (*); the beginning palynologist might wish to begin by learning these forms.

ANIMAL KINGDOM

PLANT KINGDOM

Algae

---

*Forms commonly encountered in air and sediment samples.

Fungi

Mosses

Spores of Extinct Vascular Plants

Angiosperms

INDEX AND PICTURED GLOSSARY

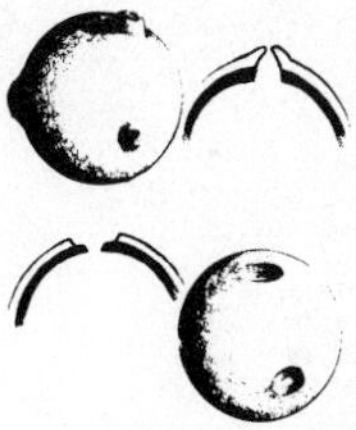

A

Figure 453

B

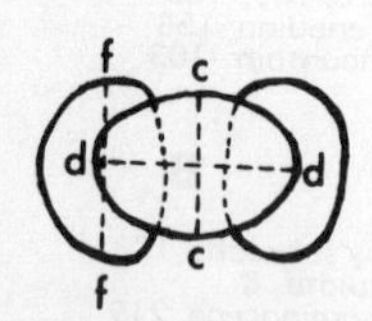

Figure 454

C

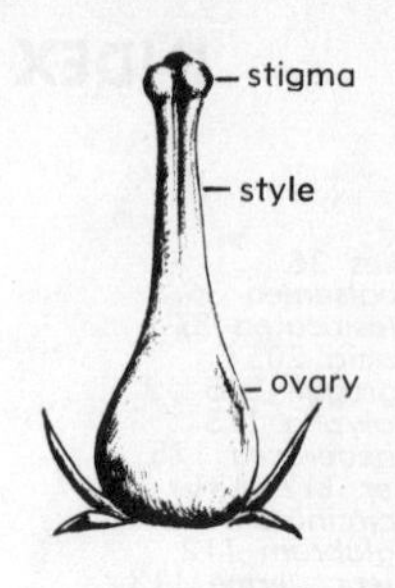

Figure 455

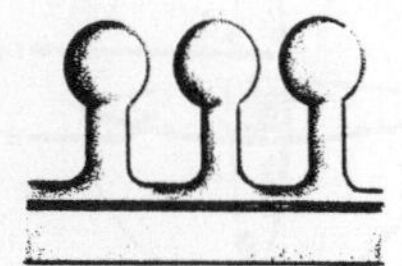

Figure 456

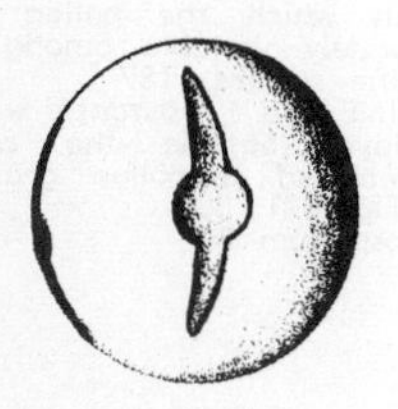

Figure 457

Figure 458

D

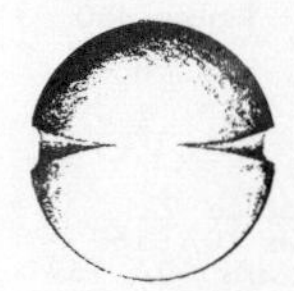

Figure 459

Figure 460

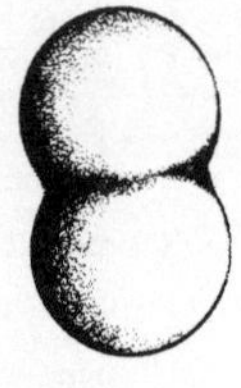

Figure 461

E

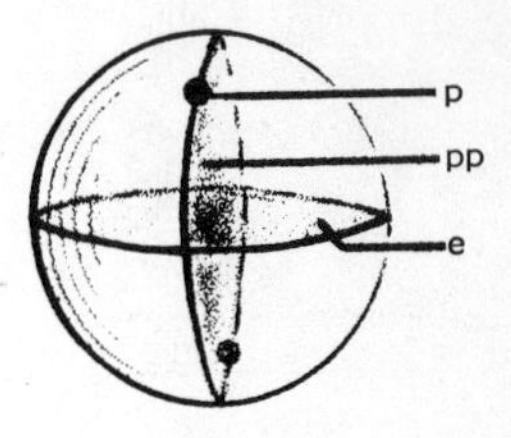

Figure 462

F

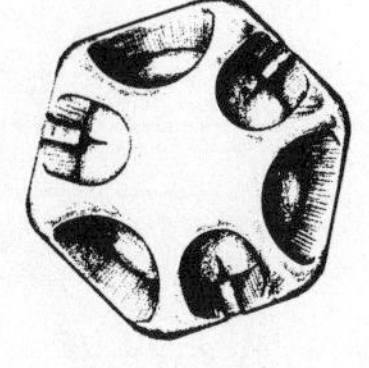

Figure 463

Figure 464

G

Figure 465

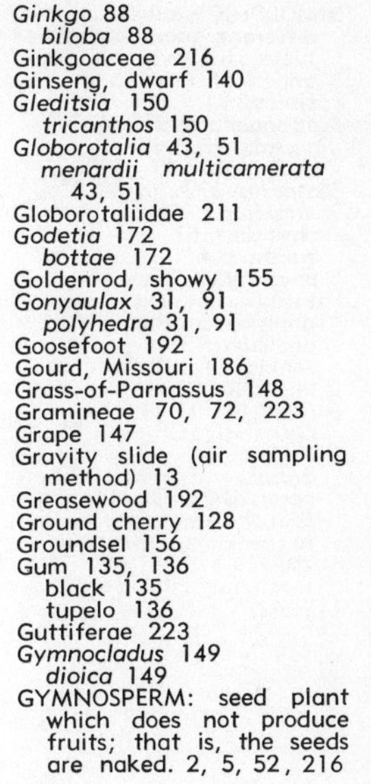

H

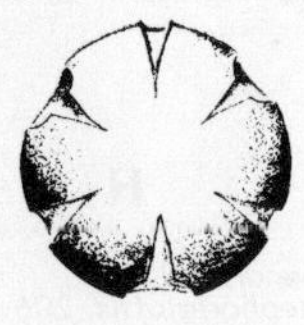
Figure 466

I

Figure 467

J

K

L

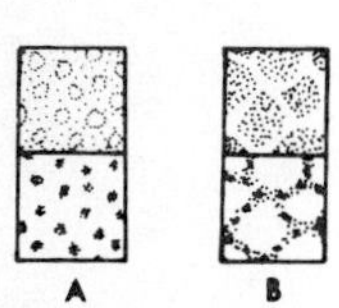

Figure 468

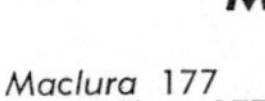

M

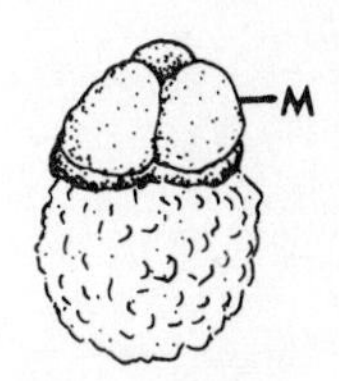

Figure 469

Fig. 470

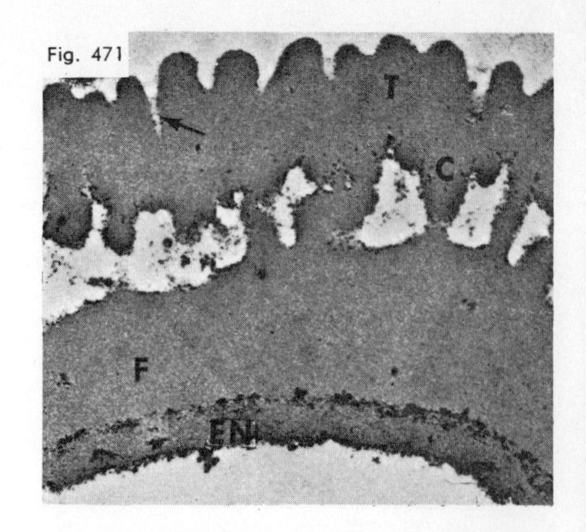

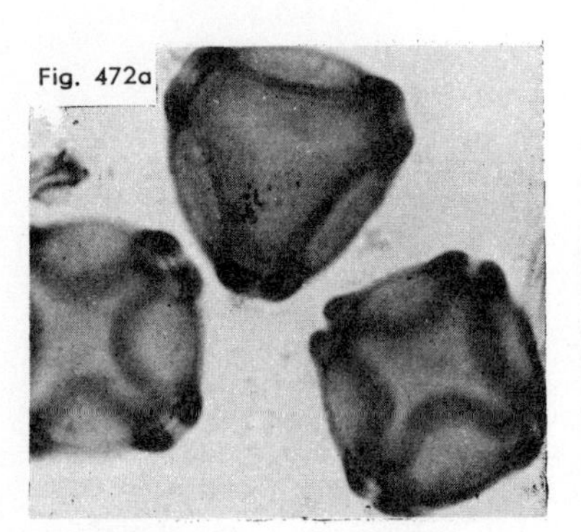

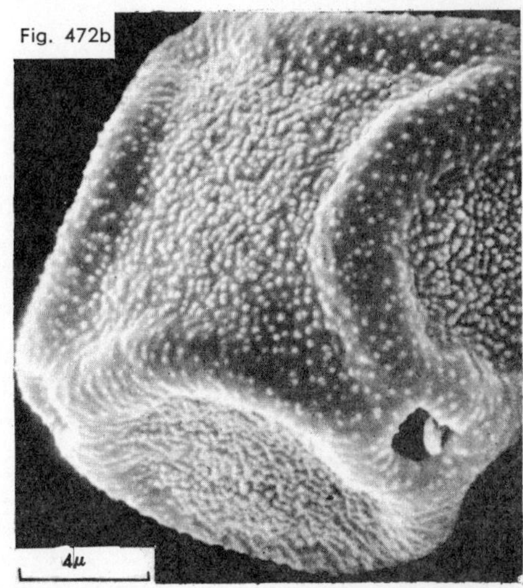

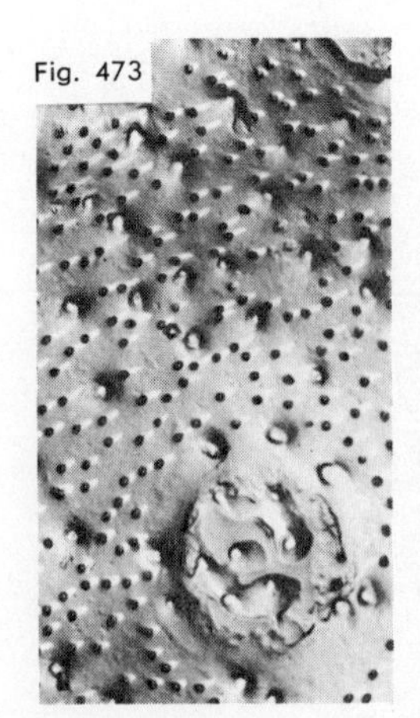

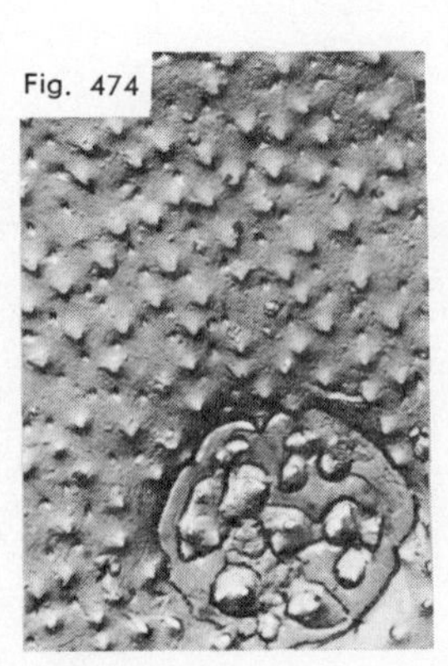

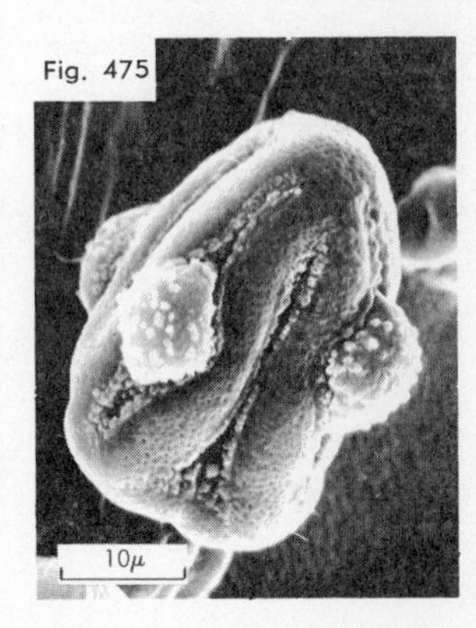

Fig. 470. Electron photomicrograph (enlarged 6,200 times) of a thin transverse section through a mature pollen grain of *Helleborus foetidus* (hellebore), a member of the buttercup family. The large nucleus, with its dense dark nucleolus, is in the center surrounded by the cytoplasm. The cellulose intine and the endexine compose the dark inner wall layer. The ektexine is clearly composed of an inner foot layer, radial, rod-like, columellae, and the outer, incomplete tectum.

From: Echlin, Patrick. 1968. *Pollen,* page 85. © Scientific American, April, 1968. Used by permission of the publishers.

Fig. 471. Electron photomicrograph (enlarged 22,000 times) of section of acetylated exine of *Photinia japonica*. The endexine (EN) is thin in comparison with the foot layer (F). Columellae (C) support the tectum (T); the arrow indicates a micropunctation in the tectum.

From: Larson, D. A., *et al.* 1962. An electron microscope study of exine stratification and fine structure. *Pollen et Spores* IV:233-246, Fig. 6.

Fig. 472A. Pollen of alder *(Alnus serrulata)* in a light photomicrograph, magnification about 1,500 X.

Fig. 472B. Acetolyzed alder pollen in a scanning electron photomicrograph magnified 5,000 times. Note the prominent arci (arcs) between the pores and the appearance of minute verrucae when the S.E.M. is used.

Figs. 472A and B are from: Martin, P. S. 1969. Pollen analysis and the scanning electron microscope. *Scanning Electron Microscopy/1969.*

Fig. 473. Carbon surface replica of *Atriplex portulacoides* pollen, shadowed with chromium and magnified X 10,000 by the electron microscope. There are many more minute holes than spinules (ratio greater than 1.6) in the pollen wall surface of members of the Chenopodiaceae.

Fig. 474. Carbon surface replica of *Amaranthus hypochondriacus* pollen, shadowed with chromium and magnified X 10,000 by the electron microscope. There are many fewer minute pores in relation to spinules (ratio less than 1.5) in the pollen wall surface of members of the Amaranthaceae when compared with the Chenopodiaceae (see Fig. 474).

Figs. 473 and 474 from: Tsukada, M. 1967. Chenopod and Amaranth Pollen: electron-microscopic identification. *Science* 157:81.

Fig. 475. Scanning electron micrograph of *Amsinckia intermedia* pollen (collected in Arizona). This heterocolpate pollen grain has eight furrows, but only four with pores; the furrows gape toward the larger end of this asymmetric grain. Clusters of microgemmae border the furrows and extend over the pore membranes.

From: Martin, P. S. and C. M. Drew. 1969. Scanning electron photomicrographs of Southwestern Pollen Grains. *Journal of the Arizona Academy of Science* 5:170 (Fig. 64A).

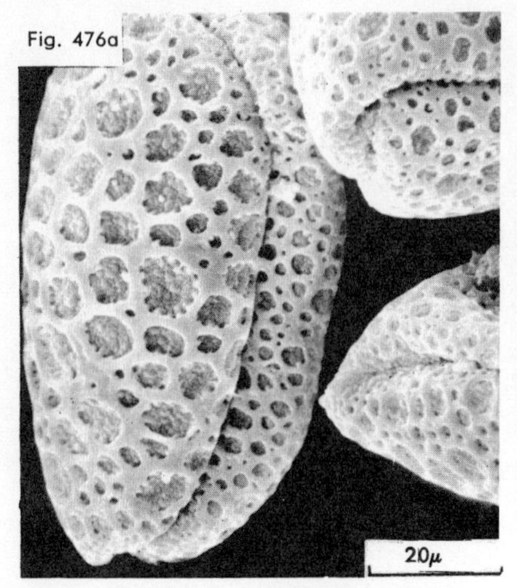
Fig. 476a
20μ

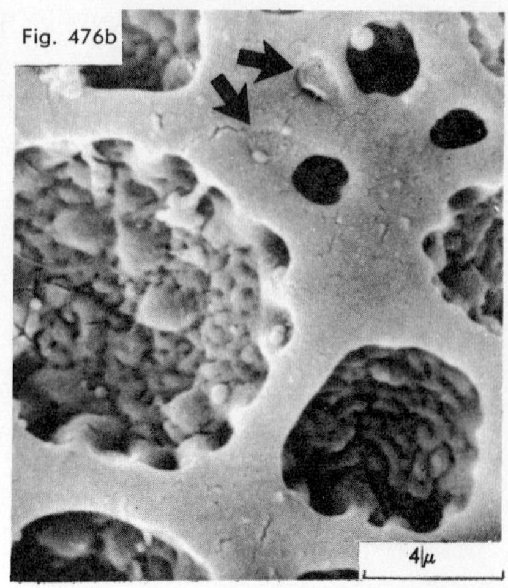
Fig. 476b
4μ

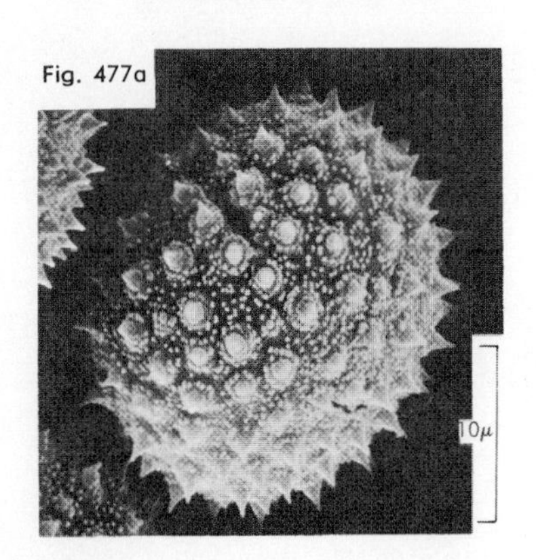
Fig. 477a
10μ

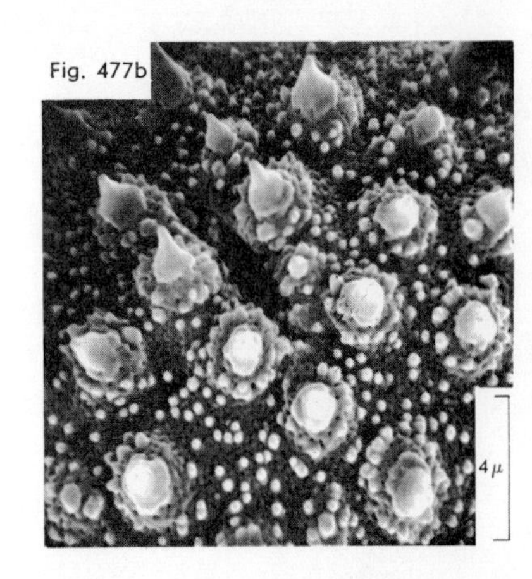
Fig. 477b
4μ

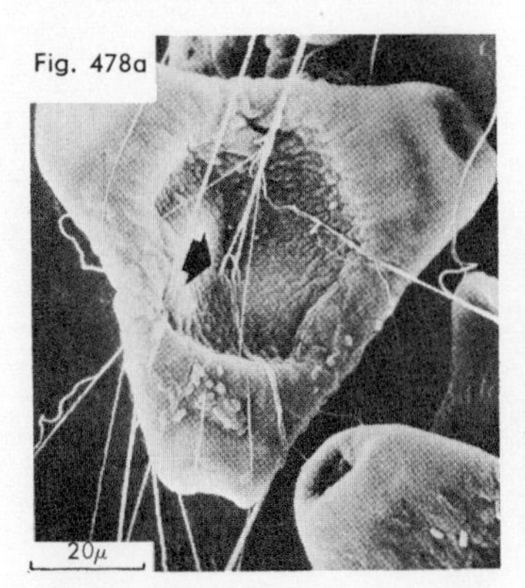
Fig. 478a
20μ

Fig. 478b
4μ

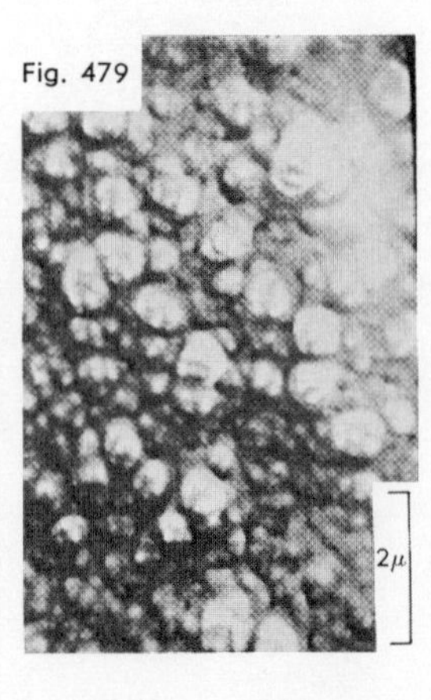
Fig. 479
2μ

Fig. 476. Scanning electron photomicrographs of pollen of the century plant *(Agave americana)*. Fig. 476A (magnified 1,000 times) reveals the reticulate surface and the single long furrow. Compare with the drawing of *Agave parryi* (Fig. 156). Fig. 476B (magnified 5,000 times) is an enlargement of a portion of the surface; note that the smooth, raised muri are supported by stout columellae. The arrows indicate the occasional small lumina which may be filled with an acid-soluble matrix; this matrix is lost in acetolysis.

From: Martin, P. S. 1969. Pollen analysis and the scanning electron microscope. *Scanning Electron Microscopy/1969:89* (Fig. 8B and 8C).

Fig. 477. Scanning electron photomicrographs of pollen of desert mallow (*Sphaeralcea, ambigua* or *emoryi*, MALVACEAE). Fig. 477A (magnified 2,000 times) shows the general morphology of this oblate, echinate grain; the apertures are three very short furrows (or pores). Compare with the drawing of *Sphaeralcea coccinea* (Fig. 359). Further magnification (5,000 times) in Fig. 477B shows the detailed ornamentation of the spines and reveals microperforations ($> 0.1 \mu$) in the surface.

From: Martin, P. S. ibid. (Fig. 23B and 23C).

Fig. 478. Scanning electron photomicrographs of pollen of the fireweed (*Epilobium angustifolium*, ONAGRACEAE, Collected in Alaska). This grain has thick-walled, strongly protruding pores; like many members of the family, the pollen grains have "viscin threads." Fig. 478A shows a general view (1,000 X), while 478B is a photograph (magnified 5,000 times) of the bases of the threads.

From: Martin, P. S. and C. M. Drew. 1969. Scanning electron photomicrographs of southwestern pollen grains. Journal of the Arizona Academy of Science 5:168 (Figs. 55A and 55C).

Fig. 479. Scanning electron photomicrograph of a small portion of the wall surface of the pollen grain of Shrub Live Oak (*Quercus turbinella*, FAGACEAE, collected in Arizona). This figure (magnified 10,000 times) reveals the surface features which give oak pollen grains the appearance of having coarse verrucae scattered over a background studded with very fine granulations. Compare with the drawing of the same species in Fig. 233.*

*Scanning electron photomicrographs (Figures 472B, 475, 476A, 476B, 477A, 477B, 478A, 478B, 479) were made possible by special arrangements with Charles M. Drew, Naval Weapons Center, China Lake, California; they were provided by Dr. R. S. Martin.

Figure 480

Figure 481

N

O

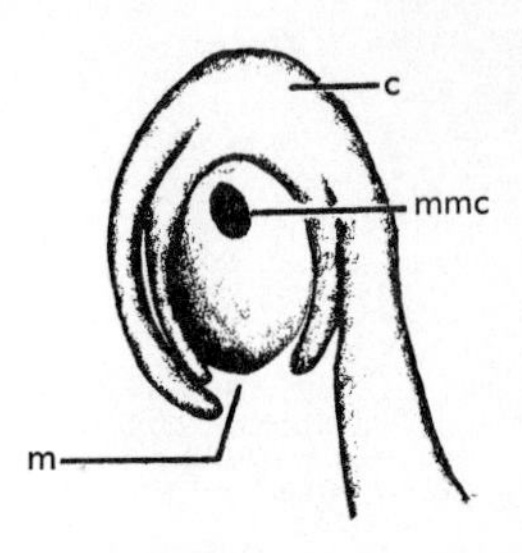

Figure 482

P

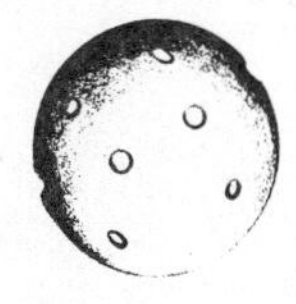

Figure 483

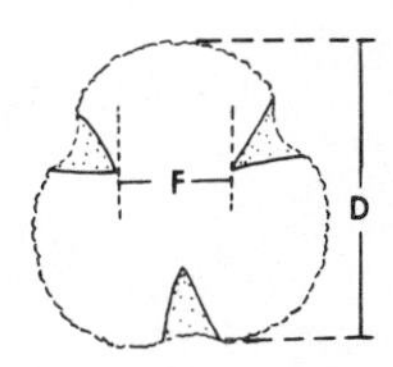

Figure 484

Figure 485

Figure 486

POLYPLICATE: refers to pollen grains with numerous surface folds. Fig. 487. 21, 39

Figure 487

PORE: a pollen aperture which is nearly round, less than twice as long as broad. Fig. 488

Figure 488

PORATE: pollen grains with pores. 6
POROID: refers to a pore or pore-like aperture.

PROPAGULE: a single-celled or multicellular unit which upon germination or cell multiplication produces a new plant.

PROTOPLAST: the protoplasmic contents of a single cell.
PROTOZOA: single-celled (or colonial) animals. 28, 31, 70, 167
PROXIMAL POLE: the position on a pollen grain or spore which was oriented nearest the center of the tetrad during development.

PSEUDOCOLPUS: a false furrow. 25

PSILATE: smooth. 8
PSILOPSID: a plant in the Psilophyta, believed to be a very primitive vascular plant.

Q

QUATERNARY: the Pleistocene Epoch or Ice Ages.

R

RETICULATE: bearing a net-like surface sculpture pattern. 8

RETICULUM: a net-like surface sculpture pattern. 35, scanning electron photomicrographs: Figs. 476a and 476b.

RHIZOME: an underground stem.

ROTATIONAL SYMMETRY: a type of symmetry which permits rotation around an axis (the polar axis in most palynomorphs) without alternation of basic pattern of symmetry. 4

S

SACCATE (vesiculate): a pollen grain bearing sacci (wings or bladders). Fig. 489.

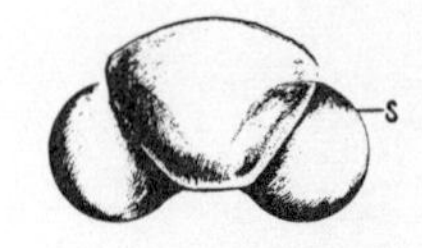

Figure 489

Figure 490

Figure 491

STRUCTURE: the structural features of the palynomorph wall, especially those characteristics due to the form and arrangement of the exine elements inside the tectum.

SUBECHINATE: with very short, small or more or less rounded spines.
SUBOBLATE: nearly oblate, subspheroidal, but tending toward an oblate shape; P/E index 0.75-0.88
SUBPROLATE: nearly prolate, subspheroidal, but tending toward a prolate shape; P/E index 1.14-1.33
SUBRETICULATE: nearly reticulate
SUBSPHEROIDAL: P/E index from 0.75-1.33. 4
SUBSTRIATE: nearly striate
SUBVERRUCATE: subdued, or nearly, verrucate.
SULCATE: the furrow-like aperture of gymnosperms and monocots. It is considered monocolpate; the sulcus is located on the distal face of the grain. H — height, W — width, L—length. Fig. 492

Figure 492

SULCOID: bearing a sulcus-like marking.

SUPRATECTATE: exine features which are peripheral to the tectum. 7, 8
SUPRATECTATE SCULPTURE: exine sculpturing external (peripheral) to the tectum. 7

SYNCOLPATE: refers to a pollen grain with fused furrows; these may be fused at the poles as in Figure 493, spirally, or in other patterns. Fig. 493. 23, 90.

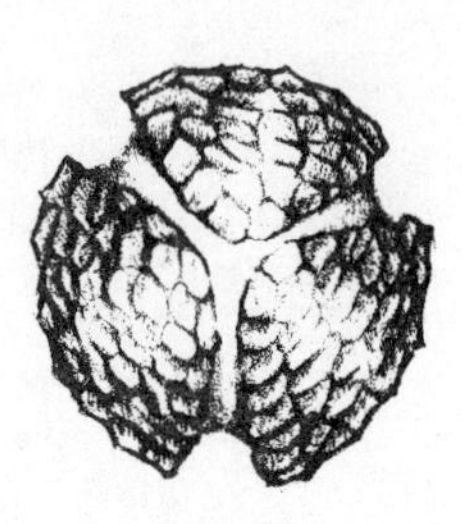
Figure 493

TECTATE: refers to an exine with a tectum. 8
TECTUM: An outer, more or less continuous layer of ektexine forming a roof over columellae which support the tectum. Fig. 471. 7
TEGILLUM: equivalent to tectum.
TELIOSPORE: One of the types of spores in the life cycle of rust fungi; frequently a dyad. Upon germination the teliospore usually produces basidia and basidiospores. 52
TERTIARY: The geologic era extending from 70 million years to 1 million years ago.
TEST (cyst): The wall (often chitinous or siliceous) of certain Protozoa. 28
TESTACEOUS: Bearing a test
TETRAD: A group of four spores or pollen grains. Several arrangement patterns are possible: A—square; B — tetrahedral; C—rhomboidal (see Figure 494).

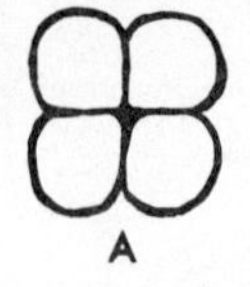

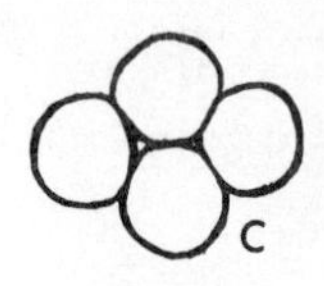

Figure 494

THECA: The wall of certain dinoflagellates.

TRANSVERSE FURROW: In pollen grains with colporate apertures, if the pore is transversely elongated (more than twice as long as wide) it is considered a transverse furrow.

TRICHOTOMOCOLPORATE: A pollen type in which three furrows are joined at one pole.
TRICOLPATE: A pollen type with three equatorial furrows. Figure 495. 24, 94.

Figure 495

TRICOLPORATE: A pollen type with three equatorial furrows, each containing a pore. Figure 496. 25, 123.

Figure 496

Trifolium 99, 151
  pratense 99, 151
TRILETE: spore having a three-pronged scar in the wall. Fig. 497. 5, 21, 22

Figure 497

Triletes 42, 43
  mamillaris 43
  triangulatis 42
Trillium 89
  flexipes 89
Triodanis 182
  coloradensis 182
TRIPORATE: A pollen type with three equatorial pores. Fig. 498. 26, 168

Figure 498

TRIRADIATE (scar): a three-pointed tetrad scar. 4
Tsuga 34
  canadensis 34
Tuberculariaceae 214
Tulip tree 84
Tupelo 136
Twin leaf 111
Typha 69, 203
  latifolia 203
Typhaceae 70, 230

U

Ulmaceae 168, 230
Ulmus 181
  rubra 181
Ulota 54
  crispa 54
Umbelliferae 123, 127, 231
Umbrella plant 150
Uromyces 71, 72, 182
  appendiculatus 72
  fabae 71, 182
Urtica 175
  procera 175
Urticaceae 168, 231
Ustilaginaceae 213
Ustilago 63
  maydis 63
Utricularia 120, 164
  minor 120, 164

V

Vaccinium 201
  corymbosum 201
VASCULAR PLANTS: plants in which xylem and phloem (vascular conductive tissue) is present.
Venus flytrap 197
Venus' looking glass 181
Verbascum 146
  virgatum 146
Verbena 104, 135
  ciliata 104, 135
Verbenaceae 94, 231
Vernonia 166
  altissima 166
  lindheimeri 166
VERRUCAE: Small, broad based surface elements which are as broad or broader than high and at least 1μ wide. Fig. 7, 8.
Verrucate 8
Vervain 104, 135
VESICULATE (saccate): pollen grains (primarily in the Pinaceae) with two air bladders or wings. 21, 34
VESTIBULE: a cavity (V) at the pore opening; formed by an inner extension of the endexine. Fig. 499.

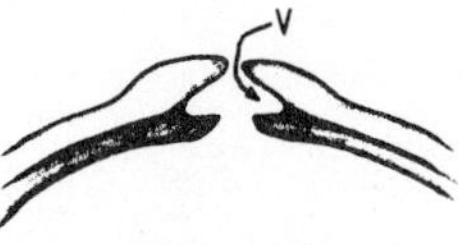

Figure 499

Vetch (hairy) 151
Viburnum 138, 139
  acerifolium 138
  lentago 139
Vicia 151
  villosa 151
Viola 105, 129, 130
  novae-angliae 130
  sagittata 105, 129
Violaceae 94, 123, 231
Violet 105, 129, 130
  arrow-leaved 129
  New England 130
VISCIN THREADS: thread-like hairs on pollen of many species in the Onagraceae. Figs. 478A & 478B.
Vitaceae 123, 231
Vitis 147
  linsecomii 147

W

Walnut 189
  black 189
Washington 105
  thorn 105
Water elm 183
Waterleaf 143
Water lily 78 see also yellow pond lily
Water milfoil 184
Watershield 83
White-buttons 92
Wild sago 87
Willow 96, 149
  black 149
Wodehouse technique for reference pollen 10
Woodbine 147
Woodrush 198
Wood sorrel (large) 102
Woodwardia 77
  virginica 77
Wool grass (bulrush) 188
Wormwood 108, 153

X

Xanthoxylum 142
  americanum 142

Y

Yeast, black 196
Yellow cress 97
Yellow pond lily 77
  see also water lily
Yerba-mansa 85
Yew (florida) 68, 80
Yucca 84
  mohavensis 84

Z

Zamia 87
  floridana 87
Zauschneria 174
  californica 174
Zea 74
  mays 74
Zlivisporis 46
  blanensis 46
Zosteraceae 231
Zygophyllaceae 185, 231
ZYGOSPORE: a dormant, thick-walled zygote, usually produced among certain algae (e.g. Spirogyra) and fungi (e.g. Rhizopus). 52

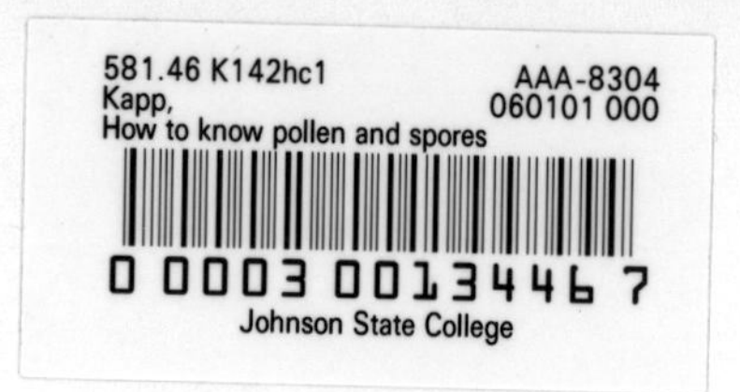
581.46 K142hc1
AAA-8304
Kapp,
060101 000
How to know pollen and spores
0 0003 0013446 7
Johnson State College